Making SEX

Making SEX

BODY AND GENDER FROM THE GREEKS TO FREUD

THOMAS LAQUEUR

HARVARD UNIVERSITY PRESS
Cambridge, Massachusetts, and London, England

First Harvard University Press paperback edition, 1992

Library of Congress Cataloging-in-Publication Data

Laqueur, Thomas Walter
Making sex : body and gender from the Greeks to Freud /
Thomas Laqueur.
p.   cm.
Includes bibliographical references.
ISBN 0-674-54349-1 (alk. paper) (cloth)
ISBN 0-674-54355-6 (paper)
1. Sex role—History.
2. Sex differences—Social aspects—History.
3. Sex differences (Psychology)—Social aspects—History.
4. Sex (Psychology).
I. Title.
HQ1075.L37   1990
305.3'09—dc20   90-35220
CIP

Designed by Gwen Frankfeldt

For Gail and Hannah

Preface

This book began without my knowing it in 1977 when I was on leave at St. Antony's College, Oxford, doing research for what was to be a history of the life cycle. I was reading seventeenth-century midwifery manuals—in search of materials on how birth was organized—but found instead advice to women on how to become pregnant in the first place. Midwives and doctors seemed to believe that female orgasm was among the conditions for successful generation, and they offered various suggestions on how it might be achieved. Orgasm was assumed to be a routine, more or less indispensable part of conception. This surprised me. Experience must have shown that pregnancy often takes place without it; moreover, as a nineteenth-century historian I was accustomed to doctors debating whether women had orgasms at all. By the period I knew best, what had been an ordinary, if explosive, corporeal occurrence had become a major problem of moral physiology.

My life-cycle project slowly slipped away. I got married; we had a child; I spent a year in medical school in 1981–1982. Precisely how these changes in my life allowed this book to take me over is still not entirely clear, but they did. (Its relevant intellectual origins are more obvious: a group of friends started *Representations*; I taught a graduate seminar on the body and the body social in nineteenth-century literature with Catherine Gallagher; I encountered feminist literary and historical scholarship; my almost daily companion in the rational recreation of drinking cappuccino, Peter Brown, was working on his book about the body and society in late antiquity.) At first the question of disappearing orgasm was the focus of my research, and what follows still bears some marks of its origins in that preoccupation. But gradually the *summa voluptas* was assimilated into the larger question of the relationship between the body and sexual difference and, indeed, the nature of sexual difference generally.

There might appear to be no problem here. It seems perfectly obvious that biology defines the sexes—what else could sex mean? Hence historians can have nothing much to say on the matter. To have a penis or not says it all in most circumstances, and one might for good measure add as many other differences as one chooses: women menstruate and lactate, men do not; women have a womb that bears children, and men lack both this organ and this capacity. I do not dispute any of these facts, although if pushed very hard they are not quite so conclusive as one might think. (A man is presumably still a man without a penis, and scientific efforts to fix sex definitively, as in the Olympic Committee's testing of the chromosomal configuration of buccal cavity cells, leads to ludicrous results.)

More to the point, though, no particular understanding of sexual difference historically follows from undisputed facts about bodies. I discovered early on that the erasure of female pleasure from medical accounts of conception took place roughly at the same time as the female body came to be understood no longer as a lesser version of the male's (a one-sex model) but as its incommensurable opposite (a two-sex model). Orgasms that had been common property were now divided. Organs that had been seen as interior versions of what the male had outside—the vagina as penis, the uterus as scrotum—were by the eighteenth century construed as of an entirely different nature. Similarly, physiological processes—menstruation or lactation—that had been seen as part of a common economy of fluids came to be understood as specific to women alone.

Some of these changes might be understood as the results of scientific progress—menstruation is not the same thing as hemorrhoidal bleeding—but the chronology of discoveries did not line up with reconceptions of the sexual body. Moreover, chronology itself soon crumbled and I was faced with the startling conclusion that a two-sex and a one-sex model had always been available to those who thought about difference and that there was no scientific way to choose between them. The former might indeed have come into prominence during the Enlightenment, but one sex did not disappear. In fact, the more I put pressure on the historical record, the less clear the sexual divide became; the more the body was pressed into service as the foundation for sex, the less solid the boundaries became. With Freud the process reaches its most crystalline indeterminacy. What began with a history of female sexual pleasure and its attempted erasure has become instead the story of how sex, as much as gender, is made.

A book that deals with so broad a range of time and materials as this one owes a multitude of debts. In the first place I could not have written it—both because the required scholarship was not in place and because the subject would not have been taken seriously—without the intellectual revolution wrought by feminism since World War II and especially during the past twenty years. My work is in some sense an elaboration of Simone de Beauvoir's claim that women are the second sex. It could also not have been written without the sustenance of my intellectual community at Berkeley and elsewhere. My colleagues on *Representations,* among whom I first went semipublic on this topic back in 1983, have offered advice, encouragement, criticism, and good company. Several of my friends and colleagues have not only read and offered detailed criticism of my manuscript but discussed it with me tirelessly in its many, many avatars over the years: Peter Brown, Carol Clover, Catherine Gallagher, Stephen Greenblatt, Thomas Metcalf, Randolph Starn, Irv Scheiner, and Reggie Zelnik. Wendy Lesser would not read it all, but she talked me through many drafts, published part of Chapter 1 in the *Threepenny Review,* and consistently represented the views of the general reader. My colleague David Keightley, leader of the Yuppie Bikers, has heard lots about sex over the miles and offered the perspective of ancient China. Marjorie Beale, Mario Biagioli, Natalie Zemon Davis, Evelyn Fox-Keller, Isabel Hull, and Roy Porter provided detailed comments on the manuscript in its penultimate form and greatly helped me to refine my arguments and the book's architecture.

The graduate-student History and Gender Group at Berkeley also read a draft and, although I have not accepted its suggestion that I bare my innermost feelings about the polymorphous perverse and erotic desire, I have profited greatly from the astute suggestions and numerous references provided by Lisa Cody, Paul Friedland, Nasser Hussain, and Vanessa Schwartz. And then, of course, a book that covers so many topics over so long a period is beholden to specialists: David Cohen, Leslie Jones, and Gregory Vlastos offered tough criticism, only some of which I accepted, on Chapter 2. Susanna Barrows, André Burguiere, William

Bouwsma, Caroline Bynum, Joan Cadden, Roger Chartier, Alain Corbin, Laura Englestein, Lynn Hunt, Sarah Blaffer Hrdy, Susan Kent, Jack Lesch, Emily Martin, Regina Morantz-Sanchez, Joan Scott, Nancy Vickers, and Judith Walkowitz have been immensely generous with references and advice. My research assistants since the early 1980s—Mary McGarry, Jonathan Clark, Eric Steinle, Ramona Curry, Jan Matlock, Catherine Kudlick, Russ Geoffrey, M.D., Alice Bullard, and Dean Bell—made it possible for me to read and begin to understand a wide range of sources. Alexander Nehamas not only answered many questions about Greek words but offered the support of an old friend and the limpid intelligence of a philosopher. My editor Lindsay Waters at Harvard University Press saw a book when none was there; he read early drafts with intelligent care and rightly forced a reluctant author back to the drawing board. Patricia Williams became my editor by adoption—she was on the spot in Berkeley—and, in addition to timely hand holding, helped me enormously in understanding what had to be done to turn what I thought was the final draft into the present book. Joyce Backman was a dream of a manuscript editor: funny, erudite, and careful.

I dedicate this book to my wife Gail Saliterman, who typed none but read most of it, and to my eight-year-old daughter Hannah, who recently pointed out that I have been working on it all her life. In ways too deep to articulate, they made my work possible.

Contents

1 Of Language and the Flesh   1

2 Destiny Is Anatomy   25

3 New Science, One Flesh   63

4 Representing Sex   114

5 Discovery of the Sexes   149

6 Sex Socialized   193

Notes   245

Credits   303

Index   305

Making SEX

ONE

Of Language and the Flesh

The first thing that strikes the careless observer is
that women are unlike men. They are "the opposite
sex" (though why "opposite" I do not know; what
is the "neighboring sex"?). But the fundamental
thing is that women are more like men than anything else in the world.

DOROTHY L. SAYERS
"THE HUMAN-NOT-QUITE-HUMAN"

An interpretive chasm separates two interpretations, fifty years apart, of
the same story of death and desire told by an eighteenth-century physician obsessed with the problem of distinguishing real from apparent
death.[1]

The story begins when a young aristocrat whose family circumstances
forced him into religious orders came one day to a country inn. He found
the innkeepers overwhelmed with grief at the death of their only daughter, a girl of great beauty. She was not to be buried until the next day, and
the bereaved parents asked the young monk to keep watch over her body
through the night. This he did, and more. Reports of her beauty had
piqued his curiosity. He pulled back the shroud and, instead of finding
the corpse "disfigured by the horrors of death," found its features still
gracefully animated. The young man lost all restraint, forgot his vows,
and took "the same liberties with the dead that the sacraments of marriage would have permitted in life." Ashamed of what he had done, the
hapless necrophilic monk departed hastily in the morning without wait-
ing for the scheduled interment.

When time for burial came, indeed just as the coffin bearing the dead
girl was being lowered into the ground, someone felt movement coming
from the inside. The lid was torn off; the girl began to stir and soon
recovered from what proved not to have been real death at all but only a

coma. Needless to say, the parents were overjoyed to have their daughter back, although their pleasure was severely diminished by the discovery that she was pregnant and, moreover, could give no satisfactory account of how she had come to be that way. In their embarrassment, the innkeepers consigned the daughter to a convent as soon as her baby was born.

Soon business brought the young aristocrat, oblivious of the consequences of his passion but far richer and no longer in holy orders because he had come into his inheritance, back to the scene of his crime. Once again he found the innkeepers in a state of consternation and quickly understood his part in causing their new misfortune. He hastened to the convent and found the object of his necrophilic desire more beautiful alive than dead. He asked for her hand and with the sacrament of marriage legitimized their child.

The moral that Jacques-Jean Bruhier asks his readers to draw from this story is that only scientific tests can make certain that a person is really dead and that even very intimate contact with a body leaves room for mistakes. But Bruhier's contemporary, the noted surgeon Antoine Louis, came to a very different conclusion, one more germane to the subject of this book, when he analyzed the case in 1752.[2] Based on the evidence that Bruhier himself offered, Louis argues, no one could have doubted that the girl was not dead: she did not, as the young monk testified, look dead and moreover who knows if she did not give some "demonstrative signs" in proof of her liveliness, signs that any eighteenth-century doctor or even layperson would have expected in the circumstances.

Bruhier earlier on in his book had cited numerous instances of seemingly dead young women who were revived and saved from untimely burial by amorous embraces; sexual ecstasy, "dying" in eighteenth-century parlance, turned out for some to be the path to life. Love, that "wonderful satisfactory *Death* and . . . voluntary Separation of Soul and Body," as an English physician called it, guarded the gates of the tomb.[3] But in this case it would have seemed extremely unlikely to an eighteenth-century observer that the innkeepers' daughter could have conceived a child without moving and thereby betraying her death.[4] Any medical book or one of the scores of popular midwifery, health, or marriage manuals circulating in all the languages of Europe reported it as a commonplace that "when the seed issues in the act of generation [from both men and women] there at the same time arises an extra-ordinary titillation and delight in all members of the body."[5] Without orgasm, another widely circulated text announced, "the fair sex [would] neither desire nuptial embraces, nor have pleasure in them, nor conceive by them."[6]

The girl *must* have shuddered, just a bit. If not her rosy cheeks then the tremors of venereal orgasm would have given her away. Bruhier's story was thus one of fraud and not of apparent death; the innkeepers' daughter and the monk simply conspired, Louis concludes, to escape culpability by feigning coma until the last possible moment before burial.

In 1836 the tale was told again, but now with a new twist. This time, the reality of the girl's deathlike comatose state was not questioned. On the contrary, her becoming pregnant under these conditions was cited by Dr. Michael Ryan as one among many other cases of intercourse with insensible women to prove that orgasm was irrelevant to conception. (In one story, for example, an ostler confesses that he came to an inn and had sex with, and made pregnant, a girl who was so dead asleep before the fire that he was long gone before she awoke.) Not only need a woman not feel pleasure to conceive; she need not even be conscious.[7]

Near the end of the Enlightenment, in the period between these two rehearsals of the tale of the innkeepers' daughter, medical science and those who relied on it ceased to regard the female orgasm as relevant to generation. Conception, it was held, could take place secretly, with no telltale shivers or signs of arousal; the ancient wisdom that "apart from pleasure nothing of mortal kind comes into existence" was uprooted.[8] Previously a sign of the generative process, deeply embedded in the bodies of men and women, a feeling whose existence was no more open to debate than was the warm, pleasurable glow that usually accompanies a good meal, orgasm was relegated to the realm of mere sensation, to the periphery of human physiology—accidental, expendable, a contingent bonus of the reproductive act.

This reorientation applied in principle to the sexual functioning of both men and women. But no one writing on such matters ever so much as entertained the idea that male passions and pleasures in general did not exist or that orgasm did not accompany ejaculation during coition. Not so for women. The newly "discovered" contingency of delight opened up the possibility of female passivity and "passionlessness."[9] The purported independence of generation from pleasure created the space in which women's sexual nature could be redefined, debated, denied, or qualified. And so it was of course. Endlessly.

The old valences were overturned. The commonplace of much contemporary psychology—that men want sex while women want relationships—is the precise inversion of pre-Enlightenment notions that, extending back to antiquity, equated friendship with men and fleshliness with women. Women, whose desires knew no bounds in the old scheme of things, and whose reason offered so little resistance to passion, became in some accounts creatures whose whole reproductive life might be spent anesthetized to the pleasures of the flesh. When, in the late eighteenth century, it became a possibility that "the majority of women are not much troubled with sexual feelings," the presence or absence of orgasm became a biological signpost of sexual difference.

The new conceptualization of female orgasm, however, was but one formulation of a more radical eighteenth-century reinterpretation of the female body in relation to the male. For thousands of years it had been a commonplace that women had the same genitals as men except that, as Nemesius, bishop of Emesa in the fourth century, put it: "theirs are inside the body and not outside it."[10] Galen, who in the second century A.D. developed the most powerful and resilient model of the structural, though not spatial, identity of the male and female reproductive organs, demonstrated at length that women were essentially men in whom a lack of vital heat—of perfection—had resulted in the retention, inside, of structures that in the male are visible without. Indeed, doggerel verse of the early nineteenth century still sings of these hoary homologies long after they had disappeared from learned texts:

> though they of different sexes be,
> Yet on the whole they are the same as we,
> For those that have the strictest searchers been,
> Find women are but men turned outside in.[11]

In this world the vagina is imagined as an interior penis, the labia as foreskin, the uterus as scrotum, and the ovaries as testicles. The learned Galen could cite the dissections of the Alexandrian anatomist Herophilus, in the third century B.C., to support his claim that a woman has testes with accompanying seminal ducts very much like the man's, one on each side of the uterus, the only difference being that the male's are contained in the scrotum and the female's are not.[12]

Language marks this view of sexual difference. For two millennia the ovary, an organ that by the early nineteenth century had become a synecdoche for woman, had not even a name of its own. Galen refers to it by the same word he uses for the male testes, *orcheis,* allowing context to make clear which sex he is concerned with. Herophilus had called the ovaries *didymoi* (twins), another standard Greek word for testicles, and was so caught up in the female-as-male model that he saw the Fallopian tubes—the spermatic ducts that led from each "testicle"—as growing into the neck of the bladder as do the spermatic ducts in men.[13] They very clearly do not. Galen points out this error, surprised that so careful an observer could have committed it, and yet the correction had no effect on the status of the model as a whole. Nor is there any technical term in Latin or Greek, or in the European vernaculars until around 1700, for vagina as the tube or sheath into which its opposite, the penis, fits and through which the infant is born.

But then, in or about the late eighteenth, to use Virginia Woolf's device, human sexual nature changed. On this point, at least, scholars as theoretically distant from one another as Michel Foucault, Ivan Illich, and Lawrence Stone agree.[14] By around 1800, writers of all sorts were determined to base what they insisted were fundamental differences between the male and female sexes, and thus between man and woman, on discoverable biological distinctions and to express these in a radically different rhetoric. In 1803, for example, Jacques-Louis Moreau, one of the founders of "moral anthropology," argued passionately against the nonsense written by Aristotle, Galen, and their modern followers on the subject of women in relation to men. Not only are the sexes different, but they are different in every conceivable aspect of body and soul, in every physical and moral aspect. To the physician or the naturalist, the relation of woman to man is "a series of oppositions and contrasts."[15] In place of what, in certain situations, strikes the modern imagination as an almost perverse insistence on understanding sexual difference as a matter of degree, gradations of one basic male type, there arose a shrill call to articulate sharp corporeal distinctions. Doctors claimed to be able to identify "the essential features that belong to her, that serve to distinguish her, that make her what she is":

> All parts of her body present the same differences: all express woman; the brow, the nose, the eyes, the mouth, the ears, the chin, the cheeks. If we shift our view to the inside, and with the help of the scalpel, lay bare the organs, the tissues, the fibers, we encounter everywhere . . . the same difference.[16]

Thus the old model, in which men and women were arrayed according to their degree of metaphysical perfection, their vital heat, along an axis whose telos was male, gave way by the late eighteenth century to a new model of radical dimorphism, of biological divergence. An anatomy and physiology of incommensurability replaced a metaphysics of hierarchy in the representation of woman in relation to man.

By the late nineteenth century, so it was argued, the new difference could be demonstrated not just in visible bodies but in its microscopic building blocks. Sexual difference in kind, not degree, seemed solidly grounded in nature. Patrick Geddes, a prominent professor of biology as well as a town planner and writer on a wide range of social issues, used cellular physiology to explain the "fact" that women were "more passive, conservative, sluggish and stable" than men, while men were "more active, energetic, eager, passionate, and variable." He thought that with rare exceptions—the sea horse, the occasional species of bird—males were constituted of catabolic cells, cells that put out energy. They spent income, in one of Geddes' favorite metaphors. Female cells, on the other hand, were anabolic; they stored up and conserved energy. And though he admitted that he could not fully elaborate the connection between these biological differences and the "resulting psychological and social differentiations," he nevertheless justified the respective cultural roles of men and women with breathtaking boldness. Differences may be exaggerated or lessened, but to obliterate them "it would be necessary to have all the evolution over again on a new basis. What was decided among the pre-historic Protozoa cannot be annulled by an act of Parliament."[17] Microscopic organisms wallowing in the primordial ooze determined the irreducible distinctions between the sexes and the place of each in society.

These formulations suggest a third and still more general aspect of the shift in the meaning of sexual difference. The dominant, though by no means universal, view since the eighteenth century has been that there are two stable, incommensurable, opposite sexes and that the political, economic, and cultural lives of men and women, their gender roles, are somehow based on these "facts." Biology—the stable, ahistorical, sexed body—is understood to be the epistemic foundation for prescriptive claims about the social order. Beginning dramatically in the Enlightenment, there was a seemingly endless stream of books and chapters of books whose very titles belie their commitment to this new vision of nature and culture: Roussel's *Système physique et moral de la femme*, Brachet's chapter "Etudes du physique et du moral de la femme," Thompson and Geddes' starkly uncompromising *Sex*. The physical "real" world in these accounts, and in the hundreds like them, is prior to and logically independent of the claims made in its name.

Earlier writers from the Greeks onward could obviously distinguish nature from culture, *phusis* from *nomos* (though these categories are the creation of a particular moment and had different meanings then).[18] But, as I gathered and worked through the material that forms this book, it became increasingly clear that it is very difficult to read ancient, medieval, and Renaissance texts about the body with the epistemological lens of the Enlightenment through which the physical world—the body—appears as "real," while its cultural meanings are epiphenomenal. Bodies in these texts did strange, remarkable, and to modern readers impossible things. In future generations, writes Origen, "the body would become less 'thick,' less 'coagulated,' less 'hardened,'" as the spirit warmed to God; physical bodies themselves would have been radically different before the fall, imagines Gregory of Nyssa: male and female coexisted with the image of God, and sexual differentiation came about only as the representation in the flesh of the fall from grace.[19] (In a nineteenth-century Urdu guide for ladies, based firmly in Galenic medicine, the prophet Mohammed is listed at the top of a list of exemplary women.[20] Caroline Bynum writes about women who in imitation of Christ received the stigmata or did not require food or whose flesh did not stink when putrifying.[21] There are numerous accounts of men who were said to lactate and pictures of the boy Jesus with breasts. Girls could turn into boys, and men who associated too extensively with women could lose the hardness and definition of their more perfect bodies and regress into effeminacy. Culture, in short, suffused and changed the body that to the modern sensibility seems so closed, autarchic, and outside the realm of meaning.

One might of course deny that such things happened or read them as entirely metaphorical or give individual, naturalistic explanations for otherwise bizarre occurrences: the girl chasing her swine who suddenly sprung an external penis and scrotum, reported by Montaigne and the sixteenth-century surgeon Ambroise Paré as an instance of sex change, was really suffering from androgen-dihydrostestosterone deficiency; she was really a boy all along who developed external male organs in puberty, though perhaps not as precipitously as these accounts would have it.[22] This, however, is an unconscionably external, ahistorical, and impoverished approach to a vast and complex literature about the body and culture.

I want to propose instead that in these pre-Enlightenment texts, and even some later ones, *sex,* or the body, must be understood as the epiphenomenon, while *gender,* what we would take to be a cultural category, was primary or "real." Gender—man and woman—mattered a great deal and was part of the order of things; sex was conventional, though modern terminology makes such a reordering nonsensical. At the very least, what we call sex and gender were in the "one-sex model" explicitly bound up in a circle of meanings from which escape to a supposed biological substrate—the strategy of the Enlightenment—was impossible. In the world of one sex, it was precisely when talk seemed to be most directly about the biology of two sexes that it was most embedded in the politics of gender, in culture. To be a man or a woman was to hold a social rank, a place in society, to assume a cultural role, not to *be* organically one or the other of two incommensurable sexes. Sex before the seventeenth century, in other words, was still a sociological and not an ontological category.

How did the change from what I have called a one-sex/flesh model to a two-sex/flesh model take place? Why, to take the most specific case first, did sexual arousal and its fulfillment—specifically female sexual arousal—become irrelevant to an understanding of conception? (This, it seems to me, is the initial necessary step in creating the model of the passionless female who stands in sharp biological contrast to the male.) The obvious answer would be the march of progress; science might not be able to explain sexual politics, but it could provide the basis on which to theorize. The ancients, then, were simply wrong. In the human female and in most other mammals—though not in rabbits, minks, and ferrets—ovulation is *in fact* independent of intercourse, not to speak of pleasure. Dr. Ryan was right in his interpretation of the story of the innkeepers' daughter in that unconscious women can conceive and that orgasm has nothing to do with the matter. Angus McLaren makes essentially this case when he argues that, in the late eighteenth century, "the rights of women to sexual pleasure were not enhanced, but eroded as an unexpected consequence of the elaboration of more sophisticated models of reproduction."[23] Esther Fischer-Homberger suggests that a new understanding of an independent female contribution to reproduction accompanied the devaluation of procreation. Its status declined as it became, so to speak, exclusively women's work. Thus, one might argue, new discoveries in reproductive biology came just in the nick of time; science seemed nicely in tune with the demands of culture.[24]

But in fact no such discoveries took place. Scientific advances do not entail the demotion of female orgasm. True, by the 1840s it had become clear that, at least in dogs, ovulation could occur without coition and thus presumably without orgasm. And it was immediately postulated that the human female, like the canine bitch, was a "spontaneous ovulator," producing an egg during the periodic heat that in women was known as the menses. But the available evidence for this half truth was at best slight and highly ambiguous. Ovulation, as one of the pioneer twentieth-century investigators in reproductive biology put it, "is silent and occult: neither self-observation by women nor medical study through all the centuries prior to our own era taught mankind to recognize it."[25] Indeed, standard medical-advice books recommended that to avoid conception women should have intercourse during the middle of their menstrual cycles, during days twelve through sixteen, now known as the period of *maximum* fertility. Until the 1930s, even the outlines of our modern understanding of the hormonal control of ovulation were unknown.

In short, positive advances in science seem to have had little to do with the shift in interpreting the story of the innkeepers' daughter. The reevaluation of pleasure occurred more than a century before reproductive physiology could come to its support with any kind of deserved authority. Thus the question remains why, before the nineteenth century, commentators interpreted conception without orgasm as the exception, an oddity that proved nothing, while later such cases were regarded as perfectly normal and illustrative of a general truth about reproduction.

Unlike the demise of orgasm in reproductive physiology, the more general shift in the interpretation of the male and female bodies cannot have been due, even in principle, to scientific progress. In the first place, "oppositions and contrasts" between the female and the male, if one wishes to construe them as such, have been clear since the beginning of time: the one gives birth and the other does not. Set against such momentous truths, the discovery that the ovarian artery is not, as Galen would have it, the female version of the vas deferens is of relatively minor significance. The same can be said about the "discoveries" of more recent research on the biochemical, neurological, or other natural determinants or insignia of sexual difference. As Anne Fausto-Sterling has documented, a vast amount of negative data that shows no regular differences between the sexes is simply not reported.[26] Moreover, what evidence there does exist for biological difference with a gendered behavioral result is either highly suspect for a variety of methodological reasons, or ambiguous, or proof of Dorothy Sayers' notion that men and women are very close neighbors indeed if it is proof of anything at all.

To be sure, difference and sameness, more or less recondite, are everywhere; but which ones count and for what ends is determined outside the bounds of empirical investigation. The fact that at one time the dominant discourse construed the male and female bodies as hierarchically, vertically, ordered versions of one sex and at another time as horizontally ordered opposites, as incommensurable, must depend on something other than even a great constellation of real or supposed discoveries.

Moreover, nineteenth-century advances in developmental anatomy (germ-layer theory) pointed to the common origins of both sexes in a morphologically androgynous embryo and thus not to their intrinsic difference. Indeed, the Galenic isomorphisms of male and female organs were by the 1850s rearticulated at the embryological level as homologues: the penis and the clitoris, the labia and the scrotum, the ovary and the testes, scientists discovered, shared common origins in fetal life. There was thus scientific evidence in support of the old view should it have been culturally relevant. Or, conversely, no one was much interested in looking for evidence of two distinct sexes, at the anatomical and concrete physiological differences between men and women, until such differences became politically important. It was not, for example, until 1759 that anyone bothered to reproduce a detailed female skeleton in an anatomy book to illustrate its difference from the male. Up to this time there had been one basic structure for the human body, and that structure was male.[27] And when differences were discovered they were already, in the very form of their representation, deeply marked by the power politics of gender.

Instead of being the consequence of increased specific scientific knowledge, new ways of interpreting the body were the result of two broader, analytically though not historically distinct, developments: one epistemological, the other political. By the late seventeenth century, in certain specific contexts, the body was no longer regarded as a microcosm of some larger order in which each bit of nature is positioned within layer upon layer of signification. Science no longer generated the hierarchies of analogies, the resemblances that bring the whole world into every scientific endeavor but thereby create a body of knowledge that is, as Foucault argues, at once endless and poverty-stricken.[28] Sex as it has been seen since the Enlightenment—as the biological foundation of what it is to be male and female—was made possible by this epistemic shift.

But epistemology alone does not produce two opposite sexes; it does so only in certain political circumstances. Politics, broadly understood as the competition for power, generates new ways of constituting the subject and the social realities within which humans dwell. Serious talk about sexuality is thus inevitably about the social order that it both represents and legitimates. "Society," writes Maurice Godelier, "haunts the body's sexuality."[29]

Ancient accounts of reproductive biology, still persuasive in the early eighteenth century, linked the intimate, experiential qualities of sexual delight to the social and the cosmic order. More generally, biology and human sexual experience mirrored the metaphysical reality on which, it was thought, the social order rested. The new biology, with its search for fundamental differences between the sexes, of which the tortured questioning of the very existence of women's sexual pleasure was a part, emerged at precisely the time when the foundations of the old social order were shaken once and for all.

But social and political changes are not, in themselves, explanations for the reinterpretation of bodies. The rise of evangelical religion, Enlightenment political theory, the development of new sorts of public spaces in the eighteenth century, Lockean ideas of marriage as a contract, the cataclysmic possibilities for social change wrought by the French revolution, postrevolutionary conservatism, postrevolutionary feminism, the factory system with its restructuring of the sexual division of labor, the rise of a free market economy in services or commodities, the birth of classes, singly or in combination—none of these things *caused* the making of a new sexed body. Instead, the remaking of the body is itself intrinsic to each of these developments.

This book, then, is about the making not of gender, but of sex. I have no interest in denying the reality of sex or of sexual dimorphism as an evolutionary process. But I want to show on the basis of historical evidence that almost everything one wants to *say* about sex—however sex is understood—already has in it a claim about gender. Sex, in both the one-sex and the two-sex worlds, is situational; it is explicable only within the context of battles over gender and power.

To a great extent my book and feminist scholarship in general are inextricably caught in the tensions of this formulation: between language on the one hand and extralinguistic reality on the other; between nature and culture; between "biological sex" and the endless social and political markers of difference.[30] We remain poised between the body as that extraordinarily fragile, feeling, and transient mass of flesh with which we are all familiar—too familiar—and the body that is so hopelessly bound to its cultural meanings as to elude unmediated access.

The analytical distinction between sex and gender gives voice to these alternatives and has always been precarious. In addition to those who would eliminate gender by arguing that so-called cultural differences are really natural, there has been a powerful tendency among feminists to empty sex of its content by arguing, conversely, that natural differences are really cultural. Already by 1975, in Gayle Rubin's classic account of how a social sex/gender system "transforms biological sexuality into products of human activity," the presence of the body is so veiled as to be almost hidden.[31] Sherry Ortner and Harriet Whitehead further erode the body's priority over language with their self-conscious use of quotation marks around "givens" in the claim that "what gender is, what men and women are . . . do not simply reflect or elaborate upon biological 'givens' but are largely products of social and cultural processes."[32] "It is also dangerous to place the body at the center of a search for female identity," reads a French feminist manifesto.[33]

But if not the body, then what? Under the influence of Foucault, various versions of deconstruction, Lacanian psychoanalysis, and poststructuralism generally, it threatens to disappear entirely.[34] (The deconstruction of stable meaning in texts can be regarded as the general case of the deconstruction of sexual difference: "what can 'identity,' even 'sexual identity,' mean in a new theoretical and scientific space where the very notion of identity is challenged?" writes Julia Kristeva.[35]) These strategies have begun to have considerable impact among historians. Gender to Joan Scott, for example, is not a category that mediates between fixed biological difference on the one hand and historically contingent social relations on the other. Rather it includes both biology and society: "a constitutive element of social relationships based on *perceived differences between the sexes* . . . a primary way of *signifying* relationships of power."[36]

But feminists do not need French philosophy to repudiate the sex/gender distinction. For quite different reasons, Catharine MacKinnon argues explicitly that gender is the division of men and women caused "by the social requirements of heterosexuality, which institutionalizes male sexual dominance and female sexual submission"; sex—which comes to the same thing—is social relations "organized so that men may dominate and women must submit."[37] "Science," Ruth Bleier argues, mistakenly views "gender attributions as *natural* categories for which biological explanations are appropriate and even necessary."[38] Thus some of the so-called sex differences in biological and sociological research turn out to be gender differences after all, and the distinction between nature and culture collapses as the former folds into the latter.

Finally, from a different philosophical perspective, Foucault has even further rendered problematic the nature of human sexuality in relation to the body. Sexuality is not, he argues, an inherent quality of the flesh that various societies extol or repress—not, as Freud would seem to have it, a biological drive that civilization channels in one direction or another. It is instead a way of fashioning the self "in the experience of the flesh," which itself is "constituted from and around certain forms of behavior." These forms, in turn, exist in relation to historically specifiable systems of knowledge, rules of what is or is not natural, and to what Foucault calls "a mode or relation between the individual and himself which enables him to recognize himself as a sexual subject amidst others." (More generally, these systems of knowledge determine what can be thought within them.) Sexuality as a singular and all-important human attribute with a specific object—the *opposite* sex—is the product of the late eighteenth century. There is nothing natural about it. Rather, like the whole world for Nietzsche (the great philosophical influence on Foucault), sexuality is "a sort of artwork."[39]

Thus, from a variety of perspectives, the comfortable notion is shaken that man is man and woman is woman and that the historian's task is to find out what they did, what they thought, and what was thought about them. That "thing," sex, about which people had beliefs seems to crumble. But the flesh, like the repressed, will not long allow itself to remain in silence. The fact that we become human in culture, Jeffrey Weeks maintains, does not give us license to ignore the body: "It is obvious that sex is something more than what society designates, or what naming makes it."[40] The body reappears even in the writings of those who would turn attention to language, power, and culture. (Foucault, for example, longs for a nonconstructed utopian space in the flesh from which to undermine "bio-power": "the rallying point for the counterattack against the deployment of sexuality ought not to be sex-desire, but bodies and pleasures."[41]

In my own life, too, the fraught chasm between representation and reality, seeing-as and seeing, remains. I spent 1980–81 in medical school and studied what was *really* there as systematically as time and circumstances permitted. Body as cultural construct met body on the dissecting table; more or less schematic anatomical illustrations—the most accurate modern science had to offer—rather hopelessly confronted the actual tangles of the human neck. For all of my awareness of how deeply our understanding of what we saw was historically contingent—the product of institutional, political, and epistemological contingencies—the flesh in its simplicity seemed always to shine through.

I remember once spending the better part of a day watching doctors and nurses trying vainly to stem the flow of blood from the ruptured esophageal varices of a middle-aged dentist, who that morning had walked into the emergency room, and to replace it pint by pint into his veins as they pumped it out of his stomach. In the late afternoon I left to hear *Don Giovanni*—I was after all only an observer and was doing the patient no good. The next morning he was dead, a fact that seemed of an entirely different order from Mozart's play on the body or the history of representation that constitutes this book. ("I know when one is dead, and when one lives. / She's dead as earth," howled Lear.)

But my acquaintance with the medical aspect of bodies goes back farther than 1981. I grew up the son of a pathologist. Most Sunday mornings as a boy I went with my father to his laboratory to watch him prepare surgical specimens for microscopic examination; he sliced up kidneys, lungs, and other organs preparatory to their being fixed in wax, stained, and mounted on slides to be "read." As he went about this delicate carving and subsequent reading, he spoke into a dictating machine about what he saw. Bodies, or in any case body parts, seemed unimpeachably real. I remember reading his autopsy protocols, stacked on the kelim-covered divan in his study, resonant with the formulas of what to me seemed like medical epic: "The body is that of a sixty-five-year-old Caucasian male in emaciated condition. It was opened with the usual Y-shaped incision." "The body is that of a well-nourished fifty-seven-year-old female. It was opened with the usual Y-shaped incision."

Three months before my father died of cancer, and only weeks before brain metastasis made it impossible for him to think, he helped me in interpreting the German gynecological literature cited in Chapters 5 and 6, some of which was by his own medical-school teachers. More to the point, he tutored me on what one could actually see, for example, in the cross section of an ovary with the naked eye or through the microscope. "Is it plausible," I would ask, "that, as nineteenth-century doctors claimed, one could count the number of ovulatory scars [the corpus albigans] and correlate them with the number of menstrual cycles?" My father was the expert on what was *really* there.

But he figures also in its deconstruction. As a recent medical-school graduate, he could not continue his studies in Nazi Germany. In 1935 he took a train to Amsterdam to ask his uncle, Ernst Laqueur, who was professor of pharmacology there, what he ought to do next.[42] Some difficulties with a German official made my father decide not to go back to Hamburg at all. Ernst Laqueur presumably secured for him the position at Leiden that he was to hold for the next year or so. I knew little of what he did there, and nothing of what he published until I went through his papers after he died. (This was well after I had completed much of the research for this book.) In his desk I found a bundle of his offprints; the earliest one, except for his "Inaugural Dissertation," is entitled "Weitere Untersuchungen uber den Uterus masculinus unter dem Einfluss verschiedener Hormone" (Further Studies of the Influence of Various Hormones on the Masculine Uterus).[43]

I had already written about how Freud the doctor severed familiar connections between the manifest evidence of bodies and the opposition between the sexes. I had read Sarah Kofman on the power of anatomy to "confuse those who think of the sexes as opposing species."[44] But my father's contribution to the confusion was a complete revelation, genuinely uncanny. It was hidden and yet so much of the home—*heimlich* but also *unheimlich*—the veiled and secret made visible, an eerie, ghostly reminder that somehow this book and I go back a long way.[45]

There are less personal reasons as well for wanting to maintain in my writing a distinction between the body and the body as discursively constituted, between seeing and seeing-as. In some measure these reasons are ethical or political and grow out of the different obligations that arise for the observer from seeing (or touching) and from representing. It is also disingenuous to write a history of sexual difference, or difference generally, without acknowledging the shameful correspondence between particular forms of suffering and particular forms of the body, however the body is understood. The fact that pain and injustice are gendered and correspond to corporeal signs of sex is precisely what gives importance to an account of the making of sex.

Moreover, there has clearly been progress in understanding the human body in general and reproductive anatomy and physiology in particular. Modern science and modern women are much better able to predict the cyclical likelihood of pregnancy than were their ancestors; menstruation turns out to be a different physiological process from hemorrhoidal bleeding, contrary to the prevailing wisdom well into the eighteenth century, and the testes *are* histologically different from the ovaries. Any history of a science, however much it might emphasize the role of social, political, ideological, or aesthetic factors, must recognize these undeniable successes and the commitments that made them possible.[46]

Far from denying any of this, I want to insist upon it. My particular Archimedean point, however, is not in the real transcultural body but rather in the *space* between it and its representations. I hold up the history of progress in reproductive physiology—the discovery of distinct germ products, for example—to demonstrate that these did not cause a particular understanding of sexual difference, the shift to the two-sex model. But I also suggest that theories of sexual difference influenced the course of scientific progress and the interpretation of particular experimental results. Anatomists might have seen bodies differently—they might, for example, have regarded the vagina as other than a penis—but they did not do so for essentially cultural reasons. Similarly, empirical data were ignored—evidence for conception without orgasm, for example—because they did not fit into either a scientific or a metaphysical paradigm.

Sex, like being human, is contextual. Attempts to isolate it from its discursive, socially determined milieu are as doomed to failure as the *philosophe*'s search for a truly wild child or the modern anthropologist's efforts to filter out the cultural so as to leave a residue of essential humanity. And I would go further and add that the private, enclosed, stable body that seems to lie at the basis of modern notions of sexual difference is also the product of particular, historical, cultural moments. It too, like opposite sexes, comes into and out of focus.

My general strategy in this book is to implicate biology explicitly in the interpretive dilemmas of literature and of cultural studies generally.

"Like the other sciences," writes François Jacob, winner of the 1965 Nobel Prize for medicine,

> biology today has lost its illusions. It is no longer seeking for truth. It is building its own truths. Reality is seen as an ever-unstable equilibrium. In the study of living beings, history displays a pendulum movement, swinging to and fro between the continuous and the discontinuous, between structure and function, between the identity of phenomena and the diversity of being.[47]

The instability of difference and sameness lies at the very heart of the biological enterprise, in its dependence on prior and shifting epistemological, and one could add political, grounds. (Jacob is of course not the first to make this point. Auguste Comte, the guiding spirit of nineteenth-century positivism, confessed that "there seems no sufficient reason why the use of scientific *fictions,* so common in the hands of geometers, should not be introduced into biology."[48] And Emile Durkheim, one of the giants of sociology, argued that "we buoy ourselves up with a vain hope if we believe that the best means of preparing for the coming of a new science is first patiently to accumulate all the data it will use. For we cannot know what it will require unless we have already formed some conception of it."[49] Science does not simply investigate, but itself constitutes, the difference my book explores: that of woman from man. (But not, for reasons discussed below, man from woman.)

Literature, in a similar way, constitutes the problem of sexuality and is not just its imperfect mirror. As Barbara Johnson argues, "it is literature that inhabits the very heart of what makes sexuality problematic for us speaking animals. Literature is not only a thwarted investigator but also an incorrigible perpetrator of the problem of sexuality."[50] Sexual difference thus seems to be already present in how we constitute meaning; it is already part of the logic that drives writing. Through "literature," representation generally, it is given content. Not only do attitudes toward sexual difference "generate and structure literary texts"; texts generate sexual difference.[51]

Johnson is careful to restrict the problem of sexuality to "us speaking animals," and thus to rest content that, among dumb animals and even among humans outside the symbolic realm, male is manifestly the opposite sex from female. But clarity among the beasts bespeaks only the very

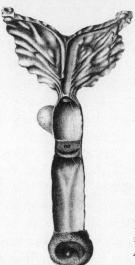

Fig. 1. Genitalia of a female elephant drawn from a fresh specimen by a nineteenth-century naturalist. From *Journal of the Academy of Natural Science,* Philadelphia, 8.4 (1881).

limited purposes for which we generally make such sexual distinctions. It matters little if the genitals of the female elephant (fig. 1) are rendered to look like a penis because the sex of elephants generally matters little to us; it is remarkable and shocking if the same trick is played on our species, as was routine in Renaissance illustrations (figs. 15–17). Moreover, as soon as animals enter some discourse outside breeding, zoo keeping, or similarly circumscribed contexts, the same sort of ambiguities arise as when we speak about humans. Then the supposedly self-evident signs of anatomy or physiology turn out to be anything but self-evident. Questions of ultimate meaning clearly go well beyond such facts. Darwin in 1861 lamented: "We do not even know in the least the final cause of sexuality; why new beings should be produced by the union of the two sexual elements, instead of by a process of parthenogenesis . . . The whole subject is as yet hidden in darkness."[52] And still today the question of why egg and sperm should be borne by different, rather than the same, hermaphroditic, creature remains open.[53]

Darkness deepens when animals enter into the orbit of culture; their sexual transparency disappears. The hare, which figures prominently in so much myth and folklore, was long thought to be capable of routine sex change from year to year and thus inherently androgynous. Or, as the more learned would have it, the male hare bears young on occasion. The hyena, another animal with prolific cultural meanings, was long thought to be hermaphroditic. The cassowary, a large, flightless, ostrich-like, and, to the anthropologist, epicene bird, becomes to the male Sambian tribesman a temperamental, wild, masculinized female who gives birth through the anus and whose feces have procreative powers; the bird becomes powerfully bisexual. Why, asks the ethnographer Gilbert Herdt, do people as astute as the Sambia "believe" in anal birth? Because anything one says, outside of very specific contexts, about the biology of sex, even among the brute beasts, is already informed by a theory of difference or sameness.[54]

Indeed, if structuralism has taught us anything it is that humans impose their sense of opposition onto a world of continuous shades of difference and similarity. No oppositional traits readily detected by an outsider explain the fact that in nearly all of North America, to use Lévi-Strauss's example, sagebrush, *Artemesia,* plays "a major part in the most diverse rituals, either by itself or associated with and at the same time, as the opposite of other plants: *Solidaga, Chrysothamnus, Gutierrezia.*" It stands for the feminine in Navaho ritual whereas *Chrysothamnus* stands for the masculine. No principle of opposition could be subtler than the tiny differences in leaf serrations that come to carry such enormous symbolic weight.[55]

It should be clear by now that I offer no answer to the question of how bodies determine what we mean by sexual difference or sameness. My claims are of two sorts. Most are negative: I make every effort to show that no historically given set of facts about "sex" entailed how sexual difference was in fact understood and represented at the time, and I use this evidence to make the more general claim that no set of facts ever entails any particular account of difference. Some claims are positive: I point to ways in which the biology of sexual difference is embedded in other cultural programs.

Chapter 2 is about the oxymoronic one-sex body. Here the boundaries between male and female are primarily political; rhetorical rather than biological claims regarding sexual difference and sexual desire are primary. It is about a body whose fluids—blood, semen, milk, and the various excrements—are fungible in that they turn into one another and whose processes—digestion and generation, menstruation and other bleeding—are not so easily distinguished or so easily assignable to one sex or another as they became after the eighteenth century. This "one flesh," the construction of a single-sexed body with its different versions attributed to at least two genders, was framed in antiquity to valorize the extraordinary cultural assertion of patriarchy, of the father, in the face of the more sensorily evident claim of the mother. The question for the classical model is not what it explicitly claims—why woman?—but the more troublesome question—why man?

Chapter 3 is the first of two chapters that examine explicitly the relationship between a model of sexual difference and scientific learning. It shows how the one-flesh model was able to incorporate new anatomical knowledge and new naturalistic forms of representation. Chapter 4 concentrates on the cultural interests that various writers had in what seems to us a manifestly counterintuitive model of sexual difference. It exposes the immense pressures on the one-sex model from the existence of two genders, from the new political claims of women, and from the claims of heterosexuality generally. I suggest through readings of legal, juridical, and literary texts that it is sustained by powerful notions of how hierarchy worked and how the body expresses its cultural meanings. At stake for the men involved in this struggle was nothing less than the suppression of the basis for a genuine, other, sex.

Chapter 5 gives an account of the breakdown of the one-sex model and the establishment of two sexes. Like Chapter 3 it maintains that these constructions were not the consequence of scientific change but rather of an epistemological and a social-political revolution. Again, the negative argument—that the scientific is not natural and given—is more forcefully put than the affirmative, in part because I am reluctant to frame my story in terms of a specific set of causes for the increasing prominence of the two-sex model. My strategy instead is to suggest, example by example, the ways in which particular struggles and rhetorical situations made men and women talk as if there were now two sexes. These contexts were of course the results of new social and political developments, but I do not draw out the connections in great detail. More detailed studies are needed to create a locally nuanced account of "Politics, Culture, and Class in the Eighteenth- and Nineteenth-Century Body."[56]

Chapter 6 functions much like Chapter 4 in that it engages the science of sex—two this time—with the demands of culture. I show specifically how cornerstones of corporeally based sexes were themselves deeply implicated in the politics of gender. But in this chapter I also present evidence for the continued life of the one-sex model. It lived on even in the midst of the most impassioned defense of two sexes, of ineradicable "organic difference . . . proved by all sound biology, by the biology of man and of the entire animal species . . . proved by the history of civilization, and the entire course of human evolution." The specter of one sex remains: the "womanliness of woman" struggles against "the anarchic assertors of the manliness of woman."[57] In some of the rhetoric of evolutionary biology, in the Marquis de Sade, in much of Freud, in slasher films, indeed in any discussion of gender, the modern invention of two distinct, immutable, and incommensurable sexes turns out to be less dominant than promised.[58] (Here I differ from Foucault, who would see one *episteme* decisively, once and for all, replacing another.) I illustrate the openness of nineteenth-century science to either a two- or a one-sex model with a discussion first of how denunciations of prostitution and masturbation reproduced an earlier discourse of the unstable individual body, open and responsive to social evil, and then of Freud's theory of clitoral sexuality in which efforts to find evidence of incommensurable sexes founders on his fundamental insight that the body does not of itself produce two sexes.

I have not written this book as an explicit attack on the current claims of sociobiology. But I hope it is taken up by those engaged in that debate. A historian can contribute little to the already existing critical analysis of particular experiments purporting to demonstrate the biological basis of gender distinctions or to lay bare the hormones and other chemicals that are meant to serve as a sort of ontological granite for observable sexual differences.[59] But I can offer material for how powerful prior notions of difference or sameness determine what one sees and reports about the body. The fact that the giants of Renaissance anatomy persisted in seeing the vagina as an internal version of the penis suggests that almost any sign of difference is dependent on an underlying theory of, or context for, deciding what counts and what does not count as evidence.

More important, though, I hope this book will persuade the reader that there is no "correct" representation of women in relation to men and that the whole science of difference is thus misconceived. It is true that there is and was considerable and often overtly misogynist bias in much biological research on women; clearly science has historically worked to "rationalize and legitimize" distinctions not only of sex but also of race and class, to the disadvantage of the powerless. But it does not follow that a more objective, richer, progressive, or even more feminist science would produce a truer picture of sexual difference in any culturally meaningful sense.[60] (This is why I do not attempt to offer a history of more or less correct, or more or less misogynistic, representations.) In other words, the claim that woman is what she is because of her uterus is no more, or less, true than the subsequent claim that she is what she is because of her ovaries. Further evidence will neither refute nor affirm these patently absurd pronouncements because at stake are not biological questions about the effects of organs or hormones but cultural, political questions regarding the nature of woman.

I return again and again in this book to a problematic, unstable female body that is either a version of or wholly different from a generally unproblematic, stable male body. As feminist scholars have made abundantly clear, it is *always* woman's sexuality that is being constituted; woman is the empty category. Woman alone seems to have "gender" since the category itself is defined as that aspect of social relations based on difference between sexes in which the standard has always been man. "How can one be an enemy of woman, whatever she may be?" as the Renaissance physician Paracelsus put it; this could never be said of man because, quite simply, "one" is male. It is probably not possible to write a history of man's body and its pleasures because the historical record was created in a cultural tradition where no such history was necessary.

But the modern reader must always be aware that recounting the history of interpreting woman's body is not to grant the male body the authority it implicitly claims. Quite the contrary. The record on which I have relied bears witness to the fundamental incoherence of stable, fixed categories of sexual dimorphism, of male and/or female. The notion, so powerful after the eighteenth century, that there had to be something outside, inside, and throughout the body which defines male as opposed to female and which provides the foundation for an attraction of opposites is entirely absent from classical or Renaissance medicine. In terms of the millennial traditions of western medicine, genitals came to matter as the marks of sexual opposition only last week. Indeed, much of the evidence suggests that the relationship between an organ as sign and the body that supposedly gives it currency is arbitrary, as indeed is the relationship between signs. The male body may always be the standard in the game of signification, but it is one whose status is undermined by its unrepentant historical inconstancy.

Although some tensions inform this book, others do not. I have given relatively little attention to conflicting ideas about the nature of woman or of human sexuality. I have not even scratched the surface of a contextual history of reproductive anatomy or physiology; even for scientific problems that I explore in some detail, the institutional and professional matrix in which they are embedded is only hurriedly sketched. There is simply too much to do in the history of biology, and too much has already been done on the condition-of-woman question or the history of ideas about sex, for any one person to master.

I want to lay claim to a different historical domain, to the broad discursive fields that underlie competing ideologies, that define the terms of conflict, and that give meaning to various debates. I am not committed to demonstrating, for example, that there is a single, dominant "idea of woman" in the Renaissance and that all others are less important. I have no interest in proving conclusively that Galen is more important than Aristotle at any one time or that a given theory of menstruation was hegemonic between 1840 and 1920. Nor will I be concerned with the gains and losses in the status of women through the ages. These are issues I must ask my readers to decide for themselves, whether the impressions they derive from these pages fit what they themselves know of the vast spans of time that I cover. My goal is to show how a biology of hierarchy in which there is only one sex, a biology of incommensurability between two sexes, and the claim that there is no publicly relevant sexual difference at all, or no sex, have constrained the interpretation of bodies and the strategies of sexual politics for some two thousand years.

Finally, I confess that I am saddened by the most obvious and persistent omission in this book: a sustained account of experience in the body. Some might argue that this is as it should be, and that a man has nothing of great interest or authenticity to say about the sexual female body as it feels and loves. But more generally I have found it impossible in all but isolated forays into literature, painting, or the occasional work of theology to imagine how such different visions of the body worked in specific contexts to shape passion, friendship, attraction, love. A colleague pointed out to me that he heard Mozart's *Così fan tutte* with new ears after reading my chapters about the Renaissance. I have felt a new poignancy in the tragicomedy of eighteenth-century disguise—the last act of *Le Nozze di Figaro,* for example—with its questioning of what it is in a person that one loves. Bodies do and do not seem to matter. I watch Shakespeare's comedies of sexual inversion with new queries, and I try to think my way back into a distant world where the attraction of deep friendship was reserved for one's like.

Further than that I have not been able to go. I regard what I have written as somehow liberating, as breaking old shackles of necessity, as opening up worlds of vision, politics, and eros. I only hope that the reader will feel the same.

TWO

Destiny Is Anatomy

Turn outward the woman's, turn inward, so to
speak, and fold double the man's [genital organs],
and you will find the same in both in every respect.

GALEN OF PERGAMUM (c.130–200)

This chapter is about the corporeal theatrics of a world where at least two
genders correspond to but one sex, where the boundaries between male
and female are of degree and not of kind, and where the reproductive
organs are but one sign among many of the body's place in a cosmic and
cultural order that transcends biology. My purpose is to give an account,
based largely on medical and philosophical literature, of how the one-sex
body was imagined; to stake out a claim that the one-sex/one-flesh model
dominated thinking about sexual difference from classical antiquity to the
end of the seventeenth century; and to suggest why the body should have
remained fixed in a field of images hoary already in Galen's time, while
the gendered self lived a nuanced history through all the immense social,
cultural, and religious changes that separate the world of Hippocrates
from the world of Newton.

Organs and the mole's eyes

Nothing could be more obvious, implied the most influential anatomist
in the western tradition, than to imagine women as men. For the dullard
who could not grasp the point immediately, Galen offers a step-by-step
thought experiment:

Think first, please, of the man's [external genitalia] turned in and extending
inward between the rectum and the bladder. If this should happen, the scrotum would necessarily take the place of the uterus with the testes lying
outside, next to it on either side.

The penis becomes the cervix and vagina, the prepuce becomes the female pudenda, and so forth on through various ducts and blood vessels. A sort of topographical parity would also guarantee the converse, that a man could be squeezed out of a woman:

> Think too, please, of . . . the uterus turned outward and projecting. Would not the testes [ovaries] then necessarily be inside it? Would it not contain them like a scrotum? Would not the neck [the cervix and vagina], hitherto concealed inside the perineum but now pendant, be made into the male member?

In fact, Galen argued, "you could not find a single male part left over that had not simply changed its position." Instead of being divided by their reproductive anatomies, the sexes are linked by a common one. Women, in other words, are inverted, and hence less perfect, men. They have exactly the same organs but in exactly the wrong places. (The wrongness of women, of course, does not follow logically from the "fact" that their organs are the same as men's, differing only in placement. The arrow of perfection *could* go either or both ways. "The silliest notion has just crossed my mind," says Mlle. de l'Espinasse in Diderot's *D'Alembert's Dream:* "Perhaps men are nothing but a freakish variety of women, or women only a freakish variety of men." Dr. Bordeu responds approvingly that the notion would have occurred to her earlier if she had known—he proceeds to give a short lecture on the subject—that "women possess all the anatomical parts that a man has.")[1]

The topographical relationships about which Galen writes so persuasively and with such apparent anatomical precision were not themselves to be understood as the basis of sexual hierarchy, but rather as a way of imagining or expressing it. Biology only records a higher truth. Thus although Galen, the professional anatomist, clearly cared about corporeal structures and their relation to the body's various functions, his interest in the plausibility of particular identifications or in maintaining the manifestly impossible implosion of man into woman and back out again, was largely a matter of rhetorical exigency.

On some occasions he was perfectly willing to argue *for* the genital oppositions he elsewhere denied: "since everything in the male is the opposite [of what it is in the female] the male member has been elongated to be most suitable for coitus and the excretion of semen" (*UP* 2.632). At other times Galen and the medical tradition that followed him were prepared to ignore entirely not only the specifically female but also the specifically reproductive quality of the female reproductive organs, not to speak of their relationship to male organs. His systematic major treatment of the uterus, for example, treated it as the archetype for a group of organs "which are especially hollow and large" and thus the locus of a generic body's "retentive faculties." The uterus was singled out not because of what we moderns might take to be its unique, and uniquely female, capacity to produce a child but because it formed the embryo in leisurely fashion, more so than a comparable organ like the stomach digested food, and was therefore "capable of demonstrating the retentive faculty most plainly."[2]

Subsequent ways of talking about the uterus reproduced these ambiguities. Isidore of Seville, the famous encyclopedist of the seventh century, for example, argued on the one hand that only women have a womb (*uterus* or *uterum*) in which they conceive and, on the other, that various authorities and "not only poets" considered the uterus to be the belly, *venter,* common to both sexes.[3] (This helps to explain why *vulva* in medieval usage usually meant vagina, from *valva,* "gateway to the belly."[4]) Isidore, moreover, assimilates this unsexed belly to other retentive organs with respect precisely to that function in which we would think it unique: during gestation, he said, the semen is formed into a body "by heat like that of the viscera."[5] A great linguistic cloud thus obscured specific genital or reproductive anatomy and left only the outlines of spaces common to both men and women.[6]

None of these topographical or lexical ambiguities would matter, however, if instead of understanding difference and sameness as matters of anatomy, the ancients regarded organs and their placement as epiphenomena of a greater world order. Then what we would regard as specifically male and female parts would not always need to have their own names, nor would the inversions Galen imagined actually have to work. Anatomy—modern sex—could in these circumstances be construed as metaphor, another name for the "reality" of woman's lesser perfection. As in Galen's elaborate comparison between the eyes of the mole and the genital organs of women, anatomy serves more as illustration of a well-known point than as evidence for its truth. It makes vivid and more palpable a hierarchy of heat and perfection that is in itself not available to the senses. (The ancients would not have claimed that one could actually feel differences in the heat of males and females.[7])

Galen's simile goes as follows. The eyes of the mole have the same structures as the eyes of other animals except that they do not allow the mole to see. They do not open, "nor do they project but are left there imperfect." So too the female genitalia "do not open" and remain an imperfect version of what they would be were they thrust out. The mole's eyes thus "remain like the eyes of other animals when these are still in the uterus" and so, to follow this logic to its conclusion, the womb, vagina, ovaries, and external pudenda remain forever as if they were still inside the womb. They cascade vertiginously back inside themselves, the vagina an eternally, precariously, unborn penis, the womb a stunted scrotum, and so forth.[8]

The reason for this curious state of affairs is the purported telos of perfection. "Now just as mankind is the most perfect of all animals, so within mankind the man is more perfect than the woman, and the reason for his perfection is his excess of heat, for heat is Nature's primary instrument" (*UP* 2.630). The mole is a more perfect animal than animals with no eyes at all, and women are more perfect than other creatures, but the unexpressed organs of both are signs of the absence of heat and consequently of perfection. The interiority of the female reproductive system could then be interpreted as the material correlative of a higher truth without its mattering a great deal whether any particular spatial transformation could be performed.

Aristotle, paradoxically for someone so deeply committed to the existence of two radically different and distinct sexes, offered the western tradition a still more austere version of the one-sex model than did Galen. As a philosopher he insisted upon two sexes, male and female. But he also insisted that the distinguishing characteristic of maleness was immaterial and, as a naturalist, chipped away at organic distinctions between the sexes so that what emerges is an account in which one flesh could be ranked, ordered, and distinguished as particular circumstances required. What we would take to be ideologically charged social constructions of gender—that males are active and females passive, males contribute the form and females the matter to generation—were for Aristotle indubitable facts, "natural" truths. What we would take to be the basic facts of sexual difference, on the other hand—that males have a penis and females a vagina, males have testicles and females ovaries, females have a womb and males do not, males produce one kind of germinal product, females another, that women menstruate and men do not—were for Aristotle contingent and philosophically not very interesting observations about particular species under certain conditions.

I do not mean to suggest by this that Aristotle was unable to tell man from woman on the basis of their bodies or that he thought it an accident that men should fulfill one set of roles and women another. Even if he did not write the *Economics* he would certainly have subscribed to the view that "the nature both of man and woman has been preordained by the will of heaven to live a common life. For they are distinguished in that the powers they possess are not applicable to purposes in all cases identical, but in some respects their functions are opposed to one another." One sex is strong and the other weak so that one may be cautious and the other brave in warding off attacks, one may go out and acquire possessions and the other stay home to preserve them, and so on.[9] In other words, both the division of labor and the specific assignment of roles are natural.

But these views do not constitute a modern account of two sexes. In the first place, there is no effort to ground social roles in nature; social categories themselves are natural and on the same explanatory level as what we would take to be physical or biological facts. Nature is not therefore to culture what sex is to gender, as in modern discussions; the biological is not, even in principle, the foundation of particular social arrangements. (Aristotle, unlike nineteenth-century commentators, did not need facts about menstruation or metabolism to locate women in the world order.) But more important, though Aristotle certainly regarded male and female bodies as specifically adapted to their particular roles, he did not regard these adaptations as the signs of sexual opposition. The qualities of each sex entailed the comparative advantage of one or the other in minding the home or fighting, just as for Galen the lesser heat of women kept the uterus inside and therefore provided a place of moderate temperature for gestation. But these adaptations were not the basis for ontological differentiation. In the flesh, therefore, the sexes were more and less perfect versions of each other. Only insofar as sex was a cipher for the nature of causality were the sexes clear, distinct, and different in kind.

Sex, for Aristotle, existed for the purpose of generation, which he regarded as the paradigmatic case of becoming, of change "in the first category of being."[10] The male represented efficient cause, the female represented material cause.

the female always provides the material, the male that which fashions it, for this is the power we say they each possess, and *this is what it is for them to be male and female* . . . While the body is from the female, it is the soul that is from the male. (*GA* 2.4.738b20–23)

the male and female principles may be put down first and foremost as the origins of generation, the former as containing the efficient cause of generation, the latter the material of it. (*GA* 2.716a5–7)

This difference in the nature of cause constitutes fully what Aristotle means by sexual opposition: "by a male animal we mean that which generates in another; by a female, that which generates in itself"; or, what comes to the same thing since for Aristotle reproductive biology was essentially a model of filiation, "female is opposed to male, and mother to father."[11]

These were momentous distinctions, as powerful and plain as that between life and death. To Aristotle being male *meant* the capacity to supply the sensitive soul without which "it is impossible for face, hand, flesh, or any other part to exist." Without the sensitive soul the body was no better than a corpse or part of a corpse (*GA* 2.5.741a8–16). The dead is made quick by the spark, by the incorporeal *sperma* (seed), of the genitor. One sex was able to concoct food to its highest, life-engendering stage, into true sperma; the other was not.

Moreover, when Aristotle discusses the capacity of the respective sexes to carry out the roles that distinguish them, he seems to want to consider bodies, and genitals in particular, as themselves opposites, indeed as making possible the efficient/material chasm itself. Males have the capacity, and females do not, to reduce "the residual secretion to a pure form," the argument runs, and "every capacity has a certain corresponding organ." It follows that "the one has the uterus, the other the male organs." (These distinctions are actually more striking in translation than in the Greek. Aristotle uses *perineos* to refer to the penis and scrotum here. He uses the same word elsewhere to refer to the area "inside the thigh and buttocks" in women. More generally he uses *aidoion* to refer to the penis, but in the plural, *aidoia,* it is the standard word for the "shameful parts," the Greek equivalent for the Latin *pudenda,* which refers to the genitals of both sexes.[12])

Nevertheless, despite these linguistic ambiguities, Aristotle does seem committed to the genital opposition of two sexes. An animal is not "male or female in virtue of the whole of itself," he insists, "but only in virtue of a certain faculty and a certain part," that is, the uterus in the female, the penis and testes in the male. The womb was the part peculiar to the female, just as the penis was distinctive of the male.[13] No slippery inversions here as in Galen. No elisions of difference or hints of one sex. "The privy part of the female is in character opposite to that of the male. In other words, the part under the pubes is hollow, and not like the male organ, protruding" (HA 1.14.493b3–4). Aristotle even adduced what he took to be experimental evidence for the fact that anatomy was the foundation of the opposing male and female "principles" of activity and passivity. A castrated male, he pointed out, assumed pretty well the form of a female or "not far short of it . . . as would be the case if a first principle is changed" (GA 1.2.716b5–12). The excision of the "ovaries" in a sow caused them to get fat and quenched their sexual appetite, while a similar operation in camels made them more aggressive and fit for war service.[14]

None of this is very surprising, since the physical appearance of the genital organs was and remains the usually reliable indicator of reproductive capacity and hence of the gender to which an infant is to be assigned.[15] But what is surprising is the alacrity with which Aristotle the naturalist blurs the distinctions of "real" bodies in order to arrive at a notion of fatherhood—the defining capacity of males—that transcends the divisions of flesh. Like Galen's, and unlike that of the dominant post-Enlightenment tradition, Aristotle's rhetoric then becomes that of one sex.

First, Aristotle's passion for the infinite variety of natural history constantly undermines the form-follows-function precision of the texts I have cited. A large penis, which one might think would render a man more manly, capable of generating in another, in fact makes him less so: "such men are less fertile than when it [the penis] is smaller because the semen, if cold, is not generative."[16] (Aristotle's biology is here playing on broader cultural themes. A large penis was thought comic in ancient Greek art and drama, appropriate to satyrs, while the preferred size was small and delicate: "little prick" (posthion) was among Aristophanes' terms of endearment. Young athletes in Athens tied down the glans with a leather string, apparently for cosmetic reasons, to make the male genitals look small and as much like the female pudenda as possible.[17]) Detail after detail further undermines the penis/male connection in Aristotle's texts: human males and stallions do indeed have proportionately large penises outside their bodies, but the male elephant's is disproportionately small—he also has no visible testes—while the dolphin has no external penis at all. (The situation is doubly confused with elephants because supposedly the female "organ opens out to a considerable extent" during intercourse (*HA* 2.1.500a33–35 and 2.1.500b6–13). Among insects, Aristotle claims, the female actually pushes her sexual organ from underneath *into* the male (*HA* 5.8.542a2ff). Indeed, the male's having a penis at all seems to depend on nothing more than the placement or indeed existence of the legs: snakes, which have no legs, and birds, whose legs are in the middle of their abdomens where the genitals ought to be, simply lack a penis entirely (*HA* 2.1.500b20–25 and *GA* 1.5.717b14–19).

As for the testes being a "first principle" in the differentiation of the sexes, little is left rhetorically of this claim when faced with specific observations and metaphors (*GA* 1.2.716b4). Aristotle demotes them in one text to the lowly task of bending certain parts of the body's piping (HA 3.1.510a13–b5). Like the weights women hang from the warp on their looms—a less than celebratory simile, which suffers from a curious mixing of genders—the testicles keep the spermatic ducts properly inclined (*GA* 1.4.717a8–b10). (Thread that is not properly held down results in a tangle; tangled seminal ducts that go back up into the body convey impotent generative material.)

These "facts" led Aristotle still further away from specific connections between opposing genitals and sex and ever deeper into the thicket of connections that constitute the one-sex model. He, like Galen five centuries later, aligned the reproductive organs with the alimentary system, common to all flesh. Animals with straight intestines are more violent in their desire for food than animals whose intestines are convoluted, Aristotle observed, and likewise those with straight ducts, creatures without testes, are "quicker in accomplishing copulation" than creatures with crooked ducts. Conversely, creatures who "have not straight intestines" are more temperate in their longing for food, just as twisted ducts prevent "desire being too violent and hasty" in animals so blessed. Testes thus end up serving the lowly but useful function of making "the movement of the spermatic secretion steadier," thus prolonging intercourse and concoction in the interest of hotter, finer sperma.[18] Aristotle makes much less of the female plumbing, but his concern to identify the ovaries as the seat of woman's specific reproductive capacity was never very serious and the one passage where he makes the case crumbles under close scrutiny.[19] Natural history, in short, works to diminish the pristine purity of testes and ovaries, penis and vagina, as signifiers of sexual opposition—of efficient versus material cause—and situates them firmly in a larger economy of the one flesh.

Moreover, when Aristotle directly confronted the question of the anatomical differences between the sexes, he unleashed a vortex of metaphor every bit as dizzying and disorienting, every bit as committed to one sex, as Galen's trope of the mole's eyes. All of the male organs, he said, are similar in the female except that she has a womb, which presumably the male does not. But Aristotle promptly assimilates the womb to the male scrotum after all: "always double just as the testes are always two in the male."[20]

This move, however, was only part of a more general conflation of male and female parts, specifically of a tendency to regard the cervix and/or vagina as an internal penis:

> The path along which the semen passes in women is of the following nature: they [women] possess a tube (*kaulos*)—like the penis of the male, but inside the body—and they breathe through this by a small duct which is placed above the place through which women urinate. This is why, when they are eager to make love, this place is not in the same state as it was before they were excited.   (*HA* 10.5.637a23–25)

The very lack of precision in this description, and especially the use of so general a term as *kaulos* for a structure that in the two-sex model would become the mark of female emptiness or lack, suggests that Aristotle's primary commitment was not to anatomy itself, and certainly not to anatomy as the foundation for opposite sexes, as much as it was to greater truths that could be impressionistically illustrated by certain features of the body.

A brief excursis on *kaulos* will help to make this case. The word refers to a hollowish tubular structure generally: the neck of the bladder or the duct of the penis or, in Homeric usage, a spear shaft or the quill of a feather (to take four charged and richly intertwined examples). In the passage I just quoted it clearly designates some part of the female anatomy though which, significantly, is unclear: the cervix [neck] of the uterus, the endo-cervical canal, the vagina, some combination of these or even the clitoris which like the penis would have been construed as hollow. But whatever *kaulos* means in this text, the part in question is spoken of elsewhere as if it functioned in women like an interior penis, a tube composed, as are both penis and vagina, of "much flesh and gristle" (*HA* 3.1.510b13).

By the time of Soranus, the second-century physician who would become the major source of the gynecological high tradition for the next fifteen centuries, the assimilation of vagina to penis through language had gone much further. "The inner part of the vagina (*tou gynaikeiou aidoiou,* the feminine private part)," Soranus said, "grows around the neck of the uterus (*kaulos,* which I take here to mean cervix) like the prepuce in males around the glans."[21] In other words, the vagina and external structures are imagined as one giant foreskin of the female interior penis whose glans is the domelike apex of the "neck of the womb." By the second century *kaulos* had also become the standard word for penis. The "protruding part" of the *aidoion* (private part) "through which flows liquid from the bladder" is called the *kaulos,* says Julius Pollux (134–192) authoritatively in his compilation of medical nomenclature.[22] Aristotle—or the pseudo-Aristotle who wrote book 10 of the *Generation of Animals*—must have imagined something like this when he wrote of the womb during orgasm violently emitting (*proiesthai*) through the cervix into the same space as the penis, i.e., into the vagina.[23] If we take this figure seriously, we must come to the extraordinary conclusion that women always have one penis—the cervix or *kaulos*—penetrating the vagina from the inside and another more potent penis, the male's, penetrating from the outside during intercourse.

There is, as G. E. R. Lloyd said, "an air of shadow boxing" about Greek debates on male and female physiology, and even a certain lunatic confusion if various claims are pushed to their limits.[24] Matters were ordinarily much clearer to the ancients, who could undoubtedly tell penis from vagina and possessed the language with which to do so. Latin and Greek, like most other tongues, generated an excess of words about sex and sexual organs as well as a great abundance of poetry and prose praising or making fun of the male or female organs, joking or cursing on the theme of what should be stuck where. I deny none of this.

But when the experts in the field sat down to write about the basis of sexual difference, they saw no need to develop a precise vocabulary of genital anatomy because if the female body was a less hot, less perfect, and hence less potent version of the canonical body, then distinct organic, much less genital, landmarks mattered far less than the metaphysical hierarchies they illustrated. Claims that the vagina was an internal penis or that the womb was a female scrotum should therefore be understood as images in the flesh of truths far better secured elsewhere. They are another way of saying, with Aristotle, that woman is to man as a wooden triangle is to a brazen one or that woman is to man as the imperfect eyes of the mole are to the more perfect eyes of other creatures.[25] Anatomy in the context of sexual difference was a representational strategy that illuminated a more stable extracorporeal reality. There existed many genders, but only one adaptable sex.

Blood, milk, fat, sperm

In the blood, semen, milk, and other fluids of the one-sex body, there is no female and no sharp boundary between the sexes. Instead, a physiology of fungible fluids and corporeal flux represents in a different register the absence of specifically genital sex. Endless mutations, a cacophonous ringing of changes, become possible where modern physiology would see distinct and often sexually specific entities.

Ancient wisdom held, for example, that sexual intercourse could alleviate conditions—mopish, sluggish behavior—caused by too much phlegm, the moist clammy humor associated with the brain: "semen is the secretion of an excrement and in its nature resembles phlegm."[26] (This already hints of the idea that conception is the male having an idea in the female body.) But more to the point here, ejaculation of one sort of fluid was thought to restore a balance caused by an excess of another sort because seminal emission, bleeding, purging, and sweating were all forms of evacuation that served to maintain the free-trade economy of fluids at a proper level. A Hippocratic account makes these physiological observations more vivid by specifying the anatomical pathways of interconversion; sperm, a foam much like the froth on the sea, was first refined out of the blood; it passed to the brain; from the brain it made its way back through the spinal marrow, the kidneys, the testicles, and into the penis.[27]

Menstrual blood, a plethora or leftover of nutrition, is as it were a local variant in this generic corporeal economy of fluids and organs. Pregnant women, who supposedly transformed otherwise superfluous food into nourishment for the fetus, and new mothers, who nursed and thus needed to convert extra blood into milk, did not have a surplus and thus did not menstruate. "After birth," says the omniscient Isidore, passing on one millennium of scholarship to the next, "whatever blood has not yet been spent in the nourishing of the womb flows by natural passage to the breasts, and whitening [hence *lac,* from the Greek *leukos* (white), Isidore says] by their virtue, receives the quality of milk."[28] So too obese women (they transformed the normal plethora into fat), dancers (they used up the plethora in exercise), and women "engaged in singing contests" (in their bodies "the material is forced to move around and is utterly consumed") did not menstruate either and were thus generally infertile.[29] The case of singers, moreover, illustrates once again the extent to which what we would take to be only metaphoric connections between organs were viewed as having causal consequences in the body as being real. Here the association is one between the throat or neck through which air flows and the neck of the womb through which the menses passes; activity in one detracts from activity in the other. (In fact, metaphorical connections between the throat and the cervix/vagina or buccal cavity and pudenda are legion in antiquity and still into the nineteenth century, as fig. 2 suggests. Put differently, a claim that is made in one case as metaphor—the emissions that both a man and a woman deposit in front of the neck of the womb are drawn up "with the aid of breath, *as with* the mouth or nostrils"—has literal implications in another: singers are less likely to menstruate.[30])

Although I have so far only described the economy of fungible fluids with respect to sperm and menstrual blood, seemingly gendered products, it in fact transcended sex and even species boundaries. True, because men were hotter and had less blood left over, they did not generally give milk. But, Aristotle reports, some men after puberty *did* produce a little milk and with consistent milking could be made to produce more (*HA* 3.20.522a19–22). Conversely, women menstruated because they were cooler than men and hence more likely at certain ages to have a surplus of nutriment. But, even so, menstruation in women was thought to have functional, nonreproductive, equivalents, which allowed it to be viewed as part of a physiology held in common with men. Thus, Hippocrates held, the onset of a nosebleed, but also of menstruation, was an indication that a fever was about to break, just as nosebleeding was a prognostic sign that blocked courses, amenorrhea, would soon resolve. Conversely,

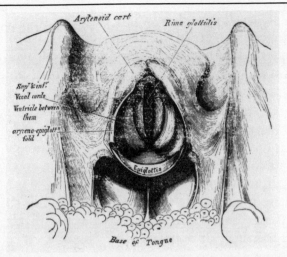

Fig. 2. Nineteenth-century illustration of a view into the aperture of the larynx which makes it look like the female external genitalia. Galen had pointed out that the uvula, which hangs down at the back of the soft palate—center view as one looks into the mouth—gives the same sort of protection to the throat that the clitoris gives to the uterus. From Max Muller, *Lectures on the Science of Language.*

a woman vomiting blood would stop if she started to menstruate.[31] The same sort of substitution works with sweat: women menstruate less in the summer and more in winter, said Soranus, because of the different amounts of evaporation that take place throughout the body in warm and cold weather. The more perspiration, the less menstrual bleeding.[32]

What matters is losing blood in relation to the fluid balance of the body, not the sex of the subject or the orifice from which it is lost. Hence, argued Araeteus the Cappadocian, if melancholy appears after "the suppression of the catemenial discharge in women," *or* after "the hemorrhoidal flux in men, we must stimulate the parts to throw off their accustomed evacuation." Women, said Aristotle, do not suffer from hemorrhoids or nosebleeds as much as men do, except when their menstrual discharges are ceasing; conversely, the menstrual discharge is slight in women with hemorrhoids or varicose veins presumably because surplus blood finds egress by these means.[33]

The complex network of interconvertibility implicit in the physiology of one sex is even wider than I have suggested and encompasses flesh as well as fluid. Aristotle, for example, finds confirmation for the common residual nature of sperm and menstrual fluid in the observation that fat creatures of *both* sexes are "less spermatic" (*spermatika*) than lean ones. Since "fat also, like semen, is a residue, and is in fact concocted blood," fat men and women have less left over to be released in orgasm or as catamenia. Lean men, on the other hand, produce more semen than fat men and for the same general reason that humans produce proportionally more semen *and* more menstrual fluid than other animals: lean men do not use up nutriment for fat; humans retain, as a surplus, material that in animals goes into their horns and hair.[34]

This sort of analysis can be extended indefinitely. Fair-complexioned men and women ejaculate more copiously than darker ones, Aristotle says, without even bothering to make explicit the assumption that this is because the latter are generally more hirsute; those on a watery and pungent diet discharge more than they would on a dry bland diet (*HA* 7.2.583a10–14). Both men and women are tired after ejaculation, not because the quantity of material emitted is so great but because of its quality: it is made from the purest part of the blood, from the essence of life (*GA* 1.18.725b6–7).

If, as I have been arguing, the reproductive fluids in the one-sex model were but the higher stages in the concoction of food—much like the lighter-weight products in the fractional distillation of crude oil—then the male and female seed cannot be imagined as sexually specific, morphologically distinct, entities, which is how they would come to be understood after the discovery of little creatures in the semen and of what was presumed to be the mammalian egg in the late seventeenth century.[35] Instead, the substances ejaculated by the "two sexes" in the one-sex body were hierarchically ordered versions of one another according to their supposed power.

The difference between so-called two-seed and one-seed theories—Galen versus Aristotle—is therefore not an empirical question that could be resolved by reference to observable facts. Even in Aristotle's one-seed theory, *sperma* and *catemenia* refer to greater or lesser refinements of an ungendered blood, except when they are used as ciphers for the male and female "principles."[36] What one sees, or could ever see, does not really matter except insofar as the thicker, whiter, frothier quality of the male semen is a hint that it is more powerful, more likely to act as an efficient cause, than the thinner, less pristinely white, and more watery female ejaculate or the still red, even less concocted, menstrua. Like reproductive organs, reproductive fluids turn out to be versions of each other; they are the biological articulation, in the language of a one-sex body, of the politics of two genders and ultimately of engendering.

The Hippocratic writer illustrates this point vividly and without the philosophical complexity we find in Aristotle's so-called one-seed theory. Perhaps, if we accept the views of Aline Rousselle, he even speaks for the otherwise silenced empirical wisdom of women.[37] Hippocrates argues for pangenesis, the view that each part of the body of each parent renders up some aspect of itself; that the representatives of the various parts form a reproductive fluid or seed; and that conception consists of a blending, in various proportions and strengths, of these germinal substances. Hippocrates abandons any effort to attribute strong or weak seed respectively to actual males or females. Although males must originate from stronger sperm, "the male being stronger than the female," both are capable of producing more or less strong seed. What each emits is the result not of any essential characteristic of male or female, but of an internal battle between each sort of seed: "what the woman emits is sometimes stronger, sometimes weaker; and this applies also to what the man emits."[38] Hippocrates insists on this point by repeating the claim and generalizing it to animals: "The same man does not invariably emit the strong variety of sperm, nor the weak invariably, but sometimes the one and sometimes the other; the same is true in the woman's case." This explains why any given couple produces both male and female offspring as well as stronger and weaker versions of each; likewise for the beasts.[39]

If both partners produce strong sperm, a male results; if both produce weak sperm, a female is born; and if in one partner the battle has gone to the weak and in the other to the strong, then the sex of the offspring is determined by the quantity of the sperm produced. A greater quantity of weak sperm, whether produced by the male or the female, can overwhelm a lesser quantity of strong sperm, of whatever origin, in the second round when the two meet in front of the uterus for renewed combat. Hippocrates is at pains to emphasize the fluidity of the situation and the interpenetration of male and female. The contest for supremacy between the sperm is,

> just as though one were to mix together beeswax and suet, using a larger
> quantity of the suet than of the beeswax, and melt them together over a
> fire. While the mixture is still fluid, the prevailing character of the mixture
> is not apparent: only after it solidifies can it be seen that the suet prevails
> quantitatively over the wax. And it is just the same with the male and female
> forms of the sperm.[40]

Male and female "forms" of sperm thus correspond neither to the genital configuration of their source nor to that of the new life they will create, but rather to gradations on a continuum of strong to weak.[41]

I think that, if pushed on the point, the Hippocratic writer would have to admit that there was something uniquely powerful about male seed, the fluid that comes from an actual male, because otherwise he would have no answer to the question with which two-seed theorists were plagued for millennia: if the female has such powerful seed, then why can she not engender within herself alone; who needs men? The Hippocratic texts, however, resolutely resist correlating the gender of the seed, its strength or weakness, with the sex of the creature that produced it. Instead, in their version of the one-sex economy of fluids, the more potent seed is by definition the more male, wherever it originated.

For Galen too each parent contributes something that shapes and vivifies matter, but he insists that the female parent's seed is less powerful, less "informing," than the male parent's because of the very nature of the female. To be female *means* to have weaker seed, seed incapable of engendering, not as an empirical but as a logical matter. "Forthwith, *of course,* the female *must* have smaller, less perfect testes, and the semen generated in them must be scantier, colder, and wetter (for these things too follow *of necessity* from the deficient heat)" (*UP* 2.631). Thus, in contrast to Hippocrates, Galen holds that the quality of the respective seeds themselves follows from the hierarchy of the sexes. Man's seed is always thicker and hotter than a woman's for the same reason that the penis is extruded and not, like the uterus and the mole's eyes, left undeveloped inside the body: humans are the most perfect animal, and man is more perfect than woman because of an "excess of heat." In opposition, however, to what he took to be Aristotle's view, Galen insisted that women did produce semen, a true generative seed. If this were not the case, he asks rhetorically, why would they have testicles, which they manifestly do? And if they had no testicles (*orcheis*) they would not have the desire for intercourse, which they manifestly have.[42] In other words, the female seed, like woman herself, "is not very far short of being perfectly warm" (*UP* 2.630).

Male and female semen, more and less refined fluids, thus stand in the same relationship to blood that penis and vagina stand to genital anatomy, extruded and still-inside organs. As the medieval Arabic physician Avicenna (ibn-Sina, 980–1037) puts it in his discussion of these Galenic texts, "the female seed is a kind of menstrual blood, incompletely digested and little converted, and it is not as far away from the nature of blood (*a virtute sanguinea*) as is the male seed."[43] He assimilates digestion and reproduction, food, blood, and seed into a single general economy of fluids driven by heat. The female in the one-sex model lacks the capacity, the vital heat, to convert food to the very highest level: sperm. But she comes close.

Aristotle and the Aristotelian "one-seed" tradition, with its radical distinction between the male and female generative materials (*gonimos*), would seem to make the Galenic intermediate position impossible and would thus also seem to provide a basis in the body for two biologically distinct and incommensurable sexes, much in the way that egg and sperm would come to function in theories like Geddes' in the nineteenth century. Males, in Aristotle's account, produce *sperma,* which is the efficient cause in generation, and females do not. Females provide instead the *catamenia,* which is the material cause and thus of an entirely different nature. But this *a priori* formal distinction entirely exhausts what Aristotle means by *sperma* and *catamenia.* Just as the bodies of males and females fail to provide fixed anatomical correlatives for his theory of generative causality, so too the reproductive fluids "in the world" do not sustain a radical two-sex account of sexual difference. Nor would Aristotle want them to.

Obviously Aristotle and his contemporaries could tell semen from menstrual blood. Men and sanguineous male animals, they knew, generally emitted a visible, palpable substance that was white because it was foam composed of invisible bubbles and thick because it was a compound of water mixed with breath (*pneuma*), the tool through which the male principle worked. Although Aristotle usually referred to this stuff as sperma, its distinguishing characteristics were not in principle aspects of the seed itself.[44] The ejaculate, he makes absolutely explicit, was but the vehicle for the efficient cause, for the sperma, which worked its magic like an invisible streak of lightning. As experience proved, it ran out of or evaporated from the vagina; it no more entered into the catemenia, into what would become the body of the embryo, than any active agent enters into passive matter when one thing is made from two. After all, no part of the carpenter merges with the bed he crafts, nor does the swordsmith's art enter the sword he is fashioning, nor does rennet or fig juice become part of the milk they curdle into cheese. Indeed the efficient cause, the artisanal, informing principle, can apparently be carried on the breeze alone, as with the Cretan mares who are "wind impregnated."[45]

All of Aristotle's metaphors discount a physically present ejaculate; sperma as artisan works in a flash, more like a genie than like a shoemaker who sticks to his last. His images bring us back to the constellation of phlegm/brain/sperm: conception is for the male to have an idea, an artistic or artisanal conception, in the brain-uterus of the female.[46]

But the female, the material, contribution to generation is only slightly more material and thus recognizable by the physical properties of menstrual blood. Aristotle is at pains to point out that catamenia, the menstrual residue itself, is not to be equated with the actual blood that one sees: "the greater part of the menstrual flow is useless, being fluid" (*GA* 2.4.739a9). But he leaves the relationship between the catamenia, wherein the sperma works its magic, and anything visible—the "useless" menstrual discharge or the fluid that moistens the vagina during intercourse—unexplored largely because it does not matter in a world in which claims about the body serve primarily as illustrations of a variety of higher truths.[47] His dominant image is of a hierarchy of blood: "The secretion of the male and the menses of the female are of a sanguinous nature."[48] Semen from men who have coitus too often reverts to its earlier bloody state; semen in boys and often in older men is, like the catamenia, unable to impart movement to matter.[49] For Aristotle, therefore, and for the long tradition founded in his thought, the generative substances are interconvertible elements in the economy of a single-sex body whose higher form is male. As physiological fluids they are not distinctive and different in kind, but the lighter shades of biological chiaroscuro drawn in blood.[50]

All of this evidence suggests that in the construction of the one-sex body the borders between blood, semen, other residues and food, between the organs of reproduction and other organs, between the heat of passion and the heat of life, were indistinct and, to the modern person, almost unimaginably—indeed terrifyingly—porous. "Anyone who has intercourse around midnight," warns a text attributed to Constantinius Africanus, "makes a mistake." Digest (concoct) food first before straining the body to give the final concoction to the seed.[51] Fifteen hundred years after Aristotle and a thousand after Galen, Dante in the *Purgatorio* still plays on the fungibility of the body's fluids and the affinities of its heats. "Undrunk" blood, perfect like a dish (*alimento*) that is sent from the table, is redistilled by the heat of the heart, sent down to the genitals, from which "it sprays in nature's vessel, on another's blood."[52] *The Secrets of Women,* compiled from ancient lore during the later Middle Ages and still popular in the eighteenth century, speaks of the appetite for intercourse as a direct result of the buildup of residue from daily food. Menstrua refined from the blood heats up a woman's vulva through an "abundance of matter" and causes her greatly to desire coition.[53]

The fluid economy of the one-sex body thus engenders the desires and the heat through which it will be perpetuated. But more generally I hope it is becoming clear that the physiology and even the anatomy of generation are but local instances of a way of talking about the body very different from our own. Visible flesh and blood cannot be regarded as the stable "real" foundation for cultural claims about it. Indeed, the interpretive problem is understanding the purchase of "real" and the degree to which biology is only the expression of other and more pervasive truths.

Orgasm and desire

"I must now tell why a great pleasure is coupled with the exercise of the generative parts and a raging desire precedes their use," Galen wrote (*UP* 2.640). However else orgasm might be tempered to fit the cultural needs of the private and the public body, it signaled the unsocialized body's capacity to generate. A basically matter-of-fact, specifically genital urge led to a grander, systemic heating of the body until it was hot enough to concoct the seeds of new life. Serous residues, exquisitely sensitive skin, and friction were the proximal causes of sexual delight and desire; "that the race may continue incorruptible forever" was their ultimate purpose. The process of generation might differ in its nuances as the vital heats, the seeds, and the physical qualities of the substances being ejaculated differed between the sexes—but libido, as we might call it, had no sex.

There was, of course, the age-old issue of whether men or women enjoyed the pleasures of Venus more, a question posed most famously in Ovid, who offers an ambiguous answer. (Ovid's account would become a regular anecdote in the professorial repertory, told to generations of medieval and Renaissance students to spice up medical lectures.) True, Tiresias, who had experienced love as both a man and a woman, was blinded by Juno for agreeing with Jupiter that women enjoyed sex more. But his qualifications for judging already suggest the slipperiness of the question: he knew either one or the other, or both, aspects of the feminine *Venus* rather than of the masculine *amor*. And the story of his "mirror" metamorphosis from man to woman, the result of his striking two copulating serpents, and back to man by striking them again eight years later, further undermines his authority on the sexual differentiation of pleasure. Snakes famously give no outward sign of their sex; they curl around one another in coition and reflect back and forth the most ambiguous and ungendered of images. Though differing perhaps in nuance, orgasm is orgasm in the one-flesh body, Ovid's story seems to say.[54]

A common neurology of pleasure in a common anatomy, it was thought, bore witness to this fact. Galen, for example, notes that "the male penis . . . as well as the neck of the uterus and the other parts of the pudendum" are richly endowed with nerves because they need sensation during sexual intercourse and that the testes, scrotum, *and* uterus are poorly endowed because they do not. Animal dissections prove, he says, that the "genital areas," in common with the liver, spleen, and kidneys, have only small nerves while the pudenda have "more considerable ones." Even the skin of the relevant organs is more irritated by the "itch" of the flesh than would be the skin of the body's other parts. Given all these adaptations, "it is no longer to be wondered at that the pleasure inherent in the parts there and the desire that precedes it are more vehement."[55]

Aristotle too is at pains to point out that "the same part which serves for the evacuation of the fluid residue is also made by nature to serve in sexual congress, and this alike in male and female."[56] Both sperma and catamenia generate heat in the genital regions, both put pressure on the sexual organs that are prepared to respond to their stimuli, though in the case of women's parts the heat seems to serve primarily to draw in semen, like a cupping vessel, and not to spur coition (*GA* 2.4.739b10).

"Semen" in this economy of pleasure is not only a generative substance but also, through its specific action on the genitals, one of the causes of libido. It is a serous, irritating humor that produces a most demanding itch in precisely that part of the body contrived by Nature to be hypersensitive to it.[57] (Or in parts not contrived for it. The only ancient text to discuss the physical causes of passive homosexuality—the unnatural desire of the male to play the socially inferior role of woman by offering his anus for penetration—attributes it both to an excess of semen and to a congenital defect that shunts this excess to an inappropriate orifice, the anus, instead of allowing it to simply build up in the proper male organ.[58]) Needless to say, great pleasure is to be had from scratching.

Orgasm thus dovetails nicely with the economy of fluids discussed in the previous section. One of Galen's arguments for the existence of a true female seed, for example, was its link to desire: it offered "no small usefulness in inciting the female to the sexual act and in opening wide the neck of the womb during coitus" (*UP* 2.643). He might actually have meant that it works like a penis. The part in question, extending out to the "pudenda" (the cervix?, the vagina?) is, he says, sinewy and becomes straight during intercourse. He does not actually claim that the womb or vagina has an erection, but he describes the penis also as a sinewy, hollow body that becomes erect when it is filled with pneuma, with breath. And elsewhere still he develops the labia/foreskin association.[59] The medieval commentator Albertus Magnus, writing still very much in this tradition almost a millennium later, makes the link explicit: a *ventositas,* a gaseous, perhaps also liquid modification of vital heat, engorges the genital organs of both sexes.[60] Organs and orgasms thus reflect one another in a common mirror.

Meanwhile Avicenna, the influential Arabic physician, broadens the discussion of the semen/pleasure nexus by explicitly connecting the anatomy and physiology of sexual pleasure in the one-sex body. An irritation of a common human flesh, caused by the acute quality or sheer quantity of sperm—again common to both sexes—engenders a specifically genital itch (*pruritum*) in the male's spermatic vessels and in the mouth of the womb (*in ore matricis*), which is relieved only by the chafing of intercourse or its equivalent. In this process the vagina, or in any case the cervix, becomes erect like the penis and is "thrust forward up against its mouth as though moving forward through the desire of attracting sperm."[61] In the telling absence of a precise technical vocabulary, it is difficult to be sure exactly what part of a woman's genital organ is moving where; but the critical general claim, that irritation by a serous fluid loosely called sperm or semen causes women like men to experience desire and erection, is made unambiguously.

Intercourse in the one-sex body, however, is not construed primarily as a genital occasion. (Nor, of course, is desire purely the product of physical forces independent of the imagination.) The genitals, to be sure, are the most sensitive gauge of the presence of residues, the point of their release, and the immediate locus of pleasure, but coitus is a generalized friction culminating in a corporeal blaze. Intercourse and orgasm are the last stage, the whole body's final exaggerated huffing and puffing, violent, stormlike agitation in the throes of producing the seeds of life. The rubbing together of organs, or even their imagined chafing in an erotic dream, causes warmth to diffuse via the blood vessels to the rest of the body. "Friction of the penis and the movement of the whole man cause the fluid in the body to grow warm," the Hippocratic writer reports; "an irritation is set up in the womb which produces pleasure and heat in the rest of the body."[62] Then, as warmth and pleasure build up and spread, the increasingly violent movement of the body causes its finest part to be concocted into semen—a kind of foam—which bursts out with the uncontrolled power of an epileptic seizure, to use the analogy Galen borrowed from Democritus.[63] Sexual heat is an instance of the heat that makes matter live and orgasm, which signals the explosive release of the seed and the heated pneuma, mimics the creative work of Nature itself.

Although specific interpretations of the male and female orgasm might differ, certain facts were generally not in dispute: both sexes experienced a violent pleasure during intercourse that was intimately connected with successful generation; both generally emitted something; pleasure was due both to the qualities of the substance emitted and its rapid propulsion by "air"; the womb performed double duty in both emitting something and then drawing up and retaining a mixture of the two emissions. Of what deeper truths these facts spoke was much debated.

In the first place, the way orgasm felt was adduced as evidence for particular embryological theories. Pangenesists could argue as follows: "the intensity of pleasure of coition" proves that seed comes from every part of both partners because pleasure is greater if multiplied and that of orgasm is so great that it must result from something happening everywhere rather than just in a few places or in one sex only. But even if this reasoning was not universally accepted, most writers nevertheless regarded orgasm as a most weighty sign.

Why, asked an ancient text, did someone having sexual intercourse, and also a dying person, cast his or her eyes upward? Because the heat going out in an upward direction makes the eyes turn in the direction in which it itself is traveling.[64] Conversely, sexual heat is the most intense form of the heat of life and so is the sign of successful generation. The early Christian writer Tertullian, for example, grounded his heterodox theory of the soul—its material origin, its entry into the body at the moment of conception, its departure at death—on the phenomenology of orgasm:

In a single impact of both parties, the whole human frame is shaken and foams with semen, in which the damp humor of the body is joined to the hot substance of the soul . . . I cannot help asking, whether we do not, in that very heat of extreme gratification when the generative fluid is ejected, feel that somewhat of our soul has gone out from us? And do we not experience a faintness and prostration along with a dimness of sight? This, then, must be the soul producing seed, which arises from the outdrip of the soul, just as that fluid is the body-producing seed which proceeds from the drainage of the flesh.[65]

This "heat of extreme gratification," however, is open to quite different secular interpretations. Lucretius regarded it as the blaze of battle in the war of sexual passion and conception. Young men are wounded by Cupid's arrow and fall in the direction of their injuries: "blood spurts out in the direction of their wound." (In context this can only be semen, pure blood and not the blood of virginity.) Then both bodies are liquefied in rapture, and their ejaculates engage in a synecdochic version of the two bodies' combat. Offspring resemble both parents, for example, because "at their making the seeds that course through the limbs under the impulse of Venus were dashed together by the collusion of mutual passion in which neither party was master or mastered."[66]

In contrast to these positions, Aristotle wants to isolate orgasm from generation so as to protect the difference between efficient and material cause from an untidy world in which both sexes have orgasms that feel as if the same process had gone on in each of them. (As it turns out, Aristotle was right but not for the reasons he gave.) Thus for him it *has* to be "impossible to conceive without the emission of the male"; whether he feels pleasure during ejaculation is irrelevant. On the other hand women *must* be able to conceive "without experiencing the pleasure usual in such intercourse" because, by definition, conception is the work of the male emission on material in, or produced by, the body of the female. (Females usually do emit something but need not do so; there can be just enough catamenial residue resting in the womb for conception to take place but no extra that needs to be expelled.) Aristotle's argument is asymmetrical here—males must emit, women need not feel—because he wants to stick to the essentials. It makes no difference how one interprets male pleasure; he must insist, however, that female pleasure—he discusses only humans in this regard—has no implication for his theory of the separation of causes. His real interest is not in interpreting orgasm, but in *not* interpreting it.[67]

It follows from this position that Aristotle would make no effort to ground two sexes in radically different passions and pleasures. Though women clearly could, in his view, conceive without feeling anything, he regarded this as a freak occurrence that resulted when "the part chance to be in heat and the uterus to have descended," that is, when the womb and vagina were warmed by something other than the friction of intercourse and experienced their internal erection without concomitant sexual excitement. "Generally speaking," he said, "the opposite is the case"; discharge by women is accompanied by pleasure just as it is in men, and "when this is so there is a readier way for the semen of the male to be drawn into the uterus."[68]

Aristotle's many allusions to sexual pleasure are clearly not directed at distinguishing the orgasms of men and women but in keeping their similarities from being relevant. What he takes to be contingent sensations must not be construed as evidence for what he regards as metaphysical truths about generation. He denies that orgasm signals the production of generative substances even for the male; "the vehemence of pleasure in sexual intercourse," he maintains, is not at all due to the production of semen but is the result instead of "a strong friction wherefore if this intercourse is often repeated the pleasure is diminished in the persons concerned."[69] The rhetorical force of this convoluted sentence is to stress the fading of feeling that comes from repetition. Elsewhere he says that pleasure arises not just from the emission of semen but from the pneuma, the breath, with which the generative substances explode. The point is simply that the phenomenological correlative of the generative act signifies nothing about its essence: there need be no seed, no efficient cause itself, for there to be an orgasm—as in young boys and old men who are not potent but nevertheless enjoy emission.[70] Conversely, both men and women can emit their respective generative products and feel nothing, as in nocturnal wet dreams.[71]

Whatever else orgasm might be or not be, mean or not mean, in various philosophical or theological contexts, it was at the very least understood as the *summa voluptas* that normally accompanied the final blast of a body heated so hot that it expelled its generative essences or, in any case, was in a state to conceive. As such, it dwelled at the intersection of nature and civilization. On the one hand, orgasm was associated with unrestrained passion, warmth, melting, rendering, rubbing, exploding, as qualities of the individual body; aspects of the process of individual generation. On the other hand, orgasm also bore witness to the power of mortal flesh to reproduce its kind and thus assure the continuity of the body social. It and sexual pleasure generally were therefore cultural facts as well: the biology of conception was at the same time a model of filiation; the effective elimination of the distinct ontological category woman in the one-sex model and the doctrine that "like seeks like" made it difficult to explain heterosexuality upon which generation depended; the unruly body spoke of the unruly heart, of the fall from grace and weakness of the will; microcosmic creation mirrored the macrocosmic. Though the social and the corporeal cannot be disentangled, for purposes of exposition I will discuss orgasm first as the physicians confronted it—as a clinical problem of fertility or infertility—and then briefly turn in the next section to its relation to the demands of culture.

Physicians and midwives needed to know how to make men and women fertile—or more covertly, how to make them infertile—and how to tell if their therapeutic interventions were on the right track. If, as was commonplace, one believed that the body gave signs through its pleasures of the capacity to generate, then these could be read and the underlying processes manipulated to ensure or prevent conception. So, for example, Aetios of Amida, physician to Justinian who summarized for the emperor much ancient medical learning, interpreted a woman's orgasmic shudder as a prognostic sign of conception. If "in the very coitional act itself, she notes a certain tremor . . . she is pregnant." (Aetios also transmitted to the Christian world the old saw that women who are forced to have intercourse against their will are sterile while those "in love conceive very often.") A woman's shiver would not have been understood simply as a sign of her "semination"; it would register also the closing off of her womb at the appropriate time, after it had drawn up her seed mixed with that of the male.[72]

Because the womb was thought to close after its orgasmic ejaculation, correct coital rhythm between partners during intercourse was thought critical for conception. If the woman is too excited before intercourse begins, the Hippocratic writer points out, she will ejaculate prematurely; then not only will her further pleasure diminish—a conclusion clearly based on men observing themselves—but also her womb will close and she will not become pregnant. In exemplary reproductive heterosexual intercourse, then, both partners reached orgasm at the same time. Like a flame that flares when wine is sprinkled on it, the woman's heat blazes most brilliantly when the male sperm is sprayed on it, Hippocrates rhapsodized. She shivers. The womb seals itself. And the combined elements for a new life are safely contained within.[73]

Orgasm in this account is thus common to both sexes but, like anatomy and the seeds themselves, it is hierarchically ordered. The man determines the nature of woman's pleasure, which is more sustained but also, because of her lesser heat, less intense; the man feels a greater pang at the secretion of bodily fluids because a greater violence accompanies their being wrenched from his blood and flesh. Feelings mirror the cosmic order and at the same time suggest the sparkling of a candle in a mist of resinated wine.

Clinically, therefore, the problem is how to manipulate the pace of passion and the heat of the body so as to produce the desired results, conception or nonconception. Aristotle (or the pseudo-Aristotelian author of book 10) gives elaborate directions for determining in cases of barrenness which partner's coital rhythms or corporeal environment was at fault. During intercourse the woman's womb should become moist but "not often or excessively too moist," lubricated as the mouth is with saliva when we are about to eat (once again a neck-of-the-womb/throat connection).[74] More natural history: if a man ejaculates quickly and "a woman with difficulty as is often the case," this prevents conception since women do contribute "something to the semen and to generation." The observation that women and men who are barren with each other are "fertile when they meet with partners who keep pace with them during intercourse" provides this further evidence for the importance of suitable coital rhythms.[75] Fifteen hundred years later, and in the very different context of prescriptions for birth control and abortion, the tenth-century Arabic writer Rhazes suggested that "if the man discharges sooner than the woman [discharges] she will not become pregnant."[76]

Anything that might diminish coital heat could also cause infertility. Insufficient friction during intercourse, for example, could keep either partner from "seminating." Thus Avicenna argues—again this is a commonplace notion—that the smallness of a man's penis might cause a woman not to be "pleased by it . . . whereupon she does not emit sperm (*sperma*), and when she does not emit sperm a child is not made." As if to raise male anxiety still further, he warns that unsatisfied women will remain in the thrall of desire and "have recourse to rubbing, with other women (*ad fricationem cum mulieribus*), in order to achieve amongst themselves the fullness of their pleasures" and to rid themselves of the pressures of seminal residue.[77]

But even if the actual pang of a woman's orgasm was regarded as a sign without the specific physiological referent of semination, sexual pleasure or at the very least desire was still regarded as part of the general care of the body that made reproduction, and hence the immortal body of the race, possible. Control of the sexual body was, as Foucault points out in his *History of Sexuality*, an aspect of more general dietary and other corporeal disciplines. Nowhere is this aspect of the domestication of sexual heat clearer than in Soranus' *Gynecology*, which was written in the second century but which in various fragments and translations was one of the most widely cited texts until the late seventeenth century.

Soranus was not much interested in female ejaculation because he remained in doubt as to whether women actually contributed an active principle, a true seed. "It seems not to be drawn upon in generation since it is excreted externally," he concluded cautiously. He nowhere denied the everyday existence of the sharp crisis of orgasm in women, but it was not of primary clinical concern. What mattered in women as in men, Soranus thought, was "the urge and appetite for intercourse." Making the body ready for generation was like making it ready to put food to best use. The physiological affinity between generation and nutrition, eating and procreation, and in later Christian formulations between gluttony and lust, are nowhere clearer: "as it is impossible for the seed to be discharged by the male, in the same manner, without appetite it can not be conceived by the female." A woman ingesting and a woman conceiving are engaged in analogous functions; food eaten when one has no appetite is not properly digested, and seed received by a woman when she has no sexual urge is not retained.[78]

But appetite alone is clearly not enough, since lecherous women feel desire all the time but are not always fertile. The body—Soranus is writing for midwives who ministered to ladies of the Roman governing class—must be properly cultivated to prepare for the civic task of procreation. They ought to be well rested, appropriately nourished, relaxed, in good order, and hot. Just as a Roman magistrate should eat only such foods as would maintain his sound judgment, so a woman should eat appropriately before sex "to give the inner turbulence an impetus toward coition" and to be sure that her sexual urges were not diverted by hunger. She should be sober. A rubdown before intercourse would be indicated, since it "naturally aids the distribution of food, [and] also helps in the reception and retention of the seed."[79] The fungibility of fluids, the equivalences of heat, are here registered in the social discipline of the body for procreation.

The demands of culture

The one-sex body would seem to have no boundaries that could serve to define social status. There are hirsute, viral women—the virago—who are too hot to procreate and are as bold as men; and there are weak, effeminate men, too cold to procreate and perhaps even womanly in wanting to be penetrated. "You may obtain physiognomic indications of masculinity and femininity," writes an ancient authority on interpreting the face and body, "from your subject's glance, movement, and voice, and then, from among these signs, compare with one another until you determine to your satisfaction which of the two sexes prevails."[80] "Two sexes" here refers not to the clear and distinct kinds of being we might mean when we speak of opposite sexes, but rather to delicate, difficult-to-read shadings of one sex. There is, for example, no inherent gendering of desire and hence of coupling. It was in no way thought unnatural for mature men to be sexually attracted to boys. The male body, indeed, seemed equally capable of responding erotically to the sight of women as to attractive young men, which is why physicians forbade sufferers of satyriasis (abnormal sexual craving characterized by unceasing erection and genital itch) to consort with either, regardless of their respective genital formations.[81] Insofar as sexual attraction had a biological basis—as opposed to a basis in the naturalness of the social order and the imperative to keep it going—it seemed more genealogical than genital. In Aristophanes' story of the origins of men and women from two aboriginal, globular creatures who had either two male organs, two female organs, or one of each, only those who descended from the hermaphroditic form would "naturally" seek the "opposite" sex in order to achieve union. Otherwise, as Aristotle pointed out in the context of "what is natural is pleasant": like loves like, jackdaw loves jackdaw. In fact, reproductive heterosexual intercourse seems an afterthought. The original globular creatures had their genitals on the outside and "cast their seed and made children, not in one another but on the ground, like cicadas." In the new cut-up state they did nothing but longingly embrace their missing halves and thus died from hunger and idleness. Zeus hit upon the idea of relocating the genitals of one half of the new creatures, "and in doing so he invented interior reproduction, *by* men *in* women." This had the great advantage that when the new male embraced the new female, he could cast his seed into her and produce children and that when male embraced male, "they would at least have the satisfaction of intercourse, after which they could stop embracing, return to their jobs, and look after their other needs in life." Genitals are very hard to picture in the first part of this account and subsist only to make the best of a bad situation. "Love is born into every human being," the story concludes; "it tries to make one out of two and heal the wound in human nature." But what we would call the sex of that human being seems of only secondary importance.[82]

But where honor and status are at stake, desire for the same sex *is* regarded as perverse, diseased, and wholly disgusting. A great deal more was written about same-sex love between men than between women because the immediate social and political consequences of sex between men was potentially so much greater. Relatively little was directly at stake in sex between women. Yet whether between men or between women, the issue is not the identity of sex but the difference in status between partners and precisely what was done to whom. The active male, the one who penetrates in anal intercourse, or the passive female, the one who is rubbed against, did not threaten the social order. It was the weak, womanly male partner who was deeply flawed, medically and morally. His very countenance proclaimed his nature: *pathicus,* the one being penetrated; *cinaedus,* the one who engages in unnatural lust; *mollis,* the passive, effeminate one.[83] Conversely it was the *tribade,* the woman playing the role of the man, who was condemned and who, like the mollis, was said to be the victim of a wicked imagination as well as an excess and misdirection of semen.[84] The actions of the mollis and the tribade were thus unnatural not because they violated natural heterosexuality but because they played out—literally embodied—radical, culturally unacceptable reversals of power and prestige.

Similarly, when power did not matter or when a utopian sharing of political responsibility between men and women is being imagined, their respective sexual and reproductive behavior is stripped of meaning as well. Aristotle, who was immensely concerned about the sex of free men and women, recognized no sex among slaves. "A 'woman,'" as Vicky Spellman puts it, "is a female who is free; a 'man' is a male who is a citizen; a slave is a person whose sexual identity does not matter."[85] For Aristotle, in other words, slaves are without sex because their gender does not matter politically.

Plato, on at least one occasion, also dismissed a distinction between the sexes which in other circumstances is critical. When in the *Republic* he wished to make a case for the absence of essential public differences between men and women, for equal participation in governance, gymnastic exercises, and even war, he supported his claim by downplaying the difference in their reproductive capacities. If something characteristic of men or women can be found which fits one or the other for particular arts and crafts, by all means assign them accordingly. But no such distinction exists, he maintains, and what Aristotle would take to be the critical difference between bearing and begetting counts for nothing.

> But if it appears that they differ only in this respect that the female bears and the male begets, we shall say that no proof has yet been produced that the woman differs from the man for our purposes, but we shall continue to think that our guardians and their wives ought to follow the same pursuits.[86]

Begetting and bearing are not radically opposed, or even hierarchically ordered. Plato uses a decidedly unphilosophical verb for begetting, the verb *ochenein,* to mount; Aristotle uses the same verb when he says that the victor among bulls "mounts" the cow and then, "exhausted by his amourous efforts," is subsequently beaten by his opponent (*HA* 6.21.575a22). Nothing more is at stake, Plato implies, than the brutish practice of man mounting woman. The macrocosmic order is not made imminent through the sexual act; the respective roles of man and woman in generation, though different, do not constitute a decisive difference.

But within the same tradition of the one sex, and in widely varying contexts, such differences could matter a great deal and were duly registered. Sperma, for Aristotle, makes the man *and* serves as synecdoche for citizen. In a society where physical labor was the sign of inferiority, sperma eschews physical contact with the catemenia and does its work by intellection. The *kurios,* the strength of the sperma in generating new life, is the microcosmic corporeal aspect of the citizen's deliberative strength, of his superior rational power, and of his right to govern. Sperma, in other words, is like the essence of citizen. Conversely, Aristotle used the adjective *akuros* to describe both a lack of political authority, or legitimacy, and a lack of a biological capacity, an incapacity that for him defined woman. She is politically, just as she is biologically, like a boy, an impotent version of the man, an *arren agonos*. Even grander differences are inscribed on the body; the insensible differences between the sexual heat of men and women turns out to represent no less a difference than between heaven and earth. The very last stage in the heating sperma comes from the friction of the penis during intercourse (*GA* 1.5.717b24). But this is not like the heat of a blacksmith's fire, which one might feel, nor is the pneuma produced like ordinary breath.[87] It is a heat "analogous to the elements of the stars," which are "carried on a moving sphere" and are themselves not fired but create warmth in things below them.[88] Suddenly the male organ in coition is a terrestrial instance of heavenly movement, and the sexed body, whose fluids, organs, and pleasures are nuanced versions of one another, comes to illustrate the major political and cosmic ruptures of a civilization.[89]

The most culturally pervasive of these ruptures is that between father and mother, which in turn contains a host of historically specific distinctions. I want to illustrate the extent to which biology in the one-sex model was understood to be an idiom for claims about fatherhood by examining three different accounts of the nature of seed put forward by Isidore of Seville, who in the sixth and seventh centuries produced the first major medieval summary of ancient scientific learning. Although the social context of a Christian encyclopedist was of course very different from that of an Athenian philosopher or an imperial Roman doctor, the structure of Isidore's arguments is paradigmatic for what is a very long-lived tradition of understanding sexual difference.

Isidore simultaneously holds three propositions to be true: that only men have sperma, that only women have sperma, and that both have sperma. It takes no great genius to see that these would be mutually contradictory claims if they are understood as literal truths about the body. But they would be perfectly compatible if they are seen as corporeal illustrations of cultural truths purer and more fundamental than biological fact. Indeed, Isidore's entire work is predicated on the belief that the origin of words informs one about the pristine, uncorrupted, essential nature of their referants, about a reality beyond the corrupt senses.[90]

In making the first case—that only man has seed—Isidore was explaining consanguinity and, as one would expect in a society where inheritance and legitimacy passes through the father, he was at pains to emphasize the exclusive origins of the seed in the father's blood.

> Consanguinity is so called by that which from one blood, that is from the same semen of the father, is begotten. For the semen of the male is the foam of blood according to the manner of water which, when beaten against rocks, makes white foam, or just as dark wine, which poured into a cup, renders the foam white.

For a child to have a father *means* that it is "from one blood, that is from the same semen as the father"; to be a father is to produce the substance, semen, through which blood is passed on to one's successors. Generation seems to happen without women at all, and there is no hint that blood—"that by which man is animated, and is sustained, and lives," as Isidore tells us elsewhere—could in any fashion be transmitted other than through the male.[91]

But illegitimate descent presents a quite different biology. In his entry on the female genitalia, Isidore argued:

> Contrary to this child [one born from a noble father and a plebian mother] is the illegitimate (*spurius*) child who is born from a noble mother but a plebian father. Likewise illegitimate is the child born from an unknown father, a spouseless mother, just the son of spurious parents.

The reason Isidore gives for why such illegitimate children, those who do not "take the name of the father" and are called *spurius,* is that they spring from the mother alone. "The ancients," he explains, "called the female genitalia the *spurium;* just as *apo tou sporou* (from the seed); this *spurium* is from the seed." (Plutarch reported that the adjective *spurius* derived from a Sabine word for the female genitalia and was applied to illegitimate children as a term of abuse.) So, while the legitimate child is from the froth of the father, the illegitimate child is from the seed of the mother's genitals, as if the father did not exist.[92]

Finally, when Isidore is explaining why children resemble their progenitors, he is vague on the vexed question of female sperm. "Whichever of the two parents bestows the form," he says cavalierly, "the newborn are conceived after equally being mixed in the maternal and paternal seed." "Newborns resemble fathers, if the semen of the fathers is potent, and resemble mothers if the mothers' semen is potent.[93] (Both parents then have seeds that engage in repeated combat for domination every time, and in each generation a child is conceived.)

These three distinct arguments about what we might take to be the same biological material are a dramatic illustration that much of the debate about the nature of the seed and of the bodies that produce it—about the boundaries of sex in the one-sex model—are in fact not about bodies at all. They are about power, legitimacy, and fatherhood, in principle not resolvable by recourse to the senses.

Freud suggests why this should be so. Until the mid-nineteenth century, when it was discovered that the union of two different germ cells, egg and sperm, constituted conception, it was perfectly possible to hold that fathers mattered very little at all. Paternity, as in Roman law, could remain a matter of opinion and of will. Spermatozoa could be construed as parasitic stirring rods whose function, in a laboratory dish, might be fulfilled by a glass rod.[94] And while the role of fathers generally in conception was settled more than a century ago, until very recently it was impossible to prove that any particular man was father to any particular child. In these circumstances, believing in fathers is like, to use Freud's analogy, believing in the Hebrew God.

The Judaic insistence that God cannot be seen—the graven-image proscription—"means that a sensory perception was given second place to what may be called an abstract idea." This God represents "a triumph of intellectuality over sensuality (*Triumph der Geistigkeit uber die Sinnlichkeit*), or strictly speaking, an instinctual renunciation." Freud briefs precisely the same case for fathers as for God in the analysis of Aeschylus' *Oresteia* that immediately follows his discussion of the second commandment. Orestes denies that he has killed his mother by questioning whether he is related to her at all. "Am I then involved with my mother by blood-bond?" he asks. "Murderer, yes," replies the chorus, pointing out quite rightly that she bore and nursed him. But Apollo saves the day for the defense by pointing out that, appearances notwithstanding, "the mother is no parent of that which is called her child, but only nurse of the new-planted seed that grows," "a stranger." The only true parent is "he who mounts."[95]

Here in the *Oresteia* is the founding myth of the Father. "Fatherdom (*Vaterschaft*), Freud concludes, "is a supposition" and like belief in the Jewish God is "based on an inference, a premiss." Motherhood (*Mutterschaft*), like the old gods, is evident from the lowly senses alone. Fatherdom too has "proved to be a momentous step"; it also—Freud repeats the phrase but with a more decisive military emphasis—is "a conquest (*einen Sieg*) of intellectuality over sensuality." It represents a victory of the more elevated, the more refined over the less refined, the sensory, the material. It is a world-historical *Kulturforschritt*, a cultural stride forward.[96]

The one-sex model can be read, I want to suggest, as an exercise in preserving the Father, he who stands not only for order but for the very existence of civilization itself. Ancient authorities make both philosophical and empirical arguments for the self-evident greater potency of the male over the female, for the absolute necessity of the genitor. If the female's seed were as potent as the male's, "there would be two principles of motion in conflict with one another," argued Galen. If woman had as much as possible of the "principle of motion," her seed would then essentially be the male's and act as one with it when mixed. Women would be men, and nature would be unnecessarily mixing two seeds. Or, if a female seed as strong as the male's need not be mixed to cause conception, then there would be no need for men at all (*UP* 2.pp632–33). (A late medieval alternative argument holds that if woman's semen were as strong as men's, then either parthenogenesis is possible—which it is not—or woman's contribution to generation would be greater than man's because she would be providing not only an active agent but also the place for conception. This, in a hierarchical world, is *ex hypothesis* impossible.[97]) If women had seed as potent as males, they could inseminate themselves and "dispense with men," Aristotle argued. A manifest absurdity (*GA* 1.18.722b14–15).

It is empirically true, and known to be so by almost all cultures, that the male is necessary for conception. It does not of course follow that the male contribution is thereby the more powerful one, and an immense amount of effort and anxiety had to go into "proving" that this was the case. Evidence based on observation of "wind eggs" (*hupenemia*)—eggs that are seemingly produced without the power of the male but that are consequently not fertile—and of *mola*—monstrous products of the womb attributed to self-insemination—seemed to bear testimony to the hierarchical ordering of the one sex. Her sperma could not ensoul matter; his could. Perhaps the confident assertions that "there needs to be a female," that the creator would not "make half the human race imperfect and, as it were, mutilated, unless there was to be some great advantage in such a mutilation," hides the more pressing but unaskable question of whether there needs to be a male. After all, the work of generation available to the senses is wholly the work of the female.[98]

But being male and being a father, having what it takes to produce the more powerful seed, is the ascendancy of mind over the senses, of order over disorder, legitimacy over illegitimacy. Thus the inability of women to conceive within themselves becomes an instance—among many other things—of the relative weakness of her mind. Since normal conception is, in a sense, the male having an idea in the woman's body, then abnormal conception, the mola, is a conceit for her having an ill-gotten and inadequate idea of her own. Seeds of life and seeds of wisdom might well come to the same thing. Plutarch cautioned that

> great care must be taken that this sort of thing does not take place in women's minds. For if they do not receive the seed[s] (*spermata*) of good doctrines and share with their husbands in intellectual advances, they, left to themselves, conceive many untoward ideas and low designs and emotions.

Her mind and her uterus are construed as equivalent arenas for the male active principle; her person is under the rational governance and instruction of her husband for the same reason that her womb is under the sway of his sperm. Similarly, he should be able to control his own passions and manage hers while being able at the same time to "delight and gratify" her sufficiently to produce children. A man who is "going to harmonize State, Forum, and Friends" should be able to have his "household well harmonized."[99]

Christianity made the possibility of such harmony between good social order and good sexual order far more problematic than it had been in Roman antiquity. It radically restructured the meanings of sexual heat; in its campaigns against infanticide, it diminished the power of fathers; in its reorganization of religious life, it altered dramatically what it was to be male and female; in its advocacy of virginity, it proclaimed the possibility of a relationship to society and the body that most ancient doctors—Soranus was the exception—would have found injurious to the health.[100]

It is also true that Augustine, as Peter Brown has argued, discovered "the equivalent of a universal law of sexuality," which represents a shift in the whole relation of human beings to society. It might stand as a metaphor for the end of the classical age and for the remaking of community associated with the rise of Christianity.[101] One's intimate experiences of sex, in this new dispensation, were the result not of an ineluctable heating of the body but of the fall and of the estrangement of will that the fall brought. Impotence, far from being paradigmatically innocent, could be construed, even more than erection, as *the* sign of the soul's alienation from God.[102] Augustine could image intercourse in paradise in which the violence, the falling on wounds, the blood gushing, the crashing of bodies that informs an account like Lucretius', would be replaced by the image of intercourse as a gentle falling asleep in the partner's arms. Uncontrolled passion would be replaced by actions no more uncontrollable than the lifting of an arm. Indeed, everything about postlapsarian sex could thus be felt as continual reminders in the flesh of the tensions of the fundamentally flawed human condition. All of this was new with the coming of Christianity.

But Augustine's images for how "impregnation and conception" might be "an act of will, instead of by lustful cravings," were very much still of the old one-sex body found in the classical doctors. Such control of the body is conceivable, he suggested, and offered as an example people who "produce at will such musical sounds from their behind (without any stink) that they seem to be singing from that region." But the more telling case is that of a presbyter named Restitutus in the diocese Calama who, "whenever he pleased (and he was often asked to perform the feat by people who desired first-hand experience of so remarkable a phenomenon) he would withdraw himself from all sensations." He would, after some initial lamentations, lie unresponsive like a corpse. But one feature of this presbyter's trance makes it a particularly apt model for the phenomenology of intercourse in paradise. When he was burned "by the application of fire he was quite insensible to pain," until of course he emerged from his state and the normally occurring wound occasioned the usual pain.[103]

Here is a model for having the *calor genitalis* without concupiscence. But it is also a lesson in the physiology of the old Adam. Bodies, when exposed to fire, burn and except in rare circumstances, feel pain. Similarly with reproduction. Augustine did not envisage the modern body in which ovulation, conception, and even male ejaculation are known to be independent of whatever subjective feelings might accompany them. Heat and pleasure remained an ineradicable part of generation. It would be a miracle, said a fifteenth-century writer of confessionals, "to stand in the flame and not feel the heat." Intercourse, argued Pope Innocent III in a diatribe against the body, is never performed without "the itch of the flesh, the heat of passion, the stench of the flesh."[104]

Thus, after Augustine as before, the body was thought to work much as pagan medical writers had described it. Augustine's new understanding of sexuality as an inner, and ever present, sign of the will's estrangement by the fall did create an alternative arena for the generative body. As Brown says, it "opened the Christian bedchamber to the priest."[105] At the same time, it kept the door open for the doctor, the midwife, and other technicians of the old flesh.

Christian and pagan notions of the body coexisted, as did the various incompatible doctrines of the seed, of generation, and of corporeal homologies, because different communities asked different things of the flesh. Monks and knights, laity and clergy, infertile couples and prostitutes seeking abortion, confessors and theologians in myriad contexts, could continue to interpret the one-sex body as they needed to understand and manipulate it, as the facts of gender changed. It is a sign of modernity to ask for a single, consistent biology as the source and foundation of masculinity and femininity.

My purpose in this chapter has been to explain what I mean by the world of one sex: mind and body are so intimately bound that conception can be understood as having an idea, and the body is like an actor on stage, ready to take on the roles assigned it by culture. In my account sex too, and not only gender, is understood to be staged.

Since I have been unwilling to tie the one-sex model to any particular level of scientific understanding of the body, and since it seems to have persisted over millennia during which social, political, and cultural life changed dramatically, the question I raised at the beginning of this chapter should perhaps be rephrased: why did the attractions of this model fade at all? I suggested two strong explanations for its longevity. The first concerns how the body was understood in relation to culture. It was not the biological bedrock upon which a host of other characteristics were supposedly based. Indeed, the paradox of the one-sex model is that pairs of ordered contrarieties played off a single flesh in which they did not themselves inhere. Fatherhood/motherhood, male/female, man/woman, culture/nature, masculine/feminine, honorable/dishonorable, legitimate/illegitimate, hot/cold, right/left, and many other such pairs were read into a body that did not itself mark these distinctions clearly.[106] Order and hierarchy were imposed upon it from the outside. The one-sex body, because it was construed as illustrative rather than determinant, could therefore register and absorb any number of shifts in the axes and valuations of difference. Historically, differentiations of gender preceded differentiations of sex.

The second explanation for the longevity of the one-sex model links sex to power. In a public world that was overwhelmingly male, the one-sex model displayed what was already massively evident in culture more generally: *man* is the measure of all things, and woman does not exist as an ontologically distinct category. Not all males are masculine, potent, honorable, or hold power, and some women exceed some men in each of these categories. But the standard of the human body and its representations is the male body.

THREE

New Science, One Flesh

The books contain pictures of all parts inserted into
the context of the narrative, so that the dissected
body is placed, so to speak, before the eyes of those
studying the works of nature.

VESALIUS, 1543

Across a millennial chasm that saw the fall of Rome and the rise of Christianity, Galen spoke easily, in various vernacular languages, to the artisans and merchants, the midwives and barber surgeons, of Renaissance and Reformation Europe. Various Latin translations, compendia, and Arabic intermediaries transmitted the one-sex body of antiquity into the age of print. "La matrice de la femme," writes Guillaume Bouchet in one late sixteenth-century potpourri of learning, "n'est que la bourse et verge renversée de l'homme" (The matrix of the woman is nothing but the scrotum and penis of the man inverted). A German doctor of no great fame pronounced, "Wo du nun dise Mutter sampt iren anhengen besichtigst, So vergleich sie sich mit allem dem Mannlichen glied, allein das diese ausserhalb das Weiblich aber inwendig ist" (Viewing the uterus along with its appendages, it corresponds in every respect to the male member except that the latter is outside and the former inside). Or "the likeness of it [the womb] is as it were a yarde reversed or turned inward, having testicles likewise," as Henry VIII's chief surgeon says in a matter-of-fact way. There was still in the sixteenth century, as there had been in classical antiquity, only one canonical body and that body was male.[1]

The various vernaculars also replicated in new voices the Latin and Greek linguistic complex of connections between organs to which we, in our medical texts, would give precise and distinctive names. *Bourse,* for example, Bouchet's word for scrotum, referred not only to a purse or bag but also to a place where merchants and bankers assemble. As bag, purse, or sack it bridges male and female bodies handily. "Purse" could mean both scrotum *and* uterus in Renaissance English.[2] An anonymous German text declares in a commonplace simile, "the uterus is a tightly sealed vessel, similar to a coin purse (*Seckel*)."[3] The womb "shuts like a purse (*bursa*)" after it draws up the male and female ejaculate, says the Pseudo-Albertus Magnus in his immensely popular and much translated *De secretis mulierum*.[4] Scrotum also links up with womb through its more social, economic meaning. *Matrice,* Bouchet's term for uterus, as well as the English variant *matrix,* had the sense of a place where something is produced or developed, as in "mountains are the matrices of gold." There is a suggestion here of the common trope of the uterus as the most remarkable, miraculously generative organ of the body. The "matrice" is thus the place where a new life is produced while "bourse" is a place where a different, and culturally less valued, kind of productivity, an exchange, takes place. Two different kinds of bags, two different ways of making and keeping money, link organs that today have no common resonances.

The body's pleasures also remained as intimately bound with generation as they had been for Hippocrates. "Much delight accompanies the ejection of the seed, by breaking forth of the swelling spirit, and the stiffness of Nerves," says the most ubiquitous sex guide in the western tradition.[5] Through a physiology shared with man, woman "suffers both wayes," the sixteenth-century physician Lemnius points out, and feels a double pleasure: "she drawes forth the man's seed, and casts her own with it," and therefore "takes more delight, and is more recreated by it."[6]

But amid these echoes of antiquity, a new and self-consciously revisionist science was aggressively exploring the body. In 1559, for example, Columbus—not Christopher but Renaldus—claims to have discovered the clitoris. He tell his "most gentle reader" that this is "preeminently the seat of woman's delight." Like a penis, "if you touch it, you will find it rendered a little harder and oblong to such a degree that it shows itself as a sort of male member." Conquistador in an unknown land, Columbus stakes his claim: "Since no one has discerned these projections and their workings, if it is permissible to give names to things discovered by me, it should be called the love or sweetness of Venus."[7] Like Adam, he felt himself entitled to name what he found in nature: a female penis.

Columbus' account is significant on two levels. First it assumes that looking and touching will reveal radically new truths about the body. The discoverer of the clitoris had nothing but contempt for his predecessors, who either did not base their claims on dissection at all or failed to report accurately and courageously what they had seen. Mondino de' Luzzi (1275–1326), for example, the premier medieval anatomist, was made the butt of heavy irony for his perfectly commonplace though relatively novel claim that the uterus had seven cells; he "might as well have called them the porches or bedrooms."[8] Columbus' colleagues, meanwhile, attacked him with equal vigor. Gabriel Fallopius, his successor at Padua, insisted that he—Fallopius—saw the clitoris first and that everyone else was a plagiarist.[9] Kaspar Bartholin, the distinguished seventeenth-century anatomist from Copenhagen, argued in turn that both Fallopius and Columbus were being vainglorious in claiming the "invention or first Observation of this Part," since the clitoris had been known to everyone since the second century.[10]

The somewhat silly but complicated debate around who discovered the clitoris is much less interesting than the fact that all of the protagonists shared the assumption that, whoever he might be, someone could claim to have done so on the basis of looking at and dissecting the human body. A militant empiricism pervades the rhetoric of Renaissance anatomists.

Columbus' discovery would also seem to be fatal, or at the very least threatening, to the ancient representations of the one-sex body. Within the constraints of common sense, if not logical consistency, women cannot have a full-size penis within (the vagina) *and* a small homologue of the penis without (the clitoris). But Renaissance writers drew no such inference. Jane Sharp, a well-informed seventeenth-century English midwife, asserts on one page that the vagina "which is the passage for the yard, resembleth it turned inward" and, with no apparent embarrassment, reports two pages later that the clitoris is the female penis: "it will stand and fall as the yard doth and makes women lustful and take delight in copulation."[11] Perhaps these positions can be reconciled in that the vagina only resembles the penis whereas the clitoris actually is one; both maintain the one-sex model's insistence on the male as the standard. But Sharp had no interest in the question. Two seemingly contradictory accounts coexisted quite neatly, and the old isomorphism dwelt in peace with the strange new homologue from another conceptual galaxy.

Just when Columbus threatens to offer a new understanding of sexual difference, his text returns to the old track and the old tensions. Woman disappears, whether the vagina or the clitoris is construed as the female penis. Sexual delight continues to flow from the homoerotic rubbing of like on like; pleasure is decoupled from the will so that her mind does not matter. "If you rub it [the clitoris] vigorously with a penis, or touch it even with a little finger, semen swifter than air flies this way and that on account of the pleasure, even with them [women] unwilling."[12] There remains but one sex, or in any case only one kind of body.

The discovery of the clitoris and its easy absorption by the one-sex model raises the central question of this chapter. Why did competent observers, self-consciously committed to new canons of accuracy and naturalistic illustration, continue to think of reproductive anatomy and physiology in a manner that is manifestly wrong and egregiously counterintuitive to the modern sensibility? In the first place, much of what is at stake is not empirically decidable. Whether the clitoris or the vagina is a female penis, or whether women have a penis at all, or whether it matters, are not questions that further research could, in principle, answer. The history of anatomy during the Renaissance suggests that the anatomical representation of male and female is dependent on the cultural politics of representation and illusion, not on evidence about organs, ducts, or blood vessels. No image, verbal or visual, of "the facts of sexual difference" exists independently of prior claims about the meaning of such distinctions.[13]

But there are empirically decidable contentions in Columbus' report and in the one-sex model generally. The clitoris (*dulcedo amoris*) he rightly says is the primary locus of venereal pleasure in women. On the other hand, he maintains—wrongly from a modern perspective—that semen, which looks very much like the male's, flies this way and that when it is stimulated and, were it not to do so, women would not conceive.[14] These are meant to be verifiable claims with the body as proof text:

> You who happen to read these laboriously produced anatomical studies of mine know that, without these protuberances [the clitoris] which I have faithfully described to you earlier, women would neither experience delight in venereal embraces nor conceive any fetuses.

> This is truly noteworthy: testes are produced in women so that they may produce semen. Indeed I myself can bear witness that, in the dissection of female testicles, I have sometimes found semen that is white and thick and very well concocted, as all the spectators have acknowledged with one voice.[15]

The specific claim that female orgasm was necessary for conception was, moreover, known to be vulnerable since antiquity.

Aristotle had pointed out that women in some circumstances could conceive "without experiencing the pleasure usual in such intercourse" and that conversely "the two sexes could reach their goal together" and the woman still not conceive.[16] Giles of Rome, a thirteenth-century scholar who was known even in that age of prolixity as "the verbose doctor," had argued at great length, on theoretical grounds, that the so-called female seed was essentially irrelevant to conception and that female orgasm was still more irrelevant. But he also offered empirical evidence of various sorts. Women purportedly told him that they had conceived without emission and presumably orgasm. Moreover, a clinical report by no less an authority than Averroës (ibn-Rushd, 1126–1198), the Arabic philosopher and author of a major medical encyclopedia, tells of a woman who became pregnant from semen floating in a warm bath. If, as this case is meant to show, penetration itself is only incidental to fertilization, how much more irrelevant still is female sexual pleasure?[17] And two thousand years after Aristotle, William Harvey repeated the old argument (though based, he says, on the evidence of "an infinite number" or at least "not a few" cases): the "violent shaking and dissolution and spilling of humours" which frequently occurs "in women in the ecstasy of coitus" is not required for the real work of making babies.[18]

It is also hard to believe that the consumers of vernacular medical literature—a wide swath of the literate public and those who might listen to them—needed the weight of tradition and learning to tell them that female orgasm did not always accompany conception.[19] Modern studies are quite consistent in showing that one third and perhaps as many as one half of women never have orgasm from intercourse alone, and certainly nowhere near such a proportion were infertile.[20] Maybe a higher percentage were orgasmic in an age in which what is now called "foreplay" was taken as a requisite prelude to procreative intercourse, but a great deal of everyday experience must nevertheless have belied the purported link between female orgasm and conception. Yet neither the evidence of the learned nor the actual experiences of marriage overturned the old model of bodies and pleasures.

Of course, some might say: those who knew—women—did not write and those who wrote—men—did not know. But this is not so telling a point. In the first place, the Hippocratic corpus and book 10 of Aristotle's *History of Animals,* for example, may well represent the voices of women, and other works give accounts much like these. Moreover, when women beginning in the Renaissance did publish on midwifery and reproduction, their views regarding the physiology of generation were entirely mainstream: Louise Bourgeois, Jane Sharp, and Madame de la Marche all propounded the common wisdom linking pleasure, orgasm, and generation. The occasional first-person account by women addressing these intimate matters, such as the remarkable autobiography of a seventeenth-century Dutch clergyman's wife, Isabella De Moerloose, further suggests that the literature I am citing reports commonly held beliefs.[21] Despite the growing tendency of the learned tradition to distance itself from "popular errors," my sense is that doctors, lay writers, and men and women in their beds shared a broad view on how the body worked in matters of reproduction.[22] The sort of highly politicized split between women's views of their bodies and that of a medical establishment would have to await the consolidation of a science-based profession beginning in the eighteenth, but not fully in place until the late nineteenth, century.[23]

Finally, there is modern evidence to suggest that women in the past might well have had no more or no less understanding of the timing and physiology of conception than did their doctors. Certainly, if advice columns are any indication, the view that orgasm is necessary for conception lives on today; physicians, both male and female, who in the early twentieth century attempted through interviews to determine the timing of ovulation during the menstrual cycle, failed to come up with consistent answers. And anthropological evidence suggests that living women whom one can interrogate actually hold views similar to those propounded by Renaissance midwifery and health guides. Thus an informant in Suye Mura told a Japanese-speaking woman anthropologist that "she [thought] that if a woman does not reach climax, she cannot conceive because her womb remains shut."[24] The Samo of Burkino Faso give an account of semen—"sex water" discharged by both men and women—blood, milk, and menstruation that is eerily like the one that dominated the western tradition.[25]

None of this argues against the fact that there must have been much local wisdom and a florid oral tradition among women in early modern Europe, which printed sources, no matter how popular, and modern evidence, no matter how wide-ranging, can never recapture. They are forever lost to historians. Nor does it prove that ordinary people, men or women, thought very much in terms of the anatomical isomorphisms of the one-sex model. Nevertheless, it does suggest that the sort of literature on which I base these chapters—the only sort we are ever likely to have—shares the same conceptual universe of Renaissance people and even of "those who knew (women)," even if it does not speak in their voices.

Evidence bearing on the empirically testable claims of the one-sex model failed to dislodge them not because such data were silenced but because these claims were part of a far more general, intricate, and many-stranded conception of the body which no observations, singly or in combination, could directly falsify. Willard Quine suggests why this should be the case on philosophical grounds. The totality of our beliefs "is a man-made fabric which impinges on experience only along the edges." So-called knowledge, switching metaphors,

> is like a field [which] is so underdetermined by its boundary conditions, experience, that there is much latitude as to what statements to reevaluate in the light of any contrary experience. No particular experiences are linked with any particular statements in the interior of the field.[26]

The ancient account of bodies and pleasure was so deeply enmeshed in the skeins of Renaissance medical and physiological theory, in both its high and its more popular incarnations, and so bound up with a political and cultural order, that it escaped entirely any logically determining contact with the boundaries of experience or, indeed, any explicit testing at all.[27]

This is by now so standard an argument in the history and philosophy of science that it even has a name: the Quine-Duhem thesis. But it is worth making again for two reasons. The empirically testable claims of the old model, which represent and are represented by the transcendental claim that there exists but one sex, are so farfetched to the modern scientific imagination that it takes a strenuous effort to understand how reasonable people could ever have held them. It is an effort worth making, if only to unsettle the stability of our own constructions of sexual difference by exposing the props of another view and by showing that the differences that make a difference are historically determined.

Second, by making manifest the web of knowledge and rhetoric that supported the one-sex model, I am setting the stage for its challengers in the eighteenth and nineteenth centuries. If its stability can be attributed to its imbrication in other discursive modes, its collapse will not need to be explained by a single dramatic discovery or even by major social upheavals. Instead, the construction of the two-sex body can then be viewed in the myriad new, and new kinds of, connections between, and within, sexual and other discourses.

The practices of anatomy

"When you meet a human being," said Freud in his comments on "Femininity" in *New Introductory Lectures,* "the first distinction you make is 'male or female?' and you are accustomed to making the distinction with unhesitating certainty." Anatomical science at first seems to support this certainty but upon further reflections turns out to be far less authoritative: "what constitutes masculinity or femininity is an unknown characteristic anatomy cannot lay hold of." The more Renaissance anatomists dissected, looked into, and visually represented the female body, the more powerfully and convincingly they saw it to be a version of the male's.

The body speaks itself. In large measure the new science greatly strengthened the old model simply because it proclaimed so vigorously that Truth and progress lay not in texts, but in the opened and properly displayed body.[28] A rhetoric of bad-mouthing reinforced the idea that only error and misguided adherence to authority stood in the way and that with care one could *see,* among many other things, that women were inverted men. Vesalius publicly denounced the whole lot of his predecessors, including his teacher Jacobus Sylvius, for considering Galen infallible, and Columbus could write of the "by no means negligible corrections" he had to make in Vesalius to produce a dissecting guide that "will tell the truth about the human body." [29] Fallopius announced that he would refute the accounts of ancient and more modern writers and completely overturn some of their doctrines, "or at least make them totter."[30]

More important, the new, extravagantly public theatrical dissection and its visual representations advertised the conviction that the opened body was the font and touchstone of anatomical knowledge.[31] What had been hidden before—there was very little if any human dissection in antiquity and no anatomical illustration—and what had been practiced only occasionally and quietly—anatomy in medieval universities—was now made available for general consumption. One need no longer imagine Galen's topographical transformations; one could verify them by sight. As Harvey Cushing argues, the famous frontispiece to Vesalius' *De humani corporis fabrica,* the founding work of modern anatomy (fig. 3), stands as

Fig. 3. Sixteenth-century dissection scene from the frontispiece to Vesalius' epochal *De humani corporis fabrica* (1543).

a rebuke to those who only read ancient texts while barber surgeons did the dissection. Compare it, for example, to the frontispiece to Mondino's *Anathomia* (figs. 4 and 5), the medical-school standard before Vesalius. Text, in the form of the name of the book, or a reader expounding *ex cathedra* dominate the earlier pictures. The body seems almost an afterthought, lying passively within the picture's plane. The anatomist's gaze in fig. 5 lights on the cadaver's face, not on its exposed viscera, as if its humanity, not its value as dead material to be studied, demands attention. Vesalius must have imagined scenes like these when he condemned ana-

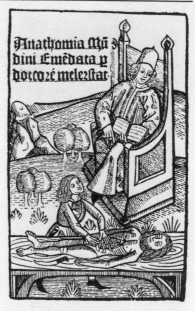

Fig. 4. Frontispiece to Johan Ketham, *Fasciculus medicinae* (Venice, 1550), a reworking of Mondino's *Anathomia*.

Fig. 5. Frontispiece to Mondino [Mundinus], *Anathomia* (1493).

tomists who "from a lofty chair arrogantly cackle like jackdaws about things they have never tried." A butcher in his meat market could teach a doctor more.[32]

By contrast, in fig. 3 the opened body is the unquestioned font of authority, enforced by the lordly skeleton that presides over the scene. Unlike the bodies in earlier representations, it comes out at us from the plane of the picture; its exposed entrails occupy dead center between the title and the bottom of the picture. An imaginary line passes down the spine of the skeleton, between its breasts and through the viscera, bisecting the image and dividing the magnificent rotunda in which the cadaver lies. Classical statues lend dignity, as they will later in the book, when the viscera are displayed in them, mediate the violence of dissection, and define the features displayed as those of a normative, median body. And, as in the frontispieces to many Renaissance anatomies, a great concourse of assorted observers looks on. This is a picture, in short, about the majestic power of science to confront, master, and represent the truths of the body in a self-consciously theatrical and public fashion.[33]

Fig. 6. Frontispiece to a 1642 Dutch edition of Vesalius' *Epitome* (1543).

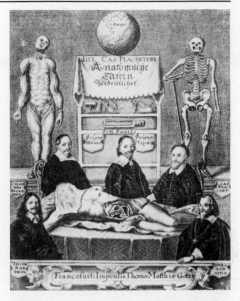

Fig. 7. Frontispiece to G. Cassario, *Anatomische Tafeln* (1656), which is a reworking of the scene in fig. 6.

The picture may seem to be, more narrowly, an assertion of male power to know the female body and hence to know and control a feminine Nature.[34] Vesalius presides here over an assemblage of men who peer into a woman's helpless, naked, and revealed body before them. The cadaver in the frontispiece (fig. 6) to a later Dutch edition of Vesalius' *Epitome,* a sort of student guide to the larger *Fabrica,* is still more shapely, her generative organs more clearly shown, her face mysteriously veiled so as to emphasize the accessibility to her body to the male gaze. Even the banner bearers are men, the sex of the skeleton evident from his cape and gravedigger's shovel.

But the politics of gender in anatomical illustration is not so simple. The frontispiece to Cassario's *Anatomische Tafeln* (fig. 7) takes the engraving used in fig. 6 and substitutes a man's body for the woman's. His face is also draped, his body is if anything more subject to domination by the instruments behind him and by the knife resting on his thigh. The young and extraordinarily eroticized cadaver being dissected in fig. 8, the frontispiece to John Riolan's text, is clearly a man though androgynously del-

Fig. 8. Frontispiece to Jean Riolan, *Les Oeuvres anatomiques* (1629). The male cadaver is if anything more erotically portrayed than either the male or female in figs. 6 and 7.

icate in his features. More generally, it simply is not true that women, sensual or not, were particularly identified with the object of anatomical study. In the frontispieces of fourteen anatomy books published between 1493 and 1658, the body being dissected is male in nine cases, female in four, and indeterminate in one. Perhaps the availability of material rather than sexual politics determined the sex of the generic cadaver.[35] In any case, the body qua body is what matters, and the programmatic point of the Renaissance anatomical frontispiece is clear: anatomists have the power to open the temple of the soul and reveal its inner mysteries (fig. 9 is paradigmatic on this point).[36]

The bodies of women must be seen in the context of two further representational strategies, both of which emphasize the theatrical display of bodies as testimony for the anatomist's claims. In the first place, even when medieval anatomies—and indeed even Renaissance books before Jacopo Berengario da Carpi's *Isagoge brevis* in 1522—were illustrated, that is, rarely, what pictures they did contain were at best superficially connected with the text, whose authority rested in the words and reputation of the author. In Berengario, however, something novel was happening. He was committed to an *anatomia sensibilis,* an anatomy of what could be seen, and illustrations were to be its printed aspect, the graphic substitute for actually seeing the structures in question and thereby vouchsafing the anatomist's words.[37] The frontispieces and the many spectacular engravings in Vesalius and subsequent works continued to invoke the authority, first, of a dramatically opened, exposed body and then, derivatively, of naturalistic representation itself.[38]

Even without words, these new illustrations were advertisements for their own truth. In them the dead act as if they were still somehow alive—not cadavers at all—and thus able to certify personally the facts that the anatomist presents and the epistemological soundness of anatomy generally. The thoroughly classical muscle man in Juan de Valverde's *Anatomia* (fig. 10) flays himself to reveal his surface structures, holding

Fig. 9. Frontispiece, after a drawing by Paolo Veronese, to Columbus, *De re anatomica* (1559).

Fig. 10. Classical figure, having flayed himself, displays both his skin and his surface musculature. From Juan de Valverde, *Anatomia del corpo umano* (1560).

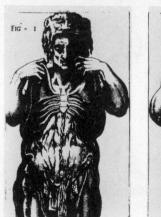

Fig. 11. Three figures in various tortured poses of revealing themselves to the readers of an anatomy text. From Valverde, *Anatomia*.

up his skin—an allusion to Michelangelo's self-portrait, part Marsias, part St. Bartholomew, from the *Last Judgment*—for extra emotional appeal.[39] Later in Valverde's book a rather self-absorbed creature calmly lifts up his belly's fat and skin to show off his abdominal fascia; for our viewing convenience, the next figure holds up still more of his fleshly clothes to reveal the omentum beneath. He gestures with his left hand and turns, as if modeling or rehearsing on stage, to ask the artist or director who hired him whether this pose or gesture will do. A third fellow needs both his hands and his teeth—they hold up the omentum—to assure us an unobstructed vista of his viscera (fig. 11). In a Belgian edition of the *Epitome* (fig. 12) an opened anatomist—no greater sacrifice in the interests of science is possible—looks heavenward as his fingers resect the ribs of a Vesalian Apollo Belvedere or perhaps himself. Various well-proportioned men in Estienne's *La Dissection des parties du corps humain,* the most lavishly produced of the pre-Vesalian anatomies, look more or less pleased, pained or pathetic, as they tear themselves apart for their viewer's somewhat minimal anatomical edification (figs. 13–14).

The art and rhetoric of Renaissance anatomies thus proclaim the authority of seeing and the power of dissection. Various stratagems for cre-

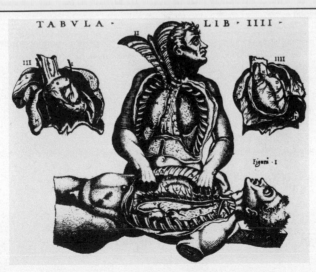

Fig. 12. One anatomized cadaver dissecting another who is represented as a fleshly version of a broken classical statue. Original also from Valverde's *Anatomia* but borrowed by a 1559 Bruges edition of Vesalius' *Epitome.*

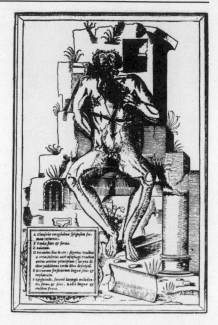

Figs. 13–14. Two male figures ripping themselves open for the edification of viewers. The "martyrdom" on the right reveals the tongue and tonsils, the one on the left the lower abdomen and genitals. From Charles Estienne, *La Dissection des parties du corps humain* (1546).

Fig. 15. A female sculpture has suddenly come alive and is leaving her pedestal to demonstrate the text's claim that the uterus is like the penis and that testicles and various vessels also correspond. From Jacopo Berengario, *Isagoge brevis* (1522).

Fig. 16. The model has left her
pedestal and gestures flamboyantly to her uterus. "You see," she
says, "how the neck of the womb
resembles a penis." From Berengario.

ating the "reality effect" make pictures stand in for bodies themselves and
witness the truths of texts that viewers are invited to construe as only one
remove from the cadaver itself. Seeing is believing the one-sex body. Or
conversely.

Believing is seeing. The new anatomy displayed, at many levels and with
unprecedented vigor, the "fact" that the vagina really is a penis, and the
uterus a scrotum.[40] Berengario makes absolutely sure that his readers do
not miss or doubt the point: "the neck of the uterus is like the penis, and
its receptacle with testicles and vessels is like the scrotum."[41] In the first
of the pictures accompanying this by now familiar assertion, a classical
statue of a decidedly feminine woman seems miraculously to have come
alive; she is in the process of throwing off her wrap and stepping carefully
down to confront the reader with proof (fig. 15). In the next one (fig.

16) she flamboyantly tosses her cloak over her head with one hand, while with the other she directs her audience's gaze to what has been removed from her open belly and placed on the pedestal from which she descended: her uterus. She—the now animated cadaver whose voice has become indistinguishable from the anatomist's—gestures epideictically and announces with obvious authority: "you see how the neck [of the uterus] . . . resembles a penis" (p. 78). Finally, a third close-up illustration

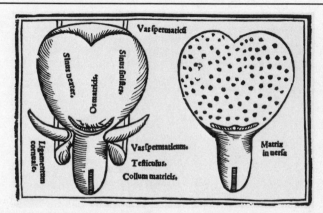

Fig. 17. The uterus and attached vessels labeled so as to make clear once again—"because a tenfold repetition is wont to please"—the correspondences between male and female organs. From Berengario.

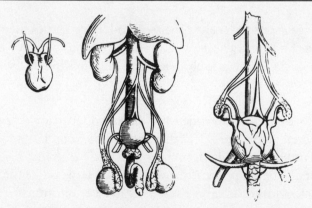

Fig. 18. Male and female organs displayed to demonstrate their correspondences. From Vesalius, *Tabulae sex* (1538).

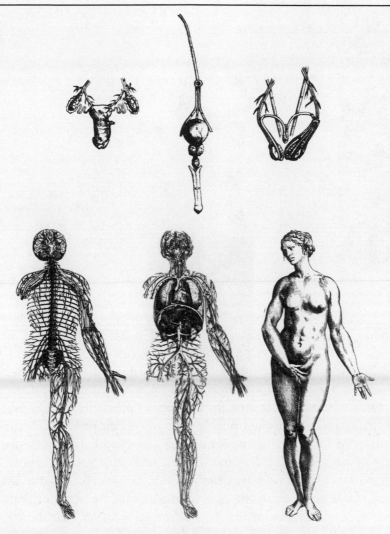

Fig. 19a–d. Top row (19a): the shorter penislike structure is the "uterus with the testes and seminal vessels"; the longer one is the male genitalia to which the student is then asked to attach the male testes. Both male and female organs were then to be glued onto fig. 19b, which in turn fit under 19c and then under 19d, a classical female nude. From Vesalius, *Epitome*.

hammers home the point visually and through labels that identify the ovaries as testicles and the Fallopian tubes as spermatic ducts (fig. 17).

Women's organs are represented as versions of man's in all three of Vesalius' immensely influential and widely plagiarized works. Among the

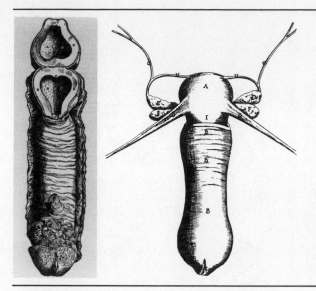

Fig. 20. (left) Vagina as penis from Vesalius, *Fabrica*.

Fig. 21. (right) The vagina and uterus from Vidus Vidius, *De anatome corporis humani* (1611)

founding images of modern anatomy is a powerful new register for the old ordering of bodies. His most reprinted image of the vagina as penis, and also the most explicit, is one of the illustrations (fig. 18) from the *Tabulae sex,* a set of cheaply printed pictures, so-called fugitive plates prepared for medical students or for lay consumption. In the *Epitome,* engravings of almost indistinguishable male and female reproductive organs are included for students to cut out and glue onto figures provided for that purpose (fig. 19).[42] But the most visually striking of Vesalius' pictures on this theme is in the *Fabrica* itself. Here (fig. 20) the uterus, vagina, and external pudenda of a young woman are not specifically arrayed, as in the *Tabulae* or the *Epitome,* to demonstrate that these structures are isomorphic with those of the male; they are just *seen as* such.

I emphasize "seeing as" because these images, and many more like them, are neither the result simply of representational conventions nor the result of error. A whole world view makes the vagina look like a penis to Renaissance observers. Of course a representational convention, a schema, is at work; Renaissance anatomical illustrators learned to depict the female genitalia from other pictures and not from nature alone (see figs. 21–24). But this does not mean that stylistic concerns kept them from seeing genital anatomy "as it really is," or as moderns see it.[43]

Nor is the strange quality of images in figs. 15–24 the result of someone's efforts to make the female body conform to some erroneous text or to distort women's genitalia so that they become a caricature of men's. The draftsman who produced fig. 21, for example, is not guilty of clandestinely substituting animal for human anatomy, as Vesalius coyly accuses Galen of doing in the *Fabrica*'s famous juxtaposition of a woodcut of a canine premaxillary bone and suture with those of a man (fig. 25). He is, moreover, innocent of what Vesalius himself did on occasion: "seeing" something that does not exist because an authority declares it to be present.[44] There are gross errors of this sort in Renaissance illustrations of the female genitalia, but they are irrelevant to the rhetorical purposes of the illustrations. In fact, if they were more accurate, they would make their point even more powerfully. If, for example, in figs. 16–17 the nonexistent "cotyledons"—the dots representing the anastomosis of veins in the uterus—were rubbed out, the suggestion of two chambers eliminated, and the vagina drawn in correct proportion to the uterus, the organs would resemble a female scrotum and penis more closely. Expung-

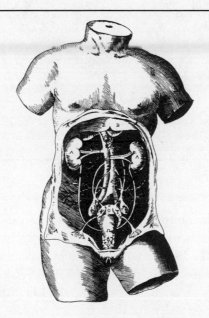

Fig. 22. The female torso, in the form of a piece of broken classical art, from which the penis-like vagina in fig. 21 was taken, following the artistic and scientific conventions of the time.

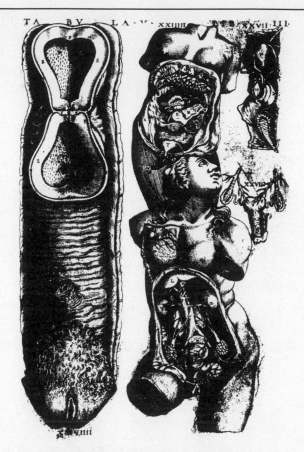

Fig. 23. This reworking of Vesalius in a 1586 edition of Valverde follows the same convention illustrated in figs. 21–22. On the left is a structure that looks like a penis; on the right are the classical female forms from which it was taken.

ing the "horns of the uterus" (GG) from John Dryander's representation of the female reproductive organs (fig. 26) or from other Renaissance illustrations (figs. 32–33 for example) would make the uterus and vagina look more, not less, like a bladder and penis; and redrawing, in the interests of accuracy, the ovarian artery and vein EE in fig. 26 so that they appear less like the epididymis, II in fig. 27, would, at worst, leave the overall effect the same.[45]

However grotesque or monstrous the woodcut of the female genitalia

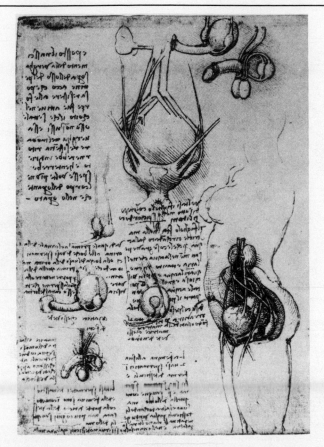

Fig. 24. Leonardo's version of the isomorphism between the womb and scrotum—upper right and lower left—is peculiar in that he renders it by making the vas deferens of the male curve around to resemble the shape of the uterus. The penis/vagina imagery is more conventional.

depicted in the *Fabrica* has appeared to some modern commentators, it is not incredible or "wrong." Its proportions are roughly those of "accurate" nineteenth-century engravings (fig. 28) and illustrations from a modern text (fig. 29), though these of course were not drawn to illustrate the isomorphism between male and female organs.[46]

Subsequent discoveries that would force changes in the labels of illustrations are of equally minor importance in the history of "seeing as." The *Zeuglin,* or testes, and the *Samadern,* seminal vesicles, did not exist, as

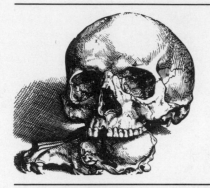

Fig. 25. "We have placed," Vesalius says in this polemical illustration from the *Fabrica*, "the skull of a dog beneath that of a man so that anyone may understand Galen's description of the bones of the upper jaw without the slightest difficulty."

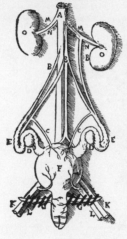

¶ Dise Figur zeyget an die innerliche gestalt eins weibs/ mit sampt den geburt glider ren / gefäß des samens/ vnnd andern bericht. A. Bedeut die großblütader / daher alle andere glider narung haben. B. Ist die weisse samadern. C C. Ader so die bermütter begreiffen / daher die frucht auch narung bekompt. D. D. Sindt weibs zeuglin. E. Da mit werden die weibs zeuglin vmbgeben/ seindt ein theyl same. vnnd ein theyl der hertza dern. F. Die bermütter gleych der blasen gestalt. G. Die ge stalt der Bermütter / daran sie dem rucken vnnd nebentu angeheffte. H. Das innerlich mundtloch der Berinütter. I. Das eusserst der Bermüt ter: die scham. K. L. Stämm odder äst der blütadern der schenckel. M. M. Harngäng vonn den Nieren. O. Bede Nieren.

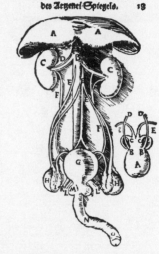

des Artzenei Spiegels.   13

Das neben klein figürlinn/ ist die blase/ mit sampt der harn vnd same adern.

Figs. 26–27. The male and female reproductive systems adapted from Vesalius' *Epitome* in Johan Dryander, *Der Gantzen Artzenei* (1542). In fig. 26 I have blocked out the nonexistent horns of the uterus to show that making a drawing like this more accurate would also make them more convincing as illustrations of the penis/vagina isomorphism. Elongating the vagina so that it is in proper proportion to the uterus would have the same effect.

Dryander's labeling claims, in both men and women; nineteenth-century histology would teach that nothing of interest follows from the observation that the uterus, labeled F in fig. 26, has the same shape as the male bladder, G in fig. 27. But these advances pale beside facts that Renaissance anatomists did know and that did nothing to discredit the whole representational convention of seeing the female genital anatomy as an interior version of the male's. The uterus bears children but the scrotum

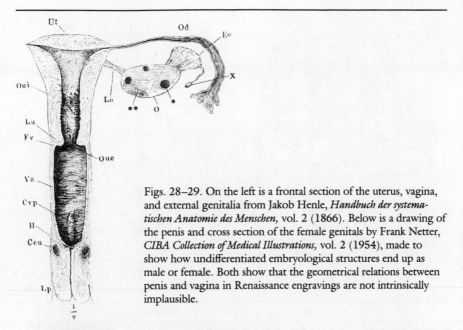

Figs. 28–29. On the left is a frontal section of the uterus, vagina, and external genitalia from Jakob Henle, *Handbuch der systematischen Anatomie des Menschen,* vol. 2 (1866). Below is a drawing of the penis and cross section of the female genitals by Frank Netter, *CIBA Collection of Medical Illustrations,* vol. 2 (1954), made to show how undifferentiated embryological structures end up as male or female. Both show that the geometrical relations between penis and vagina in Renaissance engravings are not intrinsically implausible.

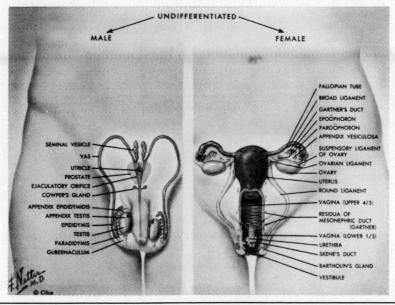

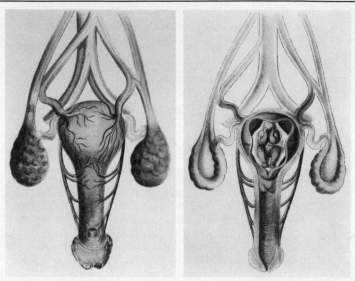

Figs. 30–31. On the left are the penislike female organs of generation from Georg Bartisch, *Kunstbuche* (1575). On the right the front of the uterus is cut away to reveal its contents.

does not; babies are delivered through the vagina and not through the penis. So what? The organ in fig. 30, for example, might be a vagina from a woman or a penis from a man. Fig. 31 relieves the suspense. It is a vagina, we now know, because what might have been either a scrotum or a uterus turns out to contain a child! The womb with its penislike extension in Walther Ryff's popular and widely translated book plays the same trick, as it becomes strangely transparent to allow readers a view of the fully formed baby within (fig. 32). A little window has been cut into the female scrotum, the uterus, in figs. 33–34, an illustration from another well-known midwifery book, to show a fully formed child, its back turned to intruders and to the penile vagina through which it will pass.

The history of the representation of the anatomical differences between man and woman is thus extraordinarily independent of the actual structures of these organs or of what was known about them. Ideology, not accuracy of observation, determined how they were seen and which differences would matter.

Seeing difference differently. Renaissance "common sense," and critical observation directed against the view of woman as man turned outside in,

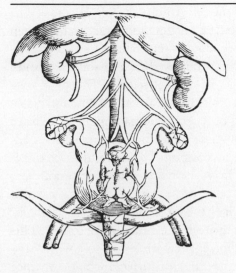

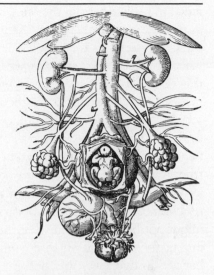

Fig. 32. The female organs of generation from Walther Ryff, *Anthomia* (1541). In this and the next illustration note that the vagina and uterus would look more like a penis and scrotum if the horns were expunged and the vagina drawn in correct proportion, that is, if they were more accurate.

Fig. 33. The female organs of generation from Jacob Rueff, *Habammenbuch* (1583), which appeared in English as the widely plagiarized and popular *The Expert Midwife* (1637). Note that the left ureter has been cut and the bladder pushed to the right from its natural position so that we might look into the window of the womb and see the child.

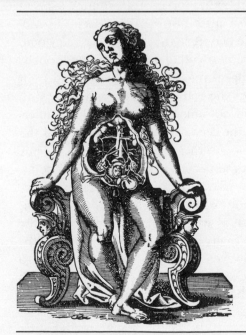

Fig. 34. The gravid uterus with its penile vagina of fig. 33 *in situ*. The bladder has been pushed left, and the child shows its profile.

failed to make a dent in the one sex-model. Arguments against the vagina as penis, for example, are to the modern imagination stranger even than the claim itself. At the simplest level, an apparent failure to find equivalences between men and women could be saved by the sort of wishful thinking that daily saves phenomena in normal science. Except in moments of revolutionary crisis, there is always a way out. Women may not seem to have a scrotum, and indeed other parts of man might be difficult to find in woman or vice versa. But these difficulties, argues Charles Estienne, can be resolved by reference to position: "You would agree this is true: if you turn a womb removed from the body inside out (quoth Galen) you will find testicles bulging out from its outer surface, by which the womb itself, by outer appearances is as a scrotum."[47] We might or might not be able to find what this anatomist claimed if we followed his instructions, but the exercise would be entirely irrelevant to a world that believes in two sexes. No pushing or pulling of surfaces would convince us to see the womb as a scrotum, any more than a topologist could make us regard a tea cup as a doughnut even if her procedures were sound, which Estienne's were not.

Conversely, perfectly sound anatomical observations adduced against the old homologies seem, from a modern perspective, so curiously peripheral—even perverse—that they serve only to cast further doubt on the whole enterprise of searching in bodies for any transcultural signs of difference. The distinguished English anatomist Helkiah Crooke argued, for example, against "any similitude betweene the bottome of the womb inverted [the cervix], and the scrotum or cod of a man," on the grounds that the skin of the "bottom of the wombe is a very thicke and tight membrane, all fleshy within" while "the cod is a rugous and thin skin." (True, but scarcely compelling, and not among the more telling differences that spring to mind between the cervix and the sack that holds the testicles.) Crooke's rejoinder to the claim that the vagina really is a penis is still more amazing. "Howsoever the necke of the wombe shall be inverted, yet it will never make the virile member," he proclaims. Why? Because "three hollow bodies cannot be made of one, but the yard consisteth of three hollow bodies" and, as we have already been told, "the necke of the womb hath but one cavity." (As figs. 35–36 make clear, Crooke is anatomically correct, however strange his argument seems to the modern sensibility.) Furthermore: "neither is the cavity of a man's yard so large and ample as that of the necke of the wombe." In short, the

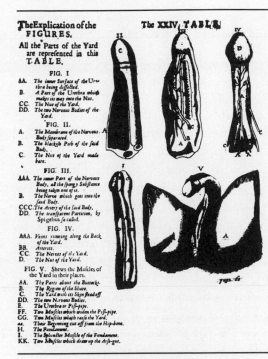

TheExplication of the
FIGURES.

All the Parts of the Yard are represented in this TABLE.

FIG. I.

AA. The inner Surface of the Urethra being dissected.
B. A Part of the Urethra which makes its way into the Nut.
CC. The Nut of the Yard.
DD. The two Nervous Bodies of the Yard.

FIG. II.

A. The Membrane of the Nervous Body separated.
B. The blackish Path of the said Body.
C. The Nut of the Yard made bare.

FIG. III.

AAA. The inner Part of the Nervous Body, all the spongy Substance being taken out of it.
B. The Nerve which goes into the said Body.
CCC. The Artery of the said Body.
DD. The transparent Partition, by Spigelius so called.

FIG. IV.

AAA. Veins running along the Back of the Yard.
BB. Arteries.
CC. The Nerves of the Yard.
D. The Nut of the Yard.

FIG. V. Shews the Muscles of the Yard in their places.

AA. The Parts about the Buttocks.
B. The Region of the Share.
C. The Yard with its Skin flead off.
DD. The two Nervous Bodies.
E. The Urethra or Piss-pipe.
FF. Two Muscles which widen the Piss-pipe.
GG. Two Muscles which raise the Yard.
aa. Their Beginning cut off from the Hip-bone.
H. The Fundament.
I. The Sphincter Muscle of the Fundament.
KK. Two Muscles which draw up the Arse-gut.

Fig. 35. Table 24 from Kaspar Bartholin, *Anatomy* (1668), showing "the parts of the yard." The drawing on the lower left shows the corpus spongiosum penis through which the urethra passes. In the drawing upper left, this passage is left intact and one of the two corpora cavernosa penis, the "nervous bodies" that were thought to produce erection, is excised: three hollows in all.

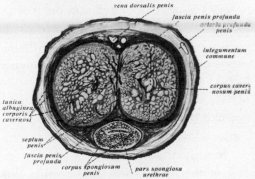

Fig. 36. Cross section of the penis from a modern atlas showing that indeed the penis does have three hollows, as Crooke said.

penis is not a vagina either because it is thrice hollow or because it is not hollow enough.[48]

But for others the hollowness test figured on the opposite side—in support of the Galenic isomorphisms—or at worst as irrelevant:

Whatever you see as a kind of opening in the entrance to the vulva [vagina] in women, such indeed is found in the foreskin of the male pudenda, like a kind of outgrowth hollow inside. The only difference between them is that this hollowness in much greater in woman than in the man.[49]

At work here is a sensibility radically different from that of doctors in the world of two sexes.

Even when the broader cultural context of the one-sex model was clear to a critic of the Galenic isomorphisms, a web of significance kept the attack narrowly focused and harmless to overarching structures. Bartholin, for example, understood Galenic sexual politics perfectly. "We must not," he argued, "think with Galen . . . and others, that these female genital parts differ from those of Men only in Situation," because to do so would be to fall prey to an ideological plot "hatched by those who accounted a Woman to be only an imperfect Man." Its perpetrators, in talking about how the woman's "coldness of temper" kept female organs inside, were simply articulating their prejudices in the language of science. (One would like to know how and why Bartholin developed so political and so astute a critique.) But, quite apart from politics, Bartholin criticized Galen and his followers for not getting their story straight. Was the "neck of the womb" or the clitoris the female penis; was the womb the female scrotum, or was at least part of it her version of the "nut of the yard"? And the spermatic preparatory vessels, he pointed out, differed in number, origin, and function in men and women, and the male has a prostate, which the female does not have.[50] Finally, illustrations hammered home the point. The clitoris is clearly rendered as *the* female penis while the womb and the vagina are portrayed in an unambiguously unpenile fashion (fig. 37).

But despite these well-developed and thoroughly articulated criticisms, Bartholin seemed incapable of transcending the ancient images he explicitly rejected. The orifice, or inner mouth of the womb (the cervix), he explained, functions "like the Hole of the Nut of the Yard," so that "no hurtful thing may enter in." The "neck of the womb"—note the use of the conventional term for the vagina—"becomes longer or shorter, broader or narrower, and swells sundry ways according to the lust of the woman." Its substance "is of a hard and nervous flesh, and somewhat spongy, like the Yard." The vagina, in other words, became once again in his imagination a penis. But the clitoris too, like the vagina, was also like the penis. It is "the female yard or prick," because it "resembles a man's yard in situation, substance, composition, repletion with spirits, and erec-

Fig. 37. Table 28 from Bartholin's *Anatomy* in which the vagina (I) is shown with its wall open and folded back so as to emphasize its hollowness. The external pudenda are no longer represented to look like the foreskin of the penis, and the clitoris (VI and VII) is clearly rendered as the female penis. These images were stolen by Venette and reprinted in his *Art of Conjugal Love* and its many translations.

tion" and because it "hath somewhat like the nut and foreskin of a Man's Yard."[51] Clearly Bartholin was caught up in a way of looking that kept him tied to the images of one sex. Indeed, the more he looked, the more he saw and the more muddled the picture became for him, with not one but two female penises to accommodate.

It did not, moreover, escape Renaissance observers that Galen's topological inversions led to ludicrous results. Again, nothing followed. The one-sex model absorbed yet another category of simile. Jacques Duval, a prominent seventeenth-century physician, for example, tried Galen's thought experiment and concluded quite rightly that "If you imagine the vulva (*vulve*) completely turned inside out . . . you will have to envisage a large-mouthed bottle hanging from a woman, a bottle whose mouth rather than base would be attached to the body."[52]

This bottle then "would bear no resemblance to what you had set out to imagine." To some, however, a bottle shaped like the vagina and womb hanging by its mouth *did* resemble a penis or scrotum enough to serve as the basis for a descriptive metaphor. William Harvey, discoverer of the blood's circulation, described a prolapsed uterus as "so rough and wrinkled as to take on the appearance of scrotum"; it hangs down, he said a few paragraphs later, "like the scrotum of a bull."[53]

Rabelais, in describing how Gargantua was dressed, also elided the distinction between the womb or, as in George Gascoigne's verse quoted below, a childbearing cradle, on the one hand, and the codpiece containing the penis and scrotum on the other.[54] True, the orange-sized emeralds on Gargantua's codpiece are said to be appropriate because "this fruit has an erective virtue." But then the pouch begins to appear as a finely embroidered and bejeweled horn of plenty, like that given by Rhea to the nymphs who nursed Jupiter. It is, the narrator says, while promising more in his forthcoming *On the Dignity of Codpieces,* "always brave, sappy, and moist, always green, always flourishing, always fructifying, full of humours, full of flowers, full of fruit, full of every delight."[55] The codpiece seems, in short, to have been transformed into the womb, which is not so odd given the ancient notion of the uterus as a belly and the late medieval sense of cod as a belly or bag. (Chaucer's Pardoner in *The Canterbury Tales* proclaims: "O wombe! O bely! O stynkyng cod.")

Moreover, the womb that to Duval seemed like a bottle hanging by its neck, and thus not a good candidate for the penis inverted, is the precise form of the codpiece, an obvious phallic sign in clothing whose visual representations are at the same time often decidedly unphallic (figs. 38–39). The codpiece tended to be, like Duval's bottle, broader at the end than at the base, blunt not sharp, decorated with ribbonlike braids. In the portrait of an unknown young aristocrat (fig. 40), it remains ambiguous whether the flower of betrothal he holds is an allusion to the hoped-for generative power of his penis or of the uterine structure in which it is coddled.[56] The codpiece indeed seems to bear a remarkable resemblance not just to a prolapsed uterus but to a swaddled child.

And this of course completes the circle back to Galen, to the womb as

Figs. 38–39. Jacobo Pontormo, *Albadiere* (1529–30). The codpiece in these pictures (close up on right) very much resembles Jacques Duval's bottle.

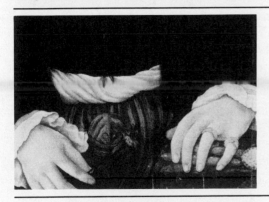

Fig. 40. Detail of *Portrait of a Young Man Before a Broad Landscape,* anonymous German painting of the 1530s, in which the codpiece is a sort of bundle for the penis. The boy holds the flower in his right hand; the bloom is to the right of his penis in the picture.

unborn penis, and to the Renaissance trope to the male organ as infant. Here is Gascoigne's "The Lullaby of a Lover":

> Eke Lullaby my loving boye,
> My little Robyn take thy rest . . .
> With lullaby now take your leave,
> With Lullaby your dreams deceive,
> And when you rise with waking eye,
> Remember then this Lullaby.[57]

Duval's argument thus turns in on itself and in a curious way makes the case against which it was directed. Seeing opposition in organs before the eighteenth century was far more problematic than would seem possible later.

The language of difference and sameness. I want to shift now from images to words. The absence of a precise anatomical nomenclature for the female genitals, and for the reproductive system generally, is the linguistic equivalent of the propensity to *see* the female body *as* a version of the male. Both testify not to the blindness, inattention, or muddleheadedness of Renaissance anatomists, but to the absence of an imperative to create incommensurable categories of biological male and female through images or words. Language constrained the seeing of opposites and sustained the male body as the canonical human form. And, conversely, the fact that one saw only one sex made even words for female parts ultimately refer to male organs. There was in an important sense no female reproductive anatomy, and hence modern terms that refer to it—vagina, uterus, vulva, labia, Fallopian tubes, clitoris—cannot quite find their Renaissance equivalents. (I think anatomy, more than physics, provides the paradigmatic case of Thomas Kuhn's argument that one cannot translate between theories across the chasm of revolution.)

There has, of course, always been in most languages a vast metaphoric elaboration of terms for organs and functions that are risqué or shameful. (When adolescent boys talk today about "getting a piece of ass," they are not referring to the anus.) Until the late seventeenth century, however, it is often impossible to determine, in medical texts, to which part of the female reproductive anatomy a particular term applies.[58]

"It does not matter," says Columbus with more insight that he was perhaps aware of, "whether you call it [the womb] matrix, uterus, or vulva."[59] And it does not seem to matter where one part stops and the other starts. He does want to distinguish the true cervix—the "mouth of the womb (*os matricis*)," which from the outside "offers to your eyes . . . the image of a tenchfish or a dog newly brought to light," which in intercourse is "dilated with extreme pleasure," and which is "open during that time in which the woman emits seed"—from what we would call the vagina, "that part into which the penis (*mentula*) is inserted, *as it were*, into a sheath (*vagina*).[60] (Note the metaphoric use of "vagina," the standard Latin word for scabbard, which was otherwise never used for the part to which it applies today.) But he offers no other term for "our" vagina, describes the labia minor as "protuberances (*processus*), emerging from the uterus near that opening which is called the mouth of the womb," and calls the clitoris, whose erectile and erotogenic qualities he is in the process of extolling, "this same part of the uterus (*hanc eadem uteri partem*).[61] The precision Columbus sought to introduce by calling the cervix the true "mouth of the womb" vanishes as the vaginal opening becomes the mouth of the womb and the clitoris one of its parts. The language simply did not exist, or need to exist, for distinguishing male from female organs. This same sort of tension is evident in other anatomists. Fallopius is anxious to differentiate the cervix proper from the vagina, but has no more specific name for it than "female pudenda," a part of a general "hollow" (*sinus*). The Fallopian tubes, as he describes them, are not the tubes that convey eggs from the ovaries to the womb, but twin protuberances of sinews (*neruei*), which do penetrate the peritoneum, are hollow, and do not have an opening into the uterus. Fallopius remained committed to the male-centered system and, despite his revolutionary rhetoric, assumed the commonplace that "all parts that are in men are present in women."[62] Indeed if they were not, women might not be human.

Gaspard Bauhin (1560–1624), professor of anatomy and botany in Basel, sought to clear up the nomenclature, but with equal lack of success. The drive to see all genital organs with reference to man is too deeply embedded in language. "Everything pertaining to the female genitalia is comprehended in the term 'of nature' (*phuseos*)," he declares, but then informs his readers that some ancient writers called the male genitalia *phuseos* as well. Among the words for the labia he cites is the Greek *mutocheila*, meaning snout, with its obvious phallic connection, or more explicitly translated, "penile lips."[63] This in turn fits the usual conflation of labia with foreskin that goes back at least to the tenth-century Arabic writer who points out that the interior of the vagina—a curious description—"possesses prolongations of skin called the lips," which are "the analogue of the prepuce in men and has as its function protection of the matrix against cold air."[64] According to Mondino, the labia guard the "the neck of the womb" in the same way that "the skin of the prepuce guardeth the penis," which is why "Haly Abbas calleth them *praputia matricis* [prepuce of the uterus, of the vagina?]".[65] Berengario simply uses the word *nymphae* to refer to both the foreskin of the penis and the foreskin of the vagina, the labia minora.[66] (And when a new female penis appears, the labia become its foreskin as well. So John Pechy, a popular English writer during the Restoration, describes the "wrinkled membranous production cloath the clitoris [not the vagina] like a foreskin."[67])

Much of the controversy around who discovered the clitoris arises out of just such a blurring of metaphorical and linguistic boundaries, the consequence of a model of sexual difference in which unambiguous names for the female genitals do not matter. I will offer only one example here. When Thomas Vicary, writing in 1548 before Columbus published, reports that the vulva "hath in the middest a Lazartus pannicle, which is called in Latin *Tentigo*," the reference would seem to be unambiguous. Moreover, *tentigo* in early seventeenth-century English means "a tenseness or lust; an attack of priapism; an erection." There is even less question that the structure in question is the female penis, the clitoris. But when Vicary reports on the functions of this part, its "two utilities," he seems to be discussing an entirely different organ. There is no mention of pleasure. "The first [utility] is that by it goeth forth the urine, or else it should be shed throughout al the Vulva: The seconde is, that when a woman does set hir thies abrode, it altereth the ayre that commeth to the Matrix for to temper the heate." What the name led us to expect, a female penis, turn out to be a pair of workaday flaps, a dual-purpose female foreskin.[68] But whatever Vicary means, it is impossible to translate across the chasm that divides this world from ours.

A web of words, like the constellation of images discussed in the previous sections, was redolent with a theory of sexual difference and thus sustained the one-sex model against more general testing. There was in both texts and images a quality of obsessive insistence, a constant circling around, always back to the male as standard. An almost defensive quality suggests that the politics of gender off the page might well have engendered the textual insistence that there really were no women after all.

The truth of the one-sex model

As I said, parts of the one-flesh model were in principle open to empirical verification and hence also to falsification. But it remained untested, not only for the reasons mentioned above but also because it was woven into a whole fabric of interpretation, clinical practice, and everyday experience that protected it from exposure to what we would construe as contrary evidence.

Orgasm and conception. It is scarcely surprising that men and women should think that there was a phenomenological correlative to so awesome and mysterious a process as generation. (Orgasm remains even today linked to conception in the imaginations of many people.) On the other hand, counterevidence must have been readily at hand that women frequently conceived without it. For a number of reasons, however, the old view survived. Systematic evidence on the subject is very difficult to gather and, even if women had been asked, it is more than likely that they would have answered what tradition dictated. They would have misremembered the night of conception or misreported their feelings because it is all too easy to dismiss a nonorgasmic conception as an anomaly or, many months later, simply to have forgotten the circumstances of conception, especially when to do otherwise would have been to fly in the face of accepted wisdom. Experience, in short, is reported and remembered so as to be congruent with dominant paradigms.

On a more technical level, it was not difficult to refute, or push to the margins, unwelcome facts. Aristotle, for example, was easy game. His own dictum that "nature never makes anything without a purpose and never leaves out what is necessary" was routinely turned on him.[69] Since women have organs that resemble the male testicles, and since they obviously experience sexual orgasm—"ye shall observe the same delight and concussion as in males"—there seemed no reason to deny them as active a role in human generation as men. "Why should we suppose Nature, beyond her custome, should abound superfluidities and useless parts," asks the progressive Oxford physician Nathaniel Highmore rhetorically.[70] Or, as Lemnius put it in 1557, in a simile that would have resonance in an increasingly commercial society, a woman's womb is not simply "hired by men, as merchant ships are to be fraited by them." And even if—as he denied—female semen had no other purpose "but only to excite, move and stir the woman to pleasure," it would be immensely important because without the "vehement and ardent lust and appetite" for carnal union, neither man nor woman would follow God's injunction to multiply and be fruitful. Thus the fact that women had gonads like men, that they had sexual desires, that they generally produced fluid during intercourse, and presumably showed signs of "delight and concussion," all confirmed the orgasm/conception link that Aristotle, at least in his philosophical persona, had sought to deny.[71]

To be sure, the fluid women produced did not look like the male ejaculate, but that was precisely what was to be expected. In the first place, a thing did not have to look like something else in order to be it, as in the bread and wine at communion. More prosaically, the Galenic model of hierarchically ordered sexes would have predicted differences in the quality of the two. Patriarchy itself was predicated on the fact that when, "by the labour and chafing of the testikles or stones," blood is turned into sperm, the man's would be "hote, white and thicke" while the woman's would be "thinner, colder, and feebler."[72]

The heat (orgasm) conception nexus was also deeply entwined in medical practice and theory generally. As we have seen, the one-flesh-model, and the role of orgasm in it, is represented in the bodily economy of fluids generally and redounds throughout the entire structure of Galenic-Hippocratic medicine. The experience of patients would have supported it, if only out of the universal tendency of people to believe in, even as they ridicule, the efficacy of their healers.

But heat, and orgasm specifically, was integral to the more mundane therapeutics of infertility, amenorrhea, and related conditions, not to speak of sexual dysfunctions whose physiological causes are the same as theirs. A physician, surgeon, midwife, wisewoman or other healer consulted regarding any of these, and especially barrenness, would immediately have suspected some caloric pathology. And since the statistical analysis of conception has evolved only very recently, and since doing nothing therapeutically has a remarkable chance of success in curing infertility, it seems probable that almost any advice Renaissance healers happened to give their patients regarding sexual heat and pleasure must have appeared to work often enough to confirm the model on which it was based.[73]

Even suspected anatomical defects might be regarded as damaging because of their effect on pleasure. If, as was thought, the generative body during coitus "shakes out" the semen, then irregularities in the actual physical contact between bodies would be among the first possibilities investigated by doctors in patients who consulted them for infertility.[74] If the penis fails to rub properly, either or both partners might fail to have an orgasm and hence to produce seed. Fallopius argues that a malformed foreskin needs to be corrected less for cosmetic reasons then because a penis without one is not "naturally lubricated"; "lubricity" is necessary for sexual pleasure and "when the pleasure is greater, the woman emits seed and suitable material for the formation of the foetus and for the production of membranes."[75] No foreskin, less friction, no female orgasm, sterility. Too short a penis could have the same result for the same reason: inability to satisfy the woman. (Avicenna was the authority on this point.) And so too could an excessively large member by diminishing female pleasure, though one sixteenth-century German doctor is skeptical: "Perhaps you have not heard too many complaints about the penis being too long," he says; "I say unto you, the longer a weed grows, the better."[76]

But genital heat, from the rubbing genitals, was in fact construed as part of the larger caloric economy, just as semen was part of a more general traffic in fungible fluids. Thus the excess heat that was thought to cause nocturnal emissions or premature ejaculation might be assuaged by cutting back on spicy foods, suppressing "images of a desired woman," or not sleeping on one's back too long (because sleeping on one's back led to warmer kidneys, which increased the production of excrement generally and therefore also of semen).[77]

These were serious matters. In a society in which one in five children died before the age of one, and even prosperous families could consider themselves fortunate if they reproduced themselves, any waste of semen was a matter of the most poignant seriousness. A French physician tells of a man who came to see him in March 1694 because "whenever he was inclined to approach his wife, the emission followed the erection so fast, that he had no ability to penetrate. This hindered him from having children; and, as he had but one left, was afraid of being left without any at all." De la Motte prescribed cooling medicines and suggested that his patient abstain from wines, ragouts, and other heating foods. His condition improved, but his wife remained barren "though very young."[78]

The problem of too much heat in women was also part of any Renaissance differential diagnosis of the causes of infertility. Excessive desire; curly, dark, and plentiful hair (in men hair was a sign of virility, bravery, and of the vital heat that arose in adolescence and distinguished them finally from women); a short or absent menses (the hot body burned off the excess materials that in normal women were eliminated in the monthly courses), and so forth, all indicated a problem of excessive warmth that would burn up the seed. Cooling drugs were called for in these situations.[79]

Insufficient heat, however, loomed far larger in the literature than did its surplus. The absence of sexual desire in men, but with minor adjustments also in women, could be cured by rubbing the loins with calorific drugs or through lascivious talk; other drugs, coquetry, and more talk could cure a "defect of spirit," the inability to have an erection when desire itself was sufficient. In women, adversity and indisposition "to the pleasures of the lawful sheets," especially when accompanied by a slow pulse, little thirst, thin urine, "no pleasure and delight" during coition, scant pubic hair, and similar signs were diagnostically important indicators of excessive coolness in their testicles and thus of insufficient heat to concoct their seed. As Jacob Rueff put it in discussing the problem of frigidity, "the fruitfulness of man and wife may be hindered very much for want of desire to be acquainted with Venus."[80]

Desire then was a sign of warmth and orgasm a sign of its sufficiency to ensure "generation in the time of copulation." To produce sufficient heat in women, talk and teasing were regarded as a good beginning.[81] They "ought be prepared for sweet embraces with lascivious words mixed with lascivious kisses," because if "the man is quicke and the woman too slow, there is not a concourse of both seeds at the same instant as the rules of conception require."[82] (Men are invariably presumed to be more quickly aroused than women.) Ambroise Paré, the foremost surgeon of his day, opens his widely translated account of generation by emphasizing the importance of flirtation, caressing, and excitement. (The audience for his advice is clearly male.) In his account, men had literally to coax the seed out of women. When a husband comes into his wife's chamber, "he must entertain her with all kinde of dalliance, wanton behaviour, and allurement to venery." If he finds her "to be slow, and more cold, he must cherish, embrace, and tickle her"; he should "creepe" into the "field of nature," intermix "wanton kisses with wanton words and speeches," and caress her "secret parts and dugs [nipples] until she is afire and "enflamed in venery." Rhythm and timing are all-important, he counsels, and if the two seeds are to come together, the man must be aware that his partner is not "all that quick in getting to that point" as he; and he must not leave the woman too soon after her orgasm "lest aire strike the open womb" and cool the seeds so recently sown.[83]

If all this failed, the Renaissance pharmacopoeia, like earlier compilations, was full of drugs that were thought to work either directly or by sympathetic magic. Paré recommended "fomenting her secret parts with a decoction of hot herbes made with muscadine, or boiled in other good wine," or that civet or musk be rubbed into her vagina. Juniper and camomile, the heart of a male quail around the neck of a man and the heart of a female around the neck of a woman—presumably because of the lecherous character of birds generally and of quails in particular—ale hoof and pease straw, were all available to manipulate the one-sex body's heat.[84] Thus savin (juniper, readily available in gin) might be prescribed to allow an impotent man to have erections, to warm an infertile woman's genitals, and to produce an inhospitably warm womb in a Somerset prostitute who sought to end her pregnancy. The same goes for mugwort (wormwood or artemesia), calamint, spices like ginger or cinnamon, and concoctions made from various animal parts.[85]

A vast body of clinical practice and learning was thus bound up with heat, orgasm, and generation. It was and remains difficult to evaluate the efficacy of particular therapies, and it should not seem strange that the experiences of patients, unchallenged by modern survey techniques and statistical analysis, would confirm the notion that more intensely pleasurable intercourse was also more fecund.

The fungibility of fluids. The economy of fluids discussed in Chapter 2 was partly ideology—a way of talking about women as colder, less well-formed, and more protean than man—and partly a way of understanding the body generally as much less bounded and restrained than we would today. But it was also a way of organizing empirical observations, which strengthened it and the vision of sexual difference it formed.

To begin with, certain anatomical discoveries that improved upon Galenic anatomy actually seemed to confirm the basic physiology of the one-sex model, though no one would have thought such testing necessary. Vesalius, for example, correctly noted that, contrary to Galen, what we would call the left ovarian and testicular veins take their origin not from the vena cava but from the left renal vein (fig. 41). From this he concluded that while the right vein may "carry the pure blood to the testis," the left one, coming as it did from nearer the kidney, might specialize in carrying a more watery, serous blood whose "salty and acrid quality may bring about an itching for the emission of the semen." What was thought to be a significant correction of Galen thus fitted nicely with the thoroughly Galenic notion of genital puritus, of sexual feeling being at least in part the result of the corrosive qualities of certain body fluids.[86]

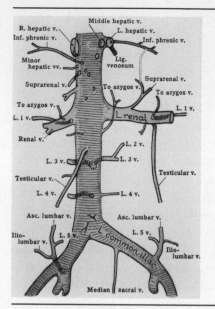

Fig. 41. This shows the left testicular vein, called the ovarian vein in women, coming off the left renal vein and not from the vena cava, the trunk running down the center of the picture.

Conversely, a finding that might have militated against the economy of fluids in the one-sex body—for example, the discovery, known already to Leonardo, that the epigastric vessels going to the breast did not originate from the uterine vessels and that therefore blood from the womb might not be so easily converted to milk and vice versa—was easily ignored. A novel bit of plumbing paled in the face of clinical and folk wisdom stretching back to Hippocrates and of the whole macrocosmic order of which such wisdom was a part.[87] "And is it not the same blood, which, having been in the womb, is now in the breasts, whitened by the vital spirit through its natural warmth?" Laurent Joubert, one of the great medical popularizers of the sixteenth century, asks rhetorically. Of course. It was common knowledge that women who were lactating usually did not menstruate, and, as Joubert said, women who had excessive menstrual flows (evidence for lots of surplus material) were also likely to have a great deal of milk once the flow stopped. (This discussion is in the context of a self-conscious effort to bring observation to bear on questions of natural history so as to get the answers right. Joubert, for example, denies the claim, made by Paré, that excess menstrual blood can produce birthmarks.[88])

Doctors continued to write as if the actual vascular pathways simply did not matter. New clinical observations seemed to confirm the view that menstruation was simply a way of ridding the body of excess and not something specific to a female organ or single route. So one doctor offered a case-by-case list of all the places and various forms blood went when it could not go out its usual place: in a Saxon woman it came from her eyes; in a nun through her ears; a woman from Stuttgart got rid of stuff by vomiting; a slave through her spittle; a woman from Trent through her bellybutton; in others from the breasts; and finally (even he thinks it "most amazing") through the index and little fingers of one Monica.[89] Christopher Wirsung, a popular German writer, argued that the menstrual flow took three separate pathways during pregnancy, even if he did not know precisely how the body effected this division: the most refined and tender was reserved for the fetus, the middle grade went "by various veins to the breasts" to be made into milk, and the coarsest remained behind to be discharged when the child is born. The route from womb to breast is clearly less relevant than the poetics of milk and blood. Someone as thoroughly up to date as the English anatomist Helkiah Crooke, who must have known that there were no connections between the vessels of the uterus and those of the chest, nevertheless argued that the breasts were uniquely well situated to "alter and labor" blood into milk because of their proximity to the heart, the "shop of heate."[90] So even if anatomy did not support the blood/milk nexus, conceptions of the heart as the body's furnace did.

Observations on the periphery of western civilization and under pathological conditions did seem to provide direct new evidence for the interconvertibility of fluids and the underlying identity, between and among men and women, of various forms of bleeding. Brazilian Indian women "never have their flowers," writes a seventeenth-century English compiler of ethnographic curiosities, because "maids of twelve years old have their sides cut by their mothers, from the armpit down unto the knee [and] some conjecture that they prevent their monthly flux in this manner." Joubert likewise thought that Brazilian women "never menstruate, no more than do female animals," while Nicholas Culpepper, the indefatigable seventeenth-century English writer and publisher, uses the fact that at least some "never have any flowers" but nevertheless are fertile as evidence for the general claim that hot women can conceive even if they do not menstruate.[91]

Conversely, in the one-sex fluid economy, strange or feminine men might lactate. Hieronymus Cardanus, court physician to the king of Denmark, says on the basis of travelers' accounts that in some places "almost all the men have great quantity of milk in their breasts."[92] (An Italian commentator cites one of Cardanus' nearer-to-home cases: "Antonio Benzo, age 34, pale, fat and scarcely bearded, had so much milk in his breasts that he could feed a baby."[93]) Men, if they were "of a cold, moist, and feminine complexion," were quite likely to have milk in their breasts thought an English doctor, a view shared by Joubert, who adds that such men are to be found primarily in the east. He gives, in addition to the evidence in Aristotle, the example of a Syrian count who nourished his child for more than six months.[94]

This is not to say that a metaphorically lactating Christ, whose blood nourishes his church as Mary's milk had nourished him, or an infant Jesus depicted with female breasts ready to spurt milk, are to be interpreted as more ethnographic examples of the sort just cited. But they do suggest that, in the world of one sex, the body was far less fixed and far less constrained by categories of biological difference than it came to be after the eighteenth century. The boundary between a more motherly, more feminine Christ lactating in religious imagery and men with milk in prosaic ethnography and clinical reports is by no means clear.[95]

Obviously the cases of amenorrhea among Indians or the more bizarre reports of lactating men need not be interpreted as confirmation of the economy of fungible fluids. The absence of the menses during lactation would today be attributed to hormonal changes and not to the conversion of surplus blood to milk. It will therefore take a certain leap of the imagination to understand how Renaissance doctors and midwives interpreted a large body of clinical material as confirmation of a very different theoretical understanding of the body. But they did; what we would imagine as distinct, sexually specific, fluids were metaphorically conflated in the one-sex model. The "irregularity" (*Gebrechen*) that "women call white stuff and doctors *menstrua alba*" was understood by a sixteenth-century German physician, for example, not as an abnormal vaginal discharge but as a fluid that "has much in common with the flow of male semen" and that arose when disordered heat, excess warmth or cold, turned the menses into something like "the male semen."[96] (The German word for regularity or law, *Regel,* which is being broken in this case is also the word for menses.)

Similarly, discharges of blood by men, occurring naturally or through phlebotomy, were interpreted not as simple instances of bleeding but as a male substitute menses in what was merely a contingently gendered economy of fluids. Men were routinely bled, usually in the spring—more often for those who exercised little—to get rid of a plethora that in women would be lost every month. Well into the eighteenth century, certain pathological bleeding in men was still likened to menstruation. Albrecht von Haller thought nosebleeds got rid of extra blood in some pubescent boys which in girls found "a more easy vent downward," and Hermann Boerhaave reported the case of a "certain merchant here at Leyden, a Man of Probity, who discharges a larger Quantity of Blood every month by the hemorrhoidal arteries than is discharged from the Uterus of the most healthy woman."[97] (This association goes back at least to Aristotle.)

Indeed, the whole matrix of medical practice connected the physiology of fluids, orgasm, conception, and heat. Cold men, less desirous, less potent, and less fecund, were more likely to suffer menstrual-like bleeding and a whole host of mental and physical ails as well; cold women were thought more likely to suffer retention of the seed or of surplus blood, amenorrhea, which in turn might have a variety of clinical sequels: depression, heaviness of limb, barrenness, green sickness, hysteria. Calorific drugs, a midwife rubbing the genitals (in the case of women), or the ardors of coition itself could warm up the cool and clammy body to normality and restore its fluid balance. The issue was warmth.

Renaissance audiences would have taken as physiologically unremarkable the case of one girl, in Robert Burton's *Anatomy of Melancholy,* who was supposedly deranged by reason of a delayed menses and who, by some stroke of good fortune—from Burton's perspective—landed in a brothel where she lay with fifteen men in a single night. The experience cured her amenorrhea and restored her sanity. On the other hand, normal or even vicarious menstruation in women was interpreted as a sign of normal body heat and sexual receptivity. The knight in George Gascoigne's *Adventures of Master F. J.* has a terrible time wooing a lady until one day she gets a torrential nose bleed. When with his help her epistaxis resolves, he finally makes it into the lady's bed.

An entire clinical tradition thus embraced the testable parts of the one-flesh model. Specific discoveries and observations—that orgasm did not always accompany conception, that there were no direct routes between uterus and breast, that the vaginal secretion of women did not look anything like the semen of men—could not, even taken together, shake ancient beliefs so deeply embedded in how men and women regarded and ministered to their bodies. And a variety of observations or putative observations, when interpreted within the constraints of the model, only confirmed its tenets.

Bodies and metaphors

Although my next chapter will consider explicitly the extraordinarily fraught relationship between the social world of two genders and the one-sex body, I do not want to end this one without briefly exploring an alternative rhetoric of difference to the anatomy of isomorphisms and the physiology of fungible fluids I have been emphasizing, one that proclaims the *unique* qualities of a woman's body and the supposed role of these corporeal attributes in determining women's health and social standing. Dr. Rondibilis in chapter 32 of Rabelais' *Tiers livre de Pantagruel,* for example, says that nature has "placed in a secret and interior place" of women's bodies "an animal, an organ, that is not in men." The seventeenth-century midwife Louise Bourgeois leaves the problem of male infertility to male doctors but argues that specifically in women it is most frequently caused by wetness of the womb, that women would be as healthy in both body and spirit as men were it not for this organ, and more generally that God created its uniquely pathogenic qualities—its tendency to wander and cause hysteria, for example—so as to prevent envy between the sexes and to lead man to pity and love woman.[98] Moreover, there is an enormous literature that relates the cold, wet humors said to dominate women's bodies to their social qualities—deceptiveness, changeability, instability—while the hot, dry humors in men supposedly account for their honor, bravery, muscle tone, and general hardness of body and spirit.

Both ways of talking, of course, unambiguously proclaim difference. Both array sexual difference on a vertical axis of hierarchy. Both acknowledge the obvious: women have a womb and men do not. Both ways of talking, to paraphrase Ian Maclean on the Aristotelian logic of sexual opposition, refer at times to an opposition "of privation," at other times to an opposition of contraries that may or may not admit intermediaries, and sometimes—I would say always—to other parts of a cognitive system, other "correlative opposites."[99]

But these ways of talking also differ in two important respects. The first is rhetorical. The anatomists, physicians, and even midwives I have cited were writing to make their readers understand the body and its fluids in a particular way. They were articulating a set of representational or semiotic claims: that the womb must be *understood* as an interior penis, that menstruation must be *understood* as women ridding themselves of a plethora which the warmer, more active bodies of men consumed in the course of everyday life. These understandings were fraught with cultural significance, but they were not expounded primarily to make points about the corporeal foundations of the social order. On the other hand, certain midwifery and medical books, by authors who wished to emphasize their specialist knowledge, as well as a vast array of books about women, for and against, treated the body as if it contained the necessary and sufficient reasons for the medical problems and behavioral characteristics with which they were specifically concerned.

The second difference (but at the same time affinity) has to do with how these two Renaissance discourses construed the body in relation to its cultural meanings. In neither is the ranking of the sexes on the great chain of being just metaphorical—nothing in this cultural system is *just* metaphor—but it is not just corporeal either. The one-flesh discourse I have been explicating seems to regard organs and the qualities of bodies generally as ways of expressing hierarchy, as elements in a network of meaning. On the other hand, the discourse on female uniqueness seems to be postulating an almost modern reductionist theory of corporeal causation, even if it does not carry the notion of incommensurable corporeal opposition as far as would post-Enlightenment writers. Yet, and this is the critical point, the metaphorical and the corporeal are so bound up with one another that the difference between the two is really one of emphasis rather than kind.

Even an apparently straightforward claim about the body like the one that Rabelais puts in the mouth of Dr. Rondibilis turns in on itself and becomes about something else as well: the womb comes once again to sound like a penis. Only women have a womb, Rondibilis says, with no hint of literary shiftiness. But the womb is "an animal," he continues, a move to metaphor and an allusion to *Timaeus* (91b-d), where Plato refers to *both* the male and female genital organs as animals prone to wander unless they are satisfied.[100] And then, in the usual Renaissance manner of piling on similes, this organ, the womb, which is said not to exist in man, becomes "un membre," a term that can of course mean simply an organ but that referred more specifically in the sixteenth century to an appendage—an arm or leg—or when used alone, as in "his member," to the penis. There was no sense in which *membre* ever referred to "her member."[101] The point here is not that Rondibilis is making a controversial claim in saying that only women have a womb; no one denied this. It is rather that once again a female organ is attracted into the metaphorical orbit of the male, not in order to make a claim about likeness but to assert that all difference is figured on the vertical scale of man.

It is also precisely in those contexts in which the womb seems most solidly the organic source of disease, as in the argument that hysteria is caused by a wandering womb, that it becomes most profoundly bound up with extracorporeal meaning. Even in classical writings it is difficult to comprehend the purchase of the claim that the womb wanders and *causes* hysteria. Herophilus in the third century B.C. discovered the uterine ligaments, and Galen merely repeated old arguments when he said that "those who are experienced in anatomy" would recognize the absurdity of a moving womb: "totally preposterous."[102] Someone must have believed literally in a rampant uterus—a folk belief perhaps—or the doctors would not have felt it necessary to keep attacking the view, and the prevalent fumigation therapies suggest that their adherents subscribed to this literal interpretation. But by the sixteenth century there was manifestly no place in the body for the womb to move to.

The new anatomy, and more specifically the widespread distribution of anatomical illustrations (such as figs. 42–44) well beyond the bounds of the learned community to midwives, barber surgeons, and laypeople, showed that not only was the uterus kept more or less in place by very broad ligaments but that the space between it and the throat was full of other organs and divided by thick membranes. Galen had already pointed out that the peritoneum covered the bladder and the uterus, but now this fact was there for anyone to see, splendidly displayed in the usual, slightly ruined classical torso.[103] The new anatomy thus made literal interpretation of a wandering womb impossible; but it did not produce a modern rhetoric of disease. Like Paracelsian iatro-chemistry, which seems to be

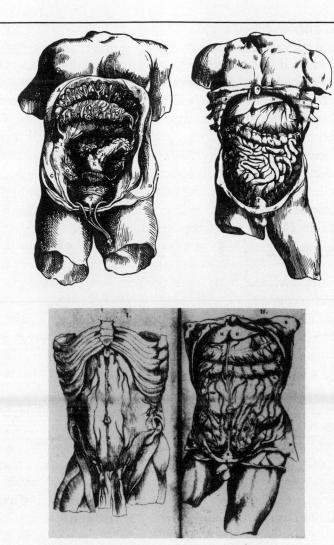

Figs. 42–44. Fig. 42, top left, shows the female torso from which the vagina in fig. 20 was removed. Vesalius tells us that the attachments of the uterus are in place but that he has removed the abdominal wall and intestines to present this view. Fig. 43 shows a male torso, a few pages before this one, opened to show the intestines still in place. Clearly this figure was meant to be be applicable to women. Two still earlier plates from the *Fabrica* (fig. 44, bottom row) showing the abdominal wall of a male torso still in place were combined and used as the opening and illustration of a leading sixteenth- and seventeenth-century midwifery manual by Raynald, *The Byrth of Mankind* (1545).

but is not a version of modern medical chemistry, the new anatomy lures us into thinking that Renaissance writers must have spoken of organs as we do, which they did not. Whatever they were debating when they pondered whether the womb wandered, it was not a discussion about the actual travels of an organ from its ligamentary anchor below, up through a foot and a half of densely packed body parts.

By the eighteenth century, this was perfectly evident. When Tobias Smollett, author of *Humphrey Clinker* as well as a surgeon and ghostwriter of Smellie's famous treatise on midwifery, ridiculed the English midwife Elizabeth Nihell for citing Plato's wandering womb, Mrs. Nihell countered that *of course* she had meant it only figuratively. Smollett, she said, had quoted her out of context to make her look bad.[104]

Though less intractable, difficulties of translation also arise when interpreting the humors. Doctors as well as laypeople in the Renaissance believed that the humorial balances of the sexes differed along the axis of hot and cold, wet and dry, that such differences had implications for anatomy as well as for behavior, and that humorial imbalance caused disease. They spoke as if there were warm or cold qualities somewhere in the body whose presence was made known by observable features; skin color, hair, temperament. On the other hand, no one believed that a quantifiable amount of some humor caused someone to be male or female. There were thought to be hot, hirsute viragos and effeminate, cold and hairless men, colder than exceptionally hot women. The claim was rather that men as a species were hotter and drier than women as a species. Nor was it claimed that one could actually feel the wetness or the coldness that distinguished women from men or that, on occasion, caused female complaints.[105] The humors were not like organs and did not play the parts organs would play in eighteenth-century nosology or social theory. Though humors were "more real" than a wandering womb and were certainly not "just metaphors" or ways of talking, they were not just corporeal attributes either.

Perhaps the most telling feature of both ways of talking about sex in the Renaissance, however, is the extent to which all talk about sex is determined contextually. In the same texts from which women are excluded and denied both separate existence and subjectivity, they enter as subjects. There they are, where most egregiously absent. Consider again Columbus' discovery of the clitoris, this time with the Latin text:

Hanc eadem uteri partem dum venerem appetunt mulieres et tanquam oes-
tro percitae, virum appetunt, ad libidinem concitae: si attinges, duriusculam
et oblongam comperies . . .

If you touch that part of the uterus while women are eager for sex and very
excited as if in a frenzy, and aroused to lust they are eager for a man, you
will find it rendered a little harder and oblong . . .

If "you" (man) touch a certain part of a woman, "you" will find it harder.
Women, in one of the few instances in which they are made the grammatical subject, are literally surrounded in the temporal clause by desire,
her desire. *Appetunt,* "are eager for," is repeated, to flank *mulieres,* women;
percitae and *concitae,* redundant predicate adjectives, attest further to *her*
sexual arousal. But then the sentence takes an unexpected turn, and the
scientifically objective, presumptively male reader is told that the part of
the female anatomy in question will become hard and oblong if touched
. . . making her semen flow "swifter than air."[106] Thus woman has entered
as a separate, desiring being in what seems to be an all-male world.

This tension is everywhere, not only in the anatomy theater but at the
Globe Theater, not only in medical texts but in the essays of Montaigne.
The cultural politics of at least two genders is never in equilibrium with
the "biology," or alternative cultural politics, of one sex. We shall see that
context determines sex in the world of two sexes as well.

FOUR

Representing Sex

Sebastian [*To Olivia*]
So comes it, lady, you have been mistook.
But nature to her bias drew in that.
You would have been contracted to a maid;
Nor are you therein, by my life, deceived:
You are betrothed both to a maid and man.

SHAKESPEARE, *TWELFTH NIGHT*

In the absence of an Archimedean point in the body that assures the stability and nature of sexual difference, one sex is, and has always been, in tension with two: stark polarities poised on the edge of chiaroscuro shadings. Specific social, political, and cultural circumstances, revealed in anecdotal moments and rhetorical contexts, favor the dominance of one or the other view, but neither is ever silent, neither is ever at rest.

We have seen that the one-sex model was deeply imbricated in layers of medical thinking whose origins stretched back to antiquity. Advances in anatomy and anatomical illustration as well as further clinical evidence, far from weakening these attachments, made the body ever more a representation of one flesh and of one corporeal economy. The considerable cultural prestige of medical learning, if not of actual practice, thus continued to weigh in on the side of one sex. But the one-sex body subsisted also, easily or not so easily, in the midst of other discourses, other political demands, other social relations, even other medical ways of speaking. It might be perfectly embedded in allegories of cosmic order, but deeply at odds with rigid gender boundaries and the social body's imperative to ensure reproductive mating.

Somehow if Olivia—played by a boy of course—is not to marry the maid with whom she has fallen in love, but the girl's twin brother Sebastian instead; if Orsino's intimacy with "Cesario" is to go beyond male bonding to marriage with Viola, "masculine usurped attire" must be thrown off and woman linked to man. Nature must be "to her bias" drawn, that is, deflected from the straight path. "Something off center, then, is implanted in nature," as Stephen Greenblatt puts it, which "deflects men and women from their ostensible desires and toward the pairings for which they are destined." But if that "something" is not the opposition of two sexes that naturally attract one another—as it came to be construed in the eighteenth century—then what is it?[1]

The answer is nothing, or at least nothing specifically and fundamentally corporeal peculiar to each sex. Having a penis does not make the man just as, to quote Feste, "cucullus non facit monachum" (the cowl does not make the monk). And yet men and women were sorted out by the configurations of their bodies—having a penis outside or inside—into their required procreative and multitudinous other gender-specific roles. The one-sex body of the doctors, profoundly dependent on cultural meanings, served both as the microcosmic screen for a macrocosmic, hierarchic order and as the more or less stable sign for an intensely gendered social order. A whole matrix of interpretive strategies and assumptions about how things come to have meaning kept the one-sex model in place, and their relative eclipse constituted the shift to an understanding of male and female as opposites. The nature of sex, I argue in this and the next two chapters, is the result not of biology but of our needs in speaking about it.

One sex and the macrocosm

We are not allowed by our Renaissance and medieval sources to forget that the word "cosmos" in both English and Greek has a double meaning. It denotes, as Angus Fletcher reminds us, both a *large-scale order* (macrocosmos) and the small-scale *sign of that order* (microcosmos). Modern science, he points out, works to reduce the metaphoric connections between various orders of the world to one, to explain man and nature, the heavens and the earth, in one neutral mathematical language and not, as in the cultural world with which we are concerned here, by adumbrating a complex structure of resemblances, creating levels upon levels of connectedness between and within the micro- and the macrocosm, engendering correspondences as the demands of meaning dictate.[2]

The new anatomy was for most purposes firmly in the old metaphorical tradition. Vesalius, for example, builds his entire account of "how nature provides for the propagation of the species" on the image of a city whose founder "does not wish to reside there," but who "still provides a plan whereby it may endure for eternity or a very long time." The human body, he begins, is necessarily subject to death and because of its very material cannot be immortal, at least not physically. All cities, even the most fortunate, have gone to ruin over the ages. But God's earthly city has endured for thousands of years, having been contrived by him "with a certain marvelous skill so that new men always succeed in place of those that waste away, and the conservation of the species becomes perpetuated."

Generation mirrors both earthly hierarchy and the wonders of creation. The male, as we might expect by now, "puts forth the most potent proportion of the principle of the fetus," but the female, having testicles and appropriate vessels as well, "adds some proportion of the primary principle," which is conceived in her womb. Pleasure, Vesalius affirms, drives humankind and indeed all animals to use their organs of generation to initiate the "miracle of nature." The creator has given them "a great desire for the uniting of bodies and a particular force of delight . . . a certain marvelous and unspeakable appetite" for their employment. The self-perpetuating macrocosmic order is, in a sense, assured by the qualities of merely mortal bodies.[3]

This constant interplay between images of the body and the world beyond it, at the same time biological and rhetorical, is so pervasive that we tend to take it for granted. Somehow the stars dictate that on certain days in April, August, and December one ought not to be bled, nor eat goose or peacock, nor take drugs (fig. 45). Heavenly bodies, one popular English tract declares, "are the forms and matrices of all Herbs . . . representing the like of every vegetable in the earth." Conversely, "every Herb is a Terrene star growing toward Heaven." From this set of correspondences followed scores of others that bring the cosmos into the body. All the species of the plant *Orchis,* for example, excite the "Venereal appetite" and aid in conception because of "their similitude of the Testicles" and because "they also have the odour of the Seed." The grapestone represents the genitals of both sexes, and wine, made of course from the grape, is therefore conducive to passion: "The Ancients, not without cause, said: Without Bacchus, Venus waxeth cold." Countless illustrations of "zodiacal man"—the male body as usual stands for generalized humanity—specify which stars correspond to which parts of the body. And between heaven and earth are countless bonds of signification.[4]

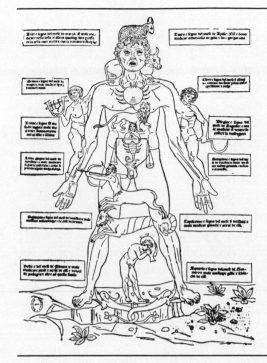

Fig. 45. Late fifteenth-century Italian zodiac man. Captions linking the zodiacal sign to organs and parts of the body also gave dietary prescriptions, directions for blood letting, and other information regarding how the heavens affected the body.

In the way that the moons of Jupiter provided for Galileo a model of the truths of Copernican astronomy, so the human body could represent the fecundity of nature and the power of the heavens. One could view the world and capture its essence by training one's instrument on Man. As the astrologer and physician John Tanner put it: "In man, as in a perspective glasse, may our Mother Earth, with her innumerable off-spring, be discovered; in him may the unruly, and restless waves of the Ocean be delineated. Nor does he only epitomize the Elementall world, but also the Celestiall."[5] I need not belabor the obvious, that the stars were thought capable of influencing human life. But I do want to draw specific attention again to the connection between generation and the cosmos, between the body and the cycles of life outside it.

Popular medical works moved vertiginously from great ontological claims to specific potions whose efficacy depends upon the macrocosmic order. Robert Bayfield's *Enchiridion medicum,* for example, begins with the Renaissance commonplace that man is an "epitome or map of the universe" and that the fall reflects a ruin upon both worlds—"upon the great world calamities and upon the little world disease and death"—and moves immediately to a kind of social mise-en-scène. The book is written, its author proclaims, for those who cannot afford the books of great men but who nevertheless need to learn medicine, God's help in time of pain. It is a treasury of palliatives for the ills encountered along the way to the destiny of all men, rich as well as poor, "to return to dust, and become as though he had never been." In the text itself the actual remedies proposed curiously mimic this movement from macro- to microcosm. To cure hysteria, for example, Bayfield suggests everything from physically heating the body in the ardor of intercourse to having a midwife rub the genitals, to applying bags of mugwart to them, to procuring "the mosse that groweth on a malefactor's scull," mixing it with the powder of that skull, and using the amalgam to alleviate seizures. The entire universe, living and dead, is brought to bear on the body in distress.[6]

The more general form of these easy movements from macro- to microcosm is in the poetics of biology itself, specifically in the language through which men and women thought about the succession of generations. This web of metaphor does not simply mirror some set of beliefs about their bodies, though it does that as well. It has a life of its own which in some measure constitutes the connections between the body and the world. That is to say, the images through which bodies and pleasures were understood in the Renaissance are less a reflection of a particular level of scientific understanding, or even of a particular philosophical orientation, than they are the expression of a whole fabric or field of knowledge. Myriad discourses echo through the body.

Thus to imagine female semen after its mixing with the male's as "expansed into filmy integuments" that surround the "new kindled deity"; to think of it as weaving a texture, "farre too fine and cunning for the fingers of Arachne," is of itself to fashion a fine network of connections. The epigenesis of the fetus is likened to godlike creation and to the making of gods, to the young Arachne who wove a picture of Europa carried across the waters by Jupiter as bull which was so realistic that "you would have thought that the bull was a live one, and that the waves were real waves," and to the humble spider, into which Arachne was changed for her hubris, spinning her webs.[7] To point out that menstruation is called *die blume* by Germans or *the flowers* by the English because "a tree in bloom is likewise regarded as capable of bearing fruit" metaphorically opens woman's bodies to all of nature.[8]

A poetry of biology similarly enables Edmund Spenser in *The Faerie*

Queene to bring the heat of the heavens into the virgin body of Chryso-gonee for the "wondrous" begetting of Belphoebe and Amoret.[9] On a hot summer's day

> In a fresh fountaine, farre from all mens vew,
> She bath'd her brest, the boyling heat t' allay;
> She bath'd with roses red, and violets blew,
> And all the sweetest flowres, that in the forrest grew.

Chrysogonee then falls asleep, naked, on the pool's bank:

> The sunne-beames upon her body play'd,
> Being through former bathing mollifide,
> And pierst into her wombe, where they embayd
> With so sweet sence and secret power unspide,
> That in her pregnant flesh they shortly fructifide.

Spenser does not claim, nor do I, that biology makes this virgin birth seem like an ordinary occurrence, that medicine naturalizes what is meant to be a wondrous virgin birth of "the wombe of Morning dew." But biology gives the metaphors of this passage resonance, and the poetry in turn envelops biology in its images. Chrysogonee's conception is not meant to seem miraculous in the sense of working through means un-known to earth. Instead Spenser writes:

> But reason teacheth that the fruitful seades
> Of all things living, through impression
> Of the sunbeames in moyst complexion,
> Doe life conceive and quickned are by kynd.

"Infinite shapes of creatures," he points out by way of example, are in-formed by the sun's rays in the mud of the Nile. These images of gener-ative heat, the body's and the sun's, are not simply expressions of now outdated scientific theories that, once reproduction is more fully under-stood, would become trivial, incomprehensible, or so implausible as to be silly.[10] But neither was biology understood only as a form of poetry: "merely" language. Rather, it is the constant back and forth, the interpre-tive dialogue between the corporeal and the linguistic, which itself con-stitutes the meanings of the body in the one-flesh model.

The absorptive powers of the whole linguistic field I have been describ-ing are nowhere more evident than in a pair of accounts of generation which encompass within a few paragraphs the grandness of creation and the tragedy of the fall, the fruitfulness of the earth and the mundane details of producing grain and baking bread. The two are distant in time and born of very different contexts, but they share the special language of corporeal openness. The first is from the extraordinary twelfth-century nun, Hildegard of Bingen. She imagines the making of Eve as the archetypal creation of new life through the power and sweetness of the sex act:

> When God created Adam, Adam experienced a sense of great love in the sleep that God instilled in him. And God gave form to that love of the man, and so woman is the man's love. And as soon as woman was formed God gave man the power of creating, that through his love—which is woman—he might procreate children.

If his love is like "a fire on blazing mountains," while hers is a small wood fire, easily quenched, hers is also "like a sweet warmth proceeding from the sun, which brings forth fruit." After the fall, their love is not so sweet, but more passionate, more violent, more human, more of this world:

> And so, because a man still feels this great sweetness in himself, and is like a stag thirsting for the fountain, he races swiftly to the woman and she to him—she like a threshing floor pounded by his many strokes and brought to heat when the grains are threshed inside her.

Within two paragraphs we move from the creation of Eve out of the sleep of Adam to ordinary human generation likened to grain coaxed into fertility through the heat of sexual ardor.[11]

A sixteenth-century German account likewise creates a matrix of metaphor in which the boundaries between the natural and the spiritual world and between the human body and the rest of creation are constantly being elided. Semen, it argues, works as a spume or froth that has the power through the movement of its spiritual, natural, and vivifying essence (*seelichen, naturlichen und lebendigen Geyst*) to create in matter a breath of air (*ein Blast*) that prepares the way for the heart. Then, like the waters parting at the creation, the two outer parts of the foam are driven to the sides, and various parts of the body arise in the space between, the spirits each producing particular parts. Thus the spiritual or psychic essence (*seelisch Geyst*) acts at the upper part of the fissure to produce the head. These extraordinary occurrences become profoundly human and mundane when we are told that a further force of nature (*naturlische Krafft*) makes a little bag (*ein Buetlin*) in which "the fruit is guarded from destruction as the bread crust protects the crumbs (*Brosam*)."[12]

These two images of bread and generation link the philosophically sophisticated notions of a great chain of being with what the Russian critic
Mikhail Bakhtin has called "the grotesque mode of representing the body
and bodily life," which "prevailed in art and creative forms of speech over
a thousand years." The model of bodies and pleasures I am explicating is
embedded in both, in the rhetoric of metaphoric resemblance and in an
image of the body whose borders with the world are porous and protean.
It will fall with their political and aesthetic collapse.[13]

By "grotesque body" Bakhtin means one "in the act of becoming" (or
dissolving), a body fecund, open, in the process of reproducing itself. The
primary organs in this act of self-creation are those that conceive new
bodies or more generally break the bounds of their host. Bakhtin identi-
fies these as the bowels and the penis, inexplicably omitting the womb.
The "main events in the life of the grotesque body" are those carried on
by these organs: ingestion, elimination from all the orifices of the body,
copulation, pregnancy, dismemberment. Conversely, Bakhtin argues, the
"logic of the grotesque image ignores the closed, smooth, and impene-
trable surface of the body." The inner body, its blood and excrement,
indeed its entire inner economy, is externally manifest. Moreover, this
image of the body is one in which particular parts—especially blood—
provide a link between generations, a bond between the death of an individual body and the continuance of the corporeal body social. Finally,
the grotesque body is "cosmic and universal." That is, the functions and
configurations of the body not only reflect the cosmic order, but are to a
great extent determined by it.[14]

Not everyone will share Bakhtin's cheerful acceptance of corporeal
openness, dismemberment, and mutilation; his blindness to the brutality
of the language directed against women; his romanticization of the role
of the carnivalesque in creating a "life of the people." For women bearing
children in particular, it must have been considerably less than joyful to
experience a world in which any perturbation of accepted order—wicked
thoughts, moral culpability, chance encounters with people or things, untimely or improperly positioned intercourse—could imprint itself disas-
trously on the flesh of their children in utero.

John Winthrop in 1638 provides an excruciating and dramatic glimpse
into this world. He reports on a child born horribly deformed to one of
the followers of the outcast Anne Hutchinson. The stillborn baby "had a
face but no head, and the ears stood upon the shoulders and were like an

ape's; it had no forehead, but over the eyes four horns, hard and sharp . . . the navel and all the belly, with the distinction of sex, were where the back should be, and the back and hips before, where the belly should have been." In short, everything about the child was as perverted as its mother's religious beliefs: front to back, animal instead of human, hard instead of soft; when it died in the mother's body two hours prior to birth, "the bed whereon the mother lay did shake, and withal there was a noisome savor," so obnoxious that women in attendance vomited and their children for the first time in their lives had convulsions. Everywhere was corruption. The midwife, suspected of being a witch, "used to give young women oil of mandrakes and other stuff to cause conception." Moreover, "coming home at this very time," the father of the "monster" was on the next Sunday "questioned in the church for divers monstrous errors." [15]

Altogether, the reproductive biology and these representations of male and female bodies are part of a specific literary mode that Bakhtin characterizes in other registers. The attacks on the grotesque which he finds in writers like Erasmus and which Norbert Elias has identified as the essence of the "civilizing process," and has associated with the rise of the absolutist state, also become attacks on the Renaissance model of sex and gender. [16] A new cultural politics will, by the eighteenth century, entail new metaphors of reproduction and new interpretations of the female body in relation to the male.

Representing one sex in a two-sex world

Talk about biological sex always threatens to collapse into theatrical gender, but it does so with special urgency and rhetorical virtuosity in the world of one sex. Elizabeth I brilliantly exploited the tensions between her masculine political body and her feminine private body in creating an erotics of court life that both engendered factions of the great men of her realm and bound them to her and to each other. She could play the alluring but inaccessible virgin queen and the warrior prince. In her famous speech to the troops at Tilbury in 1588 she proclaimed that she had "the body but of a weak and feeble woman but the heart and stomach of a king, and of a king of England too." Her rhetoric later in her life became still more reliant on masculine images. She began referring to herself more often as king, as the nation's husband rather than its virgin mother.

The nation, she said, should cast its eyes on no other *prince* as she played its Aeneas, St. George, and David. (Francis I also played on the theme of the androgyne, appearing in one painting with the head of a virago.[17] And in quite another tradition men are represented as the appropriation of female power of women by Adam, the first man, who is depicted as *really* pregnant.[18])

These sorts of slippage occur everywhere in the literature of early modern Europe. There is the fabliau in which a count cures his mother-in-law of prideful meddling in the affairs of men by claiming that her misbehavior resulted from her "balls" having descended to her loins: "You have balls like ours, and that is why your heart is so proud. I would like to feel them. If they are there, I'll have them removed." His servants stretched her out on the ground; he cut a long gash into her hip; he tugged, "removed," and displayed to his victim a huge testicle from a bull that he had earlier hidden. "After this, she thought it was real."[19] Really? And of course stories of women who actually changed sex and suddenly sprouted a penis circulated widely in both medical and other literature.

Men's bodies too could somehow come unglued. "Effeminacy" in the sixteenth century was understood as a condition of instability, a state of men who through excessive devotion to women became more like them: in one of the OED's examples, from 1589, "The king was supposed to be . . . very amorous and effeminate." Romeo, having refused to fight Tybalt, blames his softness on women:

> O Sweet Juliet,
> Thy beauty has made me effeminate
> And in my temper soft'ned valour's steel!
> (3.1.111–113)

Of course, none of these texts demands to be read as pertaining to real bodies and, therefore, to the collapse of sex into gender. And if they do, as in the case of the sex-change stories, the language of sixteenth-century texts might be readily translated into the plain naturalistic terms of modern science. Elizabeth's language is simply metaphorical; she is *like* a king or a husband but is really a queen and a maid. The fabliau plays on the commonplace that women have testicles inside, and thus the storyteller can *figure* women as becoming malelike through a slipping down of those interior balls. The mother-in law might credulously believe the bull testicle to be hers, but the count and the reader know them to be fake.

Stories of men becoming effeminate are more problematic, and it is difficult to ask of them what their authors thought "really" happened. In one sense they might be regarded as expressions of concern for the boundaries of what we would call gender roles. But this does not quite work in the textual contexts I want to consider because, if bodies were open to a wide array of astral and earthly influences, then why not open also to transgressions of gender? Bodies actually seem to slip from their sexual anchorage in the face of heterosexual sociability; being with women too much or being too devoted to them seems to lead to the blurring of what we would call sex.

As for women changing into men, naturalistic explanations are also problematic. First, they presume what ought to be questioned: that early modern men and women talked about and understood the body as we do and that their categories are readily translatable into ours. When early modern texts speak about women turning into men or receiving the stigmata or fasting for months on end—they are not doing so in neutral scientific language. To read them as such is to miss their historical specificity. Second, they presume also a fixed and modern, base-superstructure connection between gender and sex, which is again precisely what is at issue.

Instead, the texts I will consider here—those at the corporeal end of the spectrum as well as those at the metaphoric—presume a very different relationship. So-called biological sex does not provide a solid foundation for the cultural category of gender, but constantly threatens to subvert it. Foucault suggests an explanation when he argues that in the Renaissance and before there was no such thing as the one and only true sex and that a hermaphrodite could be regarded as having two, between which he/she could make a social and juridical choice. He is perhaps utopian in his political claim; gender choice was by no means so open to individual discretion, and one was not free to change in midstream. But he is right that there was no true, deep essential sex that differentiated cultural man from woman.[20] But neither were there two sexes juxtaposed in various proportions: there was but one sex whose more perfect exemplars were easily deemed males at birth and whose decidedly less perfect ones were labeled female. The modern question, about the "real" sex of a person, made no sense in this period, not because two sexes were mixed but because there was only one to pick from and it had to be shared by everyone, from the strongest warrior to the most effeminate courtier to the most aggressive virago to the gentlest maiden. Indeed, in the absence of a purportedly stable system of two sexes, strict sumptuary laws of the body attempted to stabilize gender—woman as woman and man as man—and punishments for transgression were quite severe.

In this world, the body with its one elastic sex was far freer to express theatrical gender and the anxieties thereby produced than it would be when it came to be regarded as the foundation of gender. The body is written about and drawn as if it represented the realm of gender and desire; its apparent instability marked the instability, indeed impossibility, of an all-male world with only male homoerotic desire. An open body in which sexual differences were matters of degree rather than kind confronted a world of real men and women and of the clear juridical, social, and cultural distinctions between them.

Two hundred years after the fabliaux, the all-male world of the aristocratic warrior class had waned. Courts were still overwhelmingly male, but more was required of the courtier now than military prowess and naked brutality. Political and social success depended not only on might and cunning but on the gentler skills of courtesy, dress, conversation, and all the skills of "self-fashioning."

Castiglione's *Book of the Courtier* is rampant with anxiety, expressed in the language of the body, that men engaged in such pursuits—in consorting closely with women—could become like them and, even more threateningly, that women could become like men. Much of this appears in commonplace discussion in book 3 about the worth of woman, a replay of the misogynist and antimisogynist arguments of the *querelle des femmes*.[21] But the concern that courtiers will become women also appears elsewhere in the treatise. Men can gain a "soft and womanish" countenance through overrefinement—curling their hair, plucking their brows, pampering "themselves in every point like the most wanton and dishonest women in the world." Men of this sort seem to lose the hardness and stability of male perfection and melt into unstable but protean imperfection. Becoming effeminate becomes a sort of phantasmagoric dissolution: "their members were readie to flee from one an other . . . a man woulde weene they were at that instant yielding up the ghost."[22]

Music, Castiglione's misogynist Lord Gasper proclaims, is a pastime for women and for those that have the likeness of men but not their deeds, for those who would make their minds womanish and "bring themselves in that sort to dread death." He speaks as if the body is unable to resist the pressures of blurred gender and can at any moment actually change to match its social perversion. Gasper goes so far as to suggest that heterosexuality itself can bring about man's undoing as a man. Citing Aristotle, he notes that a woman always loves the first man she has intercourse with—she after all "receiveth of the man perfection"—while a man hates his first, since "the man of the woman [receives] imperfection." By extension he hates all subsequent female lovers because "every man naturally loveth the thing that maketh him perfect, and hateth that maketh him unperfect."[23]

There is also the converse danger that thoughts or actions inappropriate to their gender could turn women into men. Lord Julian, one of Castiglione's moderates on the woman question, warns them against undertaking "manly exercises so sturdie and boisterous," against their using "swift and violent trickes [movements], or even singing or playing upon their instruments "hard and often divisions."[24] The concern here goes beyond women playing unladylike music, beyond transgressing the bounds of gender; it seems that inappropriate behaviors might really cause a change of sex. I want to strengthen this interpretation by setting Castiglione beside near contemporaneous accounts—from Michel Montaigne and from the chief surgeon to Charles IX, Ambroise Paré—of a girl whose "swift and violent movements" or other masculine activities did lead, or are reported to have led, to just the sort of sex change Julian the courtier feared.

Paré's Marie-turned-Germain story is found in a collection of clinical tales and observations: a girl, another Marie, who became Manuel when she sprouted a penis "at the time of life when girls begin their monthlies"; a young man in Reims who lived as and anatomically seemed to be a girl until the age of fourteen, when he/she, "while disporting him[/her]self and frolicking" with a chambermaid, suddenly acquired male genital parts. It is as if making love as a man suddenly gave her the organs to do it "properly." (Perhaps he was all along a man in a woman's body so that his gender, if not his sex, made the encounter, in spirit, a heterosexual one that the flesh subsequently confirmed. Or perhaps he was a woman with a homoerotic passion for a fellow servant, who was saved from sin by a last-minute sex change.) One cannot tell, and this is precisely the point. A bit more heat or acting the part of another gender can suddenly bestow a penis, which entitles its bearer to the mark of the phallus, to be designated a man.

Paré's story in which violent movement plays a major causal role—this is the one Montaigne picks up—is about Germain Garnier, christened

Marie, who was serving in the retinue of the king when the famous surgeon encountered him/her. The servant Germain was a well-built young man with a thick red beard who, until the age of fifteen (twenty-two in Montaigne's version), had lived and dressed like a girl, showing "no mark of masculinity." Then once, in the heat of puberty, the girl jumped across a ditch while chasing pigs through a wheatfield: "at that very moment the genitalia and the male rod came to be developed in him, having ruptured the ligaments by which they had been held enclosed." [25] Marie, soon to be Marie no longer, hastened home to her/his mother, who consulted physicians and surgeons, all of whom assured the somewhat shaken woman that her daughter had become her son. She took him to the bishop, who called an assembly which decided that indeed a transformation had taken place: "the shepherd received a man's name: instead of Marie . . . he was called Germain, and men's clothing was given him." (Some persisted in calling him Marie-Germain as a reminder that he had once been a girl.) Montaigne in both his *Travel Journal* and the *Essays* tells the same story, adding the observation that there was in the area still "a song commonly in the girls' mouths, in which they warn one another not to stretch their legs too wide for fear of becoming males, like Marie-Germaine." The girls' answer to the dangers of effeminacy. [26]

Paré offers the following, entirely naturalistic, explanation for Marie's transformation: the fact that "women have as much hidden within the body as men have exposed outside; leaving aside, only, that women don't have so much heat, nor the ability to push out what by the coldness of their temperament is held bound to the interior." So puberty, jumping, active sex, or something else whereby "warmth is rendered more robust" might be just be enough to break the interior-exterior barrier and produce on a "woman" the marks of a "man." Succinctly put by the learned Gaspard Bauhin: "women have changed into men" when "the heat, having been rendered more vigorous, thrusts the testes outward." But the reason heat works in this way and not in reverse—men cannot be physically transformed into women—is as much metaphysical as physiological in any modern sense. Movement is always up the great chain of being: "we therefore never find in any true story that any man ever became a woman, because Nature tends always toward what is most perfect and not, on the contrary, to perform in such a way that what is perfect should become imperfect." [27]

Paré, Montaigne, and Bauhin are of course writing in a long tradition stretching back to antiquity. They all cite Pliny, who asserts that "transformation of females into males is not an idle story" and that, in addition to various reliably reported cases, he himself "saw in Africa a person who had turned into a male on the day of marriage to a husband."[28] (There is another tale in the Greek corpus about a thirteen-year-old girl who developed a severe stomachache on the eve of her marriage and was saved from becoming a child bride when four days later she emitted a great cry and produced male genitals.) The celebrated seventeenth-century English physician and author Sir Thomas Browne concluded in his *Vulgar Errors*—an attack on a variety of false popular beliefs—that one could not deny the transition from one sex to another in hares, "it being observable in Man." Man, after all, is in an "androgynal condition."[29]

To the protagonists of the *Courtier,* or even to the count who castrated his mother-in-law in the fabliau, the lesson of Paré's stories and of the tradition going back to the Greeks is not that a woman is at any moment likely to change sex and become a man or, worse, that a man will lose his member and become a woman. Male anxiety about effeminacy or about the acquisition of masculine traits by women might find resonance in the tale of Marie-Germain but cannot have been caused, or even given credence, by the genre it represents. Real sex changes are, in other words, not the objective correlatives of imagined ones. If the only danger were such extraordinary transformations, then the terrifying erosions of sex/gender boundaries would not figure as prominently as they do in so many kinds of literature.

The problem is rather that in the imaginative world I am describing there is no "real" sex that in principle grounds and distinguishes in a reductionist fashion two genders. Gender is part of the order of things, and sex, if not entirely conventional, is not solidly corporeal either. Thus the modern way of thinking about these texts, of asking what is happening to sex as the play of genders becomes indistinct, will not work. What we call sex and gender are in the Renaissance bound up in a circle of meanings from which escape to a supposed biological substratum is impossible.

Montaigne's recounting of Germain's transformation in his essay "Of the Force of the Imagination" illustrates this point. Whatever Montaigne thinks really happened to the girl who jumped the fence, the essay resolutely obscures; it simply refuses to come to rest on the question of what is imaginative and what is real. The force of the imagination brings forth horns on the head of Cyppus, king of Italy, who had attended and assisted at a bull baiting and had "dreamed of hornes in his head." Montaigne cites Pliny's reports of having seen women turn into men on their wedding night.

Finally, just before the story of Germain, Montaigne alludes to another example—this time from Ovid—of getting a penis: "Iphis a boy, the vowes then paid,/Which he vow'd when he was a maid." [30]

This is the happy ending to the story of a girl who was born and raised as a boy, who was engaged by her father to be married to a beautiful girl, and who just in the nick of time—in response to her virtuous mother's prayers—did actually turn into a boy: her features sharpened, her strength grew, and presumably she gained a penis to match the phallus she already carried within.

Montaigne never makes clear what this myth has to do with the girl chasing her pigs in Vitry to whose transformation he next bears personal witness.[31] Nor is it clear how we are to take the following extraordinary claim, which seems to normalize what happened to Iphis and Marie on the grounds that we men may as well grant all women penises since they will get them anyway:

> It is not so great a marvel that this sort of accident is frequently met with. For if the imagination has power in such things, it is so continually and vigorously fixed on this subject that in order not to relapse so often into the same thought and sharpness of desire, it is better that once and for all it incorporates this masculine member in girls.[32]

Is it that women would like to have a penis, intensely desire a penis, and consequently will get one? Do they want one of their own, or is this a joke that plays on Montaigne's certainty that they want a man's (his) penis? Why is it better "once and for all" to give them a penis? Because they will get one anyway? The supposed real and the imaginary, the representational and the actual, phallus and penis, are hopelessly jumbled.

Perhaps Montaigne's penis is at stake. After various other quick tributes to the power of the imagination—the stigmata, the scars on King Dagobert, his friend fainting and being subsequently prone to fits after hearing about someone else with these afflictions—he settles into the only sustained topic of the essay: impotence and the power of the imagination, and of women, to cause it. Certain women of Scythia supposedly had the power to kill men who had provoked them with their looks;

others could set "us" afire only to "extinguish us"; tortoises and ostriches hatch their eggs with looks alone, "a sign they have ejackulative virtues"; women transfer marks to their children in utero; an unusual young girl from Pisa was presented to Charles of Bohemia because she was shaggy in consequence of her mother's having a picture of John the Baptist over her bed when the girl was conceived. And so on.

Perhaps there is much of Montaigne the ironist here. But the essay does not allow of certainty on the bounds of sex. His impotency—finding "himself so short"—Germain's new real penis, and incorporating "this virile part into women," who already have it within, are all part of the same discursive whirl. An intensely gendered discussion—this is a man writing about his organ—seems to float over a chasm of fabled sex in which penises come and go at the mind's command.

I want to illustrate the fluid boundaries of sex and the more rigid distinctions of gender in one more context: the court of the lascivious Francis I. It is a powerfully gendered cultural venue. This was the court in which the Diana in Cellini's famous *Nymph of Fontainebleau* was uncomfortably posed over the entrance of the palace, the object of an unmistakably male gaze and especially of the privileged gaze of the king. Here men wrote blazons and counterblazons for one another's enjoyment on the subject of women's parts, ideological constructions of the female body. The beautiful breast—ivory, rose, a fruit—poetically confronts the ugly breast—black, sagging, stinking, shapeless—in this discourse between men.[33]

And courtly anatomy was similarly gendered. The artistically magnificent, if scientifically nugatory, work of the king's physician Charles Estienne is the product of an implicitly male science. Male intellect and male hands open up bodies and reveal nature's secrets, even as the illustrations show bodies of male sex tearing themselves apart for the male viewer's edification (figs. 13–14). Estienne cautions his students to hide the face and private parts of their cadavers so as not to divert the attention of spectators.[34]

There is in all of this a powerful homoerotic quality as women seem to mediate and create bonds between men. But still the women in Estienne's anatomy text are aggressively conventional in their heterosexual appeal. The first engraving (fig. 46) from a series illustrating the female reproductive system proclaims the "voluptuous" feminine erotic qualities of its model. And why not? It is, in fact, a reworking of the Florentine Perino

Fig. 47. Perino del Vaga's engraving *Venus and Cupid* from which Estienne took his anatomical model in fig. 46.

Fig. 46. A female figure from Charles Estienne's *La Dissection des parties du corps humain* (1546) in which the abdominal wall has been resected to reveal the placenta. The anatomically relevant section has in fact been inserted into a figure borrowed for this purpose.

del Vaga's *Venus and Cupid* (fig. 47).[35] A curtain sack, which at least in northern art of the period was an icon for the womb, has been added to fig. 47 in the process of refurbishing Venus so that she might serve the scientific purpose of fig. 46. A vase has replaced the cherub. It too may represent the womb—the uterus with handles as "seminal vessels" and the bearded men as ovaries—both linguistically and because of its shape (Latin *vas*, French *vase*, container or vessel). A few surgical tools are strewn in the foreground, and a little window has been carved out of Venus' belly into which a woodcut of the placenta has been set. Looking through it we see that the goddess of love, in her new incarnation as an anatomy model, is pregnant.[36] Another engraving (fig. 48) shows her in a slightly different though no less alluring pose, reclining on luxuriant

Fig. 48. This nude from Estienne's *Dissection* shows the womb opened, the kidneys, and the major vessels. The placenta that had been revealed in fig. 46 is now lying on the footstool. Again the anatomically relevant sections have been inserted into a figure produced for another purpose.

Fig. 49. The last in the series of female nudes from Estienne's *Dissection*. This one shows the womb with its "neck" (the vagina and its folds) and its "mouth" (the external pudenda).

bolsters, this time with an engraver's window displaying her womb into which a second window has been cut. The placenta, seen from the outside in fig. 46, now lies on the table where a cupid once sat.

Finally, in the most alluring pose of the series (fig. 49), Venus seems to be writhing in ecstasy on her plush cushions. Her hand holds onto the pillow, her foot seeks the trunk for support as she balances on the edge of the bed. We need to remember that this is only the background for an anatomical drawing: her liver and intestine are in full view, her genitals brazenly exposed. But these genital organs, which in a jurisprudential context even Renaissance anatomists would regard as distinguishing male from female, turn out to be just like a man's. Estienne is thoroughly, indeed obsessively, Galenic:

so that what is inside women, likewise sticks out in males, but what is the foreskin in males is the pudendum in women. For, says Galen, whatever you see as a kind of opening in the entrance to the vulva in women, such indeed is found in the foreskin of the male pudendum.

He continues in this fashion for several more paragraphs, to be sure that his readers understand that the overtly eroticized female figures he has presented really have the same genitals as men: "we call the throat of the womb that which is the shaft of the male's penis; it is like it nearly . . . what is a small covering in the opening of the vulva, such appears as a circular outgrowth of the male genitals."[37] Even in their tiny compartment we can see both the cervix and the vulva represented as glans-like structures. The notion, so powerful after the eighteenth century, that there was something concrete and specific inside, outside, and throughout the body that defined male as opposed to female, and provided the foundation for the attraction of opposites, was absent in the Renaissance.

In one of the illustrations (fig. 50) in Estienne's book, a man—perhaps Everyman—stands on a balcony overlooking a public square strewn with debris (perhaps ruins). His head is tilted slightly upward, he looks through a glass into the distance, and he fails to notice a naked, pregnant, opened-up, and most uncomfortable woman enthroned below. Despite its appearance in an anatomy book, this picture, and the others I have discussed from Estienne, are about what happens on the surface. They are about theater, about appearances, about erotic fetishes. Writhing St. Sebastians, ripped-apart men, naked women in courtyards, and similar dramatic tableaux capture the eye, while the organs themselves whimper for attention. In short, these are anatomical pictures about gender and not about what we would call sex, or the structures in the body that mark male and female. About these they are remarkably uninformative.

The prostitute Nanna in one of Pietro Aretino's erotic dialogues delights in precisely this theatricality of sex. Obviously she is a woman, different from man but as much because of artifice as biology. A "luscious pair of buttocks"—displayed in men more than in women by the clothes of the period—is the source of her power. The "mysteries of enchantment" lie between her legs, she says, shifting ground. But then what is between her legs? A vaginal opening "so finely cleft that one could barely find the place where it was."[38] Her erotic powers are not those of sexual anatomy but of an immensely powerful erotization of surface. Gender, not sex, is what matters. The tiny, invisible, closed cleft, not the vagina

Fig. 50. A pregnant woman with open womb enthroned in a courtyard as a man on the balcony, upper left, dangles a scroll from the window and looks heavenward. From a Latin version of Estienne's *Dissection*.

and organs within, defines Nanna as desirable, and considerable art has to go into making nature be "to her bias drawn."

Sex, gender, doctors, and law

Renaissance doctors understood there to be only one sex. On the other hand, there were manifestly at least two social sexes with radically different rights and obligations, somehow corresponding to ranges or bands, higher and lower, on the corporeal scale of being. Neither sort of sex—social or biological—could be viewed as foundational or primary, although gender divisions—the categories of social sex—were certainly construed as natural. More important, though, biological sex, which we generally take to serve as the basis of gender, was just as much in the domain of culture and meaning as was gender. A penis was thus a status symbol rather than a sign of some other deeply rooted ontological essence: *real* sex. It could be construed as a certificate of sorts, like the diploma of a doctor or lawyer today, which entitled the bearer to certain rights and privileges. In this section I will explore how, in difficult cases, sex was determined so as to fit a person for clear and unambiguous categories of gender. By showing how sex was fixed at the margins, perhaps I can shed light on its cultural nature at the core and on the tensions between an unbounded one sex and gender boundaries that mattered deeply.

In the ordinary course of events, sexing was of course no problem. Creatures with an external penis were declared to be boys and were allowed all the privileges and obligations of that status; those with only an internal penis were assigned to the inferior category of girl. In a world where birth mattered desperately, sex was another ascriptive characteristic with social consequences; being of one sex or another entitled the bearer to certain social considerations, much as being of noble birth entitled one to wear ermine under sumptuary laws governing clothing. Dress, occupation, and particular objects of desire were allowed to some and not to others, depending on whether they had sufficient heat to extrude an organ. The body thus seemed to be the absolute foundation for the entire system of bipolar gender.

But sex is a shaky foundation. Changes in corporeal structures, or the discovery that things were not as they seemed at first, could push a body easily from one juridical category (female) to another (male). These categories were based on gender distinctions—active/passive, hot/cold, formed/unformed, informing/formable—of which an external or an internal penis was only the diagnostic sign. Maleness and femaleness did not reside in anything particular. Thus for hermaphrodites the question was not "what sex they are *really*," but to which gender the architecture of their bodies most readily lent itself. The concern of magistrates was less with corporeal reality—with what we would call sex—than with maintaining clear social boundaries, maintaining the categories of gender.

Hermaphrodites "are called either male or female," Columbus says, "from their superabundance, as they are more suited or are believed to be more suited for forming humans or receiving one."[39] Sex is assigned as a consequence of formative capacity; once again, to be male is to be a father, which is to be the author of life. The nearer a creature approaches "creativity," the more it is male. Conversely, Columbus notes that the difficulties in diagnosing the sex of one woman he had seen arose from her being "unable to be either *rightly active or passive*." The reason for uncertainty is presented as organic: "her penis did not exceed the length or thickness of a little finger," while "the opening of her vulva was so narrow that it scarcely left the space of the tip of a little finger."[40] And Columbus, were he before a court of law, would apply widely accepted medical criteria for deciding which organ ought to decide sex. But he does not do so here; he does not say which organ is real. This person is deemed a woman because she is socially and juridically a woman, but one who can neither "rightly" act the passive role nor play the active one that would constitute a serious violation of sexual sumptuary laws, a woman pretending to be a man, a woman dressing above her station. It is almost as if the more general early modern concern about comporting oneself above one's place, born of the breakdown of patronage networks, the insidious workings of money, and the rise of new state-sponsored positions, was transferred to the world of gender.

By the nineteenth century, behavior is irrelevant. The question of sex is biological, pure and simple, writes the leading French forensic physician Ambroise Tardieu. It is "a pure question of fact that can and ought to be resolved by the anatomical and physiological examination of the person in question." Any notion of genuine sexual ambiguity or neutrality is nonsense because sex is absolutely there in and throughout the body.[41] In the late sixteenth century, the situation was very different; a woman taking the man's role in lovemaking with another woman was assumed to be a tribade (*fricatrice*), one who illicitly assumed the active role, who did the rubbing when she ought to have been primarily the one rubbed against. She stood accused as a woman who had violated a law of gender by playing the man's part during intercourse.

Marie de Marcis came close to being burned at the stake for this transgression.[42] She was baptized with a girl's name and grew to what appeared to be normal adulthood in a village near Rouen. Her master and mistress testified that she had regular periods, and medical testimony at her trial confirmed that she was indeed what she had been gendered from birth. But she fell in love with a female servant with whom she shared a bed, revealing to her that she had a penis and was therefore a man. They sought to marry.

Instead of being publicly acknowledged as a man once she had sprung a penis, as happened to Marie-Germain in Montaigne's story, Marie de Marcis was tried for sodomy—no assumption of natural heterosexuality here—and convicted; he/she could not produce the necessary organ under the pressure of a trial. But then Dr. Jacques Duval entered the case, found the missing member by probing his/her vulva, and proved that it was not a clitoris by rubbing it until it ejaculated a thick masculine semen. (Since the emphasis in this case was on illicit penetration, attention was focused not on whether Marie had an internal penis—a vagina—but whether her candidate for an external penis entitled her to the prerogatives of penis possession.) Duval's intervention saved Marie from the stake but did not immediately entitle her to a new gender. The court ordered that she continue to wear woman's clothing until she was twenty-five—as if the transition to maleness had to be made gradually—and that she refrain from having intercourse with either sex while she continued life as a woman.

The serious concern of the judges in this case seemed to be not with underlying sex but with gender: what signs of status, what clothes, what postures could Marie legitimately assume? Despite the court's obvious concern with organs, the central question is whether someone not born to the more elevated station, someone who had lived all her life as a woman, had what it took to legitimately play a man and more generally whether a "person" is entitled to a certain place in the social order.[43]

Women playing, or becoming, men is the dominant trope. In early seventeenth-century Holland there is, for example, Henrika Schuria, a "woman of masculine demeanor who had grown weary of her sex." She dressed as a man, enlisted in the army, and passed in her new role until she was caught taking the man's part in sexual intercourse. When she returned from the wars, she was accused of "immoral lust":

> For sometimes even exposing her clitoris outside the vulva and trying not only licentious sport with other women . . . but even stroking and rubbing them . . . so that a certain widow, who burned with immoderate lusts, found her depraved longings so well satisfied that she would gladly—except for legal prohibition—have married her.[44]

Her clitoris, it was said, "equalled the length of half a finger and in its stiffness was not unlike a boy's member." Schuria was tried, convicted, and sentenced to be burned as a tribade, but a merciful judge recommended that she be "nipped in the bud, and sent into exile." She was, in other words, relieved of the organ that she supposed would allow her to leave the "sex of which she had grown weary"; but she was punished with exile, a man's sentence. (This case shows that only one of the female penis isomorphisms really counts; her internal penis has to descend, as did Marie-Germain's, if she is to be entitled to a change. An enlarged clitoris does not count.) Her partner, the widow and the woman in their transaction, was chastised in an unspecified way and allowed to remain in the city. Having played the woman, she could be assumed to be less culpable, less dangerous, and less deserving of severe punishment. There are other cases, real or imagined, like this.[45]

But there are also cases that work the other way around, of men playing women for their own advantage. In 1459, the story goes, there was born a creature who "had the kinds of both male and female," though "man's nature did prevail." But because his "disposition and portraiture of body represented a woman," he/she? was able to find work as a maid servant and in this capacity to share a bed with his master's daughter, who became pregnant by him. For setting himself out to be a woman, this "monstrous beast" of a man was burned at the stake. Just how "a man's nature did prevail" when his body "represented" a woman's is not made explicit. Nor is it clear whether the offender lived as a woman generally or only on the occasions of bedding the daughter of the house. Whether the "damsel" understood the encounter to be with a woman throughout or only initially is also left ambiguous: was she deceived into allowing this man into her bed as a woman and then accepted him sexually as a man, or did she think until near the end that she was making love with a woman? There is no doubt, however, that someone used the ambiguities of his body to live as a woman—bad enough perhaps—but then reverted to having sex as a man. He was burned, as was the false Martin Guerre, for flouting the conventions that make civilization possible.

It seemed to matter little in any of these cases what sex the protagonists felt themselves to be, what they were inside. One of the disconcerting and poignant aspects of cases like Marie de Marcis' is how little regard was paid, in the accounts themselves and in the final determination of sex, to what we would call core gender identity, the sense that infants acquire very early on of whether they are girls or boys. No one probed what gender a person thought herself or himself to be before a change occurred or an accusation was made (I use the words "sex" and "gender" interchangeably here precisely because the distinction has now broken down). As long as sign and status lined up, all was well. Or, conversely, gender as a social category was made to correspond to the sign of sex without reference to personhood. The authorities assumed that the transformation from one to another state was absolutely precipitous, like moving from being married to being unmarried. Subjects were assumed to change from being socially defined girls to being socially defined boys with no difficulty or inner turmoil. Indeed, if instantaneous conversion was not forthcoming, the full penalties of the law were.

Montaigne recounts in his *Travel Journal* the story of a group of girls in Chaumont-en-Bassigni "who plotted together a few years ago to dress up as males and thus continue their life in the world." One of them came to Vitry, where Montaigne was visiting, worked as a weaver, and made friends. *He* became engaged to a woman with whom he subsequently fell out; still earning his living at the said trade, *he* fell in love with another woman, whom *he* actually did marry and live with for four or five months, "to her satisfaction, so they say." But then the weaver was recognized by someone from home. Just as abruptly as the social sex of the protagonist changed, so Montaigne changes his use of the personal pronoun: "*she* was condemned to be hanged . . . *she* said she would rather undergo [it] than return to a girl's status." And she was hanged "for using illicit devices to supply her defect in sex."[46]

Like Iphis, the girl in this story was gendered as a boy; she was every bit as much a boy as her mythical counterpart. But unlike Ovid's character, the French girl was able to consummate her love with a woman, without recourse to a penis and without the emotional storms Iphis suffered because he lacked one. But the gods did not come to the young weaver's rescue and did not bring forth the penis that would have entitled her to continue life as a man. The fact that he felt himself a man, that he had the skills of a man, and that he had lived as one was only more evidence of his crime: he lacked the birthmark of acquired status. For this he died a woman.

This does not seem very remarkable. Renaissance doctors and laypeople differentiated between the genital organs of males and females, and those with a penis were designated men. Sex then, as today, determined status, gender. But one also has the distinct feeling that in texts like Montaigne's, somehow "there is no there," no ontological sex, only organs with assigned legal and social status. At the very moment when genitals seem to display their full, unambiguous extralinguistic reality— when the language of one sex collapses—they also assume their fullest civil status, their fullest integration into the world of meaning. Corporeal solidity is shaken when it seems most stable, and we enter the shoals of language.

I want to illustrate this point with mention of how Paolo Zacchia's *Questionum medico-legalium,* the major Renaissance medical-jurisprudential text and one of the founding works of the discipline, treats the question of assigning sex.[47] It is, Zacchia argues, first of all a matter for the doctors and not for poets, soothsayers, quacks, or others among the medically ignorant. Hermaphrodites, he insists, are not dangerous, portentous monsters or prodigious inhabitants of the lands of Prester John, but rather people with ambiguous sexual organs who raise serious legal questions. Their deformations can be classified: three primary sorts in the male hermaphrodite, one in the female. There are true hermaphrodites who have both kinds of organs, and apparent hermaphrodites in whom, for example, a prolapsed uterus or an enlarged clitoris is mistaken for a penis. All this can be satisfactorily sorted out by an experienced professional observer.[48] Zacchia spends the remaining nineteen folio pages explaining who is to be called woman and who man.

The clinical and professional tone of the *Questionum*—case histories, taxonomies, learned reviews of the literature on various points—would lead one to assume that organs will be treated as the sign of something solidly corporeal, something that thoroughly informs its subject and determines its identity. But Zacchia, like Montaigne, treats organs as if they were contingent certificates of status: "members conforming to sex are not the causes that constitute male or female or distinguish between them . . . Because it is so, the members of one sex could appear in someone of the opposite sex."[49]

Zacchia's language, most blatantly in his discussion of clitoral hypertrophy, reveals his fundamentally cultural concerns. "It should be enough now to observe" he argues, "that, in regard to women who have turned into males, in the most, this has followed a promotion (*beneficium*) of the clitoris, as several anatomists think." He does not use the obvious noun for what might have happened, *incrementum* or *amplification,* an enlargement, and writes instead of *beneficium,* a kindness or favor, especially in the political sense of an advancement or a grant that endows ecclesiastical property or a feudal right. An enlarged clitoris must not be mistaken for a promotion on the scale of being, although, as in the case of Marie-Germain, having an internal penis pop out just might. Getting a certifiable penis is getting a phallus, in Lacanian terms, but getting a large clitoris is not.[50]

Similarly, when Zacchia is discussing hermaphrodites with both sets of organs, he distinguishes, following Aristotle, the valid sex (*sexum ratum*) from the ineffectual, invalid, useless sex (*inritum*). Again the sense is political—valid or invalid testaments or laws—and not morphological. Political judgments, the claims of gender, are already contained in judgments about sex because politics is already contained in the biology of generation. Thus, when Zacchia is arguing that humans cannot have two valid sexes, he is alluding less to a biological fact than to a social or cultural fact: males inform and women bear, and it is impossible for any one creature to do both, however his/her organs might be configured. In the absence of evidence regarding actual generation, the old Pythagorean oppositions, not some alternative anatomical or physiological criteria, come into play: the organ on the right (in the case of hermaphrodites with side-by-side genitals) or the organ on top (in the case of those with organs arranged vertically along the body's axis) is the one that counts.[51]

Even when there are no genital organs visible at all, there are signs to indicate which sex is the more potent and which is less so or impotent (*potentiorem ab impotentiorem*). Again the language is at least as much political as biological: the secondary sexual characteristics to which one would have reference in lieu of genital organs are the consequence of the greater or lesser vital heat that defines man and woman. Heat, for Renaissance doctors, was of course supposed to have physical correlatives. But heat was also so inextricably bound up with the great chain of being that it is difficult to unpack its meaning from the meaning of perfection itself.

For example, women can turn into men, Zacchia maintains, but men cannot turn into women. Why? He offers a straightforward anatomical reason—no room inside a man for a penis to invert into—but this throwaway line carries little conviction. His main lines of argument are metaphysical. Generally speaking, most authorities agree that "nature always tends towards the more perfect." But more specifically, if a transformation of sex does take place, it occurs because of what men have, heat. Heat, he says, "drives forward, diffuses, dilates; it does not compress, contract, or retract." The active principle therefore works so that "members which project outwards will never recede inwards." (Male heat, in other words, obeys the laws of thermodynamics.) Men cannot become women by expulsion because, as he has already shown, this works in the opposite direction, and they cannot become women by attraction because "this, when it works properly, draws together that which is favorable for the animal," and becoming more imperfect is clearly not more favorable.[52]

Biology, in other words, is restrained by cultural norms just as much as culture is based on biology. In the one sex-world generally, and specifically in the work of Zacchia, when the talk turns—for good, everyday, practical legal reasons—to the biology of two clear and distinct foundational sexes, it becomes at the same time enmeshed in the body/gender continuum of the one-sex model. During much of the seventeenth century, to be a man or a woman was to hold a social rank, to assume a cultural role, and not to *be* organically one or the other of two sexes. Sex was still a sociological, not an ontological, category.

Imagining generation in Harvey's work

Live Modern Wonder and be read alone,
Thy brain have issue though thy Loins have none.
Let fraile Succession be the Vulgar Care,
Great Generation's Selfe is now thy Heire.[53]

The childless "Live Modern Wonder" whose brain had issue was William Harvey, the man who discovered that the blood circulated, the man credited with being the first to say that all life comes from an egg, the man who thought that conception was the having of an idea, sparked by sperm, in the womb. I close this chapter with a brief discussion of his *Disputations Touching the Generation of Animals*[54] because it is the last major story about generation and the body still deeply embedded in the political aesthetics of the one-sex model with, at the same time, its claims to epistemological authority, its experimental strategies, and its ontology of reproduction—Harvey claims to be talking, for the first time in history, about a specific germ product, the egg—cast overtly in the language of the new biology. In Harvey we can begin to glimpse what will become clearer in the next two chapters: not only that theories of sexual difference help to determine what scientists see and know but, more important, that the opposite is not the case. What scientists see and know at any given time does not circumscribe how sexual difference is understood or limit the aesthetics of its expression. Quite to the contrary, observations and the prestige of science generally lend the art of difference new weight without affecting its content.

The question of this section can be posed formally. Harvey's *On the Motion of the Heart and Blood in Animals,* like other great scientific texts, powerfully achieves closure. Cleanly, crisply, and economically it destroys two thousand years of physiology and establishes beyond a doubt that whatever else the heart is, it is a pump, and whatever else the blood does, it must circulate even if the passages through which it goes from arteries to veins, the capillaries, could not yet be demonstrated. The far longer *Disputations,* on the other hand, endlessly defers coming to an end; stories multiply but go nowhere. The book corrects a few relatively minor errors in previous accounts of the embryology of the chick, makes a strong but inconclusive case for epigenesis, suggests experimentally but does not prove the important point that fertilization is not the merging of a mass of semen with a mass of menstrual blood, and fails after desperate efforts to understand the mystery of generation.[55] Why this lack of closure?

The book's length and narrative openness are not due primarily to scientific failures that no degree of clearheadedness and lack of cultural baggage could have avoided. The fact that Harvey, lacking a microscope, could not see egg or sperm was not the reason that he could come to no closure on the issue of conception, just as the discovery of egg and sperm in the eighteenth century also could offer no convincing solution. By the last half of the nineteenth century, cell theory allowed conception to be understood as the fusion of two distinct cells, which suggested the view that recognizable males and females were the projections somehow of radically different germ products. But then the DNA revolution has once again taken sex out of conception; strands of DNA do not sustain a vision of sexual dimorphism. Molecular biology has begun to illuminate with a precision unimaginable in Harvey's day—or indeed before the late 1940s—how epigenesis works. It has not provided answers to the "mystery of life" in relation to a socially sexed world.

The peculiar narrative openness of the *Disputations* is also not the result of Harvey's particular political agenda, if for no other reason than that his perfectly conventional positions on matters of gender have deeply ambiguous and inconsistent resonances in his other work. One can argue that Harvey emphasizes the passivity of women and matter in reproduction and that this is consistent with "new scientific values based on the control of nature and women integral to the new capitalist modes of production," and more generally with the "cultural biases" or "prevailing cultural ideas of male superiority."[56] His declaration to his anatomy students—as if it were a law of nature—that "Men *woe allure make love;* female *yield condescen suffer;* the contrary *preposterous*" is certainly evidence for the spilling over of politics into science.[57] And when clinical evidence fails him on why women do not produce semen, he resorts to the genital teleology of the one-sex model: it is unthinkable that "such imperfect and inconspicuous parts" as the genital apparatus of the female could produce a semen "so concocted and so vital" as to be able to share influence with that of the male, "so concocted with quickening heat, refined in so many vessels and leaping with so much spirit."

Yet Harvey did abandon a traditional Aristotelian account of the active male who acts upon the passive female. The female "primordium," in his account of generation, was *both* a material and an efficient cause of generation.[58] The form and the matter of the fetus come from the mother whose womb, once ignited, has within it, specifically within the primordium or egg, the "spirit" or idea of new life. Indeed, Harvey's account borders on parthenogenesis and lays so much stress on the female's having the idea of new life inside her that it prompted one wag to remark that, if it were true, women should be able to conceive by just thinking about it.[59] The point, however, is not which of Harvey's stories about generation is the dominant one, but rather that there are so many stories to be told.

The William Harvey who wrote on biological and social sex relies on the authority of nature and experiment for these tales just as aggressively as the William Harvey who wrote on the blood's circulation and who is, for that work, much admired by those exploring the origins of modern science. Narratives about sex in the *Disputations* are presented as if they were self-evident in Nature, "herself the most faithful interpreter of her own secrets." (A feminine Nature is here both scientist and object.) What is obscure in one species, Nature exhibits clearly in another, and now "that the whole theater of the world" lies open, only willful sloth would make one rely on the wisdom of others: it is "sweet not only to grow weary but even to faint away" in following Nature's lead down the path she chalks till at last we are "received into her closest secrets." One could, Harvey believed, actually get at the thing itself, which was perforce more real than any image or representation of it (its *eidos*). Thus what one discovers through the senses is clearer than what one might discover in books, and it is a sign of moral degeneracy, of baseness, "to be tutored by other men's commentaries without making trial of the things themselves, especially since Nature's book is so open and so legible."[60] By extension we are invited to regard Harvey's account of generation as morally and epistemologically superior to one based on the ratiocination of Galen or on blind submission to the authority of the ancients, even of Aristotle. Harvey expounds the triumphant empiricist epistemology, the new reductionism of the new science.

To Harvey the crowning glory of his entire enterprise was his famous demonstration to Charles I that the Galenists were wrong in holding that male and female matter actually mixed at conception and that Aristotle was wrong in holding that menstrual blood was the material basis for new life. This exercise, in Harvey's view, speaks not only of the particular truth in question but of the very power of formal experimental procedures to adjudicate between theories.[61] He had shown the king a deer's uterus in the early stages of pregnancy and "made clear to him that no trace of seed or conception could be found in the hollow of the womb." When Charles communicated this news to some of his followers, they declared that Harvey had been deceived and had led the king into error. They declared that a conception forming "without any trace of the male's seed surviving," with nothing remaining in the uterus after coitus, "ranked among the *adunata,* the things impossible." To settle the matter and "in order that this finding of so great moment might be understood more clearly by posterity," the king ordered an experiment, which Harvey devised. A dozen does were isolated in Richmond Park after the rutting season and kept away from bucks after an initial mating. Harvey dissected some of them—presumably fertile, as shown by the fact that those remaining alive did become pregnant "as if by some contagion, and gave birth to their fawns in due season"—and found that there were "no remains in the uterus either of the semen of the male or female . . . nothing produced by any admixture of these fluids . . . nothing of the menstrual blood present as 'matter' as Aristotle will have it."[62]

Never mind that this experiment was deeply flawed, that by the time Harvey looked he would not have found sperm even if he had the lens necessary to see it. Never mind that the demonstration for Charles made the search for semen in the wombs of postcoital females a new research problem in an already overcrowded field. (The great Dutch anatomist Frederik Ruysch (1638–1731) is said to have gone out in the middle of the night to dissect a woman, caught and murdered by her husband in the act of adultery, only to have his discovery of semen in her uterus discounted because the room was dark and his eyes were weakened by age.) Harvey's experiment makes an important negative case. Menstrual blood does not, in fact, go into making a fetus, and the great bulk of the male ejaculate is indeed irrelevant to actual conception, although of course a sperm does materially enter an egg.[63] More important, it provides the materials out of which one can imagine the profound truth and mystery of epigenesis, of making a complex organism from unformed matter which somehow assumes the shape and characteristics of the creature it came from.

But Harvey, like his predecessors and successors, was incapable of writing about sexual reproduction outside an already gendered language, in his case that of the one-sex model. Generation by the union of two sexes must be made *to have meaning* beyond itself, involving the social realm that such a union sustains. Having argued convincingly that the hen's egg—by extension the human female's—is not produced, contra Galen, from any female outpourings at coitus, Harvey nevertheless felt compelled to render culturally significant a chicken's, and a woman's, post-coital behavior. The hen's acting "as one ravished with gentle delight," though not a sign of semination, is a sign of gratitude toward the male for his godlike act:

> She shakes herself for joy, and, as if she had now received the greatest gift, preens her feathers as if giving thanks for the blessing of issue granted by Jove the creator. The dove . . . expresses her joy in coitus in wondrous wise; she leaps and spreads her tail and with it sweeps the earth below her, and combs her feathers with her beak and settles them, as if the gift of fertility did lead to the greatest glory.[64]

Somehow the female primordium, with its mysterious capacity to form itself sequentially into an ordered body, must be ignited and given life. Somehow the unfathomable drama of generation must have its objective correlative in the social world. Enter the male. Sperm acts by "contagion" to ignite the egg. Indeed, sperm is prolific in some measure because it is "permeated with spirit by the fervency of coitus or desire and froth with the nature of spume [foam]." The heat of intercourse corresponds to no earthly blaze but to the stars, so that sperm bears, Prometheus-like, the celestial fire while fertilization itself is the male's reenactment of what

God wrought at the moment of creation. Impregnation for Harvey becomes metaphorically the igniting of women, setting them aflame as if struck by lightning. Or, in a metaphor even more evocative of the Word, of the Logos "informing" the world, it is like the formation of a conception in the brain. Here the image gets a bit more complex because the sperm alone is definitely not the idea, even though the uterus alone is the brain: "the generation of things in Nature and the generation of things in Art take place in the same way . . . Both are first moved by some conceived form which is immaterial and *is produced by conception*." The brain is "the instrument of conception" in producing art because it is the instrument of the soul, "without the intervention of matter"; meanwhile, the "uterus or egg" is the brain or the instrument of conception in Nature. But the idea in question seems to be not, as in Aristotle, the sperm alone but rather the thing "produced by conception" which generates the living work of art.

Harvey had earlier prepared the way for the uterus-as-brain metaphor. The pregnant uterus of the deer swells up, "and a most soft and pulpy substance, like that of the brain, fills the cavity." A few sentences later he writes that the interior of the uterus is so delicate and smooth that "you might think it the softness of the ventricles of the brain." Elsewhere: "the appearance or form of the chick is in the uterus or egg without any material, just as the concept of the house is in the brain of the builder." In other words, sperm might act "as if the Almighty should say, 'Let there be offspring,' and straightaway it is so," but only insofar as it allows an idea—the primordium or egg—to be generated in the uterine brain of woman.[65]

While rejecting Galen's interpretation of female orgasm as a sign of semination, Harvey saw sexual passion as deeply significant, an expression of the body's vital force. The sheer carnality of intercourse bespeaks life's energy and tragically prefigures its end: "And it is clear that parents are youthful, beautiful, perfected and live joyously no longer than they can beget eggs and fecundate them, and by the mediation of these eggs, bear their own like." Males and females, Harvey told his students in 1616, are *"never more brave sprightly blithe valiant plesant or bewtifull"* than now that coitus is about to be performed."[66]

But when "this office of life is once ended, alas!" Just as a man is sad after coition, so all animals are sad unto death when the spark, of which orgasm is the sign, is exhausted: "even he flags after long use of venery and like a soldier time expired grows weary, and the hens too, like plants, become past laying and are exhausted." Only now do we realize that Harvey's account of life's drama has been flitting between the barnyard and the bedroom. Thus for Harvey, as much as for the confirmed Galenist, the heats and passions of the body express the hierarchy of creation.

Harvey's new epistemology and substantive discoveries led right back to new versions of old stories. Generation, the body's most social function, remained beyond the reach of a nonexistent neutral language of organs and functions. Desperate to understand how it all worked, Harvey spun story after poignant story about sexual difference, always pretending that it was Nature herself who spoke.

In the eighteenth century, the voice of Nature would be heard more loudly. Meaning, it would be thought then, existed not in the echoes of macrocosm and microcosm but in the thing itself. The mechanical world picture promised truth from the material world. But a new epistemology would not shield sexual anatomy and reproduction from the demands of culture. While the one flesh did not die—it lives today in many guises— two fleshes, two new distinct and opposite sexes, would increasingly be read into the body. No longer would those who think about such matters regard woman as a lesser version of man along a vertical axis of infinite gradations, but rather as an altogether different creature along a horizontal axis whose middle ground was largely empty.

FIVE

Discovery of the Sexes

> The bicycle's triumph . . . necessitates an androgy-
> nous outfit worn by its adepts of the weaker sex . . .
> Will we never make our skirted publishers and soci-
> ologists in dresses understand that a woman is nei-
> ther equal nor inferior nor superior to a man, that
> she is a being apart, another thing, endowed with
> other functions by nature than the man with whom
> she has no business competing in public life? A
> woman exists only through her ovaries.
>
> VICTOR JOZÉ, 1895

Sometime in the eighteenth century, sex as we know it was invented. The reproductive organs went from being paradigmatic sites for displaying hierarchy, resonant throughout the cosmos, to being the foundation of incommensurable difference: "women owe their manner of being to their organs of generation, and especially to the uterus," as one eighteenth-century physician put it.[1] Here was not only an explicit repudiation of the old isomorphisms but also, and more important, a rejection of the idea that nuanced differences between organs, fluids, and physiological processes mirrored a transcendental order of perfection. Aristotle and Galen were simply mistaken in holding that female organs are a lesser form of the male's and by implication that woman is a lesser man. A woman is a woman, proclaimed the "moral anthropologist" Moreau in one of the many new efforts to derive culture from the body, everywhere and in all things, moral and physical, not just in one set of organs.[2]

Organs that had shared a name—ovaries and testicles—were now linguistically distinguished. Organs that had not been distinguished by a name of their own—the vagina, for example—were given one. Structures that had been thought common to man and woman—the skeleton and the nervous system—were differentiated so as to correspond to the cultural male and female. As the natural body itself became the gold standard of social discourse, the bodies of women—the perennial other—thus became the battleground for redefining the ancient, intimate, fundamental social relation: that of woman to man. Women's bodies in their corporeal, scientifically accessible concreteness, in the very nature of their bones, nerves, and, most important, reproductive organs, came to bear an enormous new weight of meaning. Two sexes, in other words, were invented as a new foundation for gender.

Woman's purported passionlessness was one of the many possible manifestations of this newly created sex. Female orgasm, which had been the body's signal of successful generation, was banished to the borderlands of physiology, a signifier without a signified. Previously unquestioned, the routine orgasmic culmination of intercourse became a major topic of debate. The assertion that women were passionless; or alternatively the proposition that, as biologically defined beings, they possessed to an extraordinary degree, far more than men, the capacity to control the bestial, irrational, and potentially destructive fury of sexual pleasure; and indeed the novel inquiry into the nature and quality of female pleasure and sexual allurement—all were part of a grand effort to discover the anatomical and physiological characteristics that distinguished men from women. Orgasm became a player in the game of new sexual differences.

This did not happen all at once, nor did it happen everywhere at the same time, nor was it a permanent shift. When in the 1740s the young Princess Maria Theresa was worried that she had not immediately become pregnant after her marriage to the future Hapsburg emperor, her physician responded with advice that was no different from what Soranus might have offered a Roman matron: "Ceterum censeo vulvam Sanctissimae Majestatis ante coitum esse titillandum" (Moreover, I think the vulva of Her Most Holy Majesty should be titillated before intercourse.) She bore more than a dozen children.[3] Physicians in the nineteenth and early twentieth centuries could offer little more, and even today doctors disabuse patients of beliefs as old as Hippocrates:

> Dear Dr. Donohue: I am ashamed to ask my doctor: Do you only get pregnant when you have an orgasm?
> *Answer:* Pregnancy results when sperm meets and fertilizes an egg. Orgasm has nothing to do with it.[4]

As for the one-sex model, it too lived on. In the eighteenth and nineteenth centuries, books like *Aristotle's Masterpiece* and Nicholas Venette's

The Art of Conjugal Love, or to a lesser extent the Pseudo-Albertus Magnus' *Secrets of Women,* transmitted Galenic learning to hundreds of thousands of lay readers, whatever their doctors might have thought. And in a variety of contexts, physicians themselves also spoke in the language of the one-sex model (such as those who feared that German women workers engaged in unfeminine occupations would become *Mannweiber,* male women).[5]

There are two explanations for how the two modern sexes as we imagine them were, and continue to be, invented: one is epistemological and the other is, broadly speaking, political.[6] The epistemological explanation in turn has at least two articulations. The first is part of the story in which fact comes to be more clearly distinguished from fiction, science from religion, reason from credulity. The body is the body is the body, said a new group of self-appointed experts with ever more authority, and there are only certain things it can do. Lactating monks, women who never ate and exuded sweet fragrance, sex changes at the whim of the imagination, bodies in paradise without sexual difference, monstrous births, women who bore rabbits, and so on, were the stuff of fanaticism and superstition even if they were not so far beyond the bounds of reason as to be unimaginable. Skepticism was not created in the eighteenth century, but the divide between the possible and the impossible, between body and spirit, between truth and falsehood, and thus between biological sex and theatrical gender, was greatly sharpened.

The second part of the epistemological explanation is essentially the one given by Foucault: the episteme "in which signs and similitudes were wrapped around one another in an endless spiral," in which "the relation of microcosm to macrocosm should be conceived as both the guarantee of that knowledge and the limit of its expansion," ended sometime in the late seventeenth century.[7] All the complex ways in which resemblances among bodies, and between bodies and the cosmos, confirmed a hierarchic world order were reduced to a single plane: nature. In the world of reductionist explanation, what mattered was the flat, horizontal, immovable foundation of physical fact: sex.

Or, put differently, the cultural work that had in the one-flesh model been done by gender devolved now onto sex. Aristotle did not need the facts of sexual difference to support the claim that woman was a lesser being than man; it followed from the *a priori* truth that the material cause is inferior to the efficient cause. Of course males and females were in daily life identified by their corporeal characteristics, but the assertion that in generation the male was the efficient and the female the material cause was, in principle, not physically demonstrable; it was itself a restatement of what it *meant* to be male or female. The specific nature of the ovaries or the uterus was thus only incidental to defining sexual difference. By the eighteenth century, this was no longer the case. The womb, which had been a sort of negative phallus, became the uterus—an organ whose fibers, nerves, and vasculature provided a naturalistic explanation and justification for the social status of women.

The context for the articulation of two incommensurable sexes was, however, neither a theory of knowledge nor advances in scientific knowledge. The context was politics. There were endless new struggles for power and position in the enormously enlarged public sphere of the eighteenth and particularly the postrevolutionary nineteenth centuries: between and among men and women; between and among feminists and antifeminists. When, for many reasons, a preexisting transcendental order or time-immemorial custom became a less and less plausible justification for social relations, the battleground of gender roles shifted to nature, to biological sex. Distinct sexual anatomy was adduced to support or deny all manner of claims in a variety of specific social, economic, political, cultural, or erotic contexts. (The desire of male for female and female for male was natural—hence the new slogan "opposites attract"—or it was not.) Whatever the issue, the body became decisive.

But no one account of sexual difference triumphed. It may well be the case that almost as many people believed that women by nature were equal in passion to men as believed the opposite.[8] We simply do not know how many people believed, with the eighteenth-century moral anthropologist Pierre Roussel and the nineteenth-century English feminist Elizabeth Wolstenholme, that menstruation was a contingent pathology of civilization and how many believed the opposite, that menstruation showed the power of the uterus over women's lives and hence was a natural foundation for gender difference.[9] For everyone who thought that women of color were especially responsive sexually because of the structure of their genitalia, someone else thought that their coarse nervous systems and dry mucous membranes resulted in a "want of genital sensitiveness."[10]

Studies of the micropolitics of these alternative accounts would be rewarding, but we should not lose sight of the fact that the very terms of the debates are new: difference that had been expressed with reference to gender now came to be expressed with reference to sex, to biology. There were no books written before the late seventeenth century with titles like *De la femme sous ses rapports physiologiques, morals et littéraires* or *De la puberté . . . chez la femme, au point de vue physiologue, hygiènigue et medical* that argued so explicitly for the biological foundations of the moral order. There were hundreds if not thousands of such works in which sexual differences were articulated in the centuries that followed.

Scientists did far more than offer neutral data to ideologues. They lent their prestige to the whole enterprise; they discovered or bore witness to aspects of sexual difference that had been ignored. Moreover, the politics of gender very clearly affected not only the interpretation of clinical and laboratory data but also its production.[20] On the other hand, a number of new research traditions did produce considerable knowledge about the developmental and mature anatomy of the male and female body, about the nature of ovulation and the production of sperm, about conception, menstruation, and in the 1920s and 1930s the hormonal control of reproduction generally. By the early decades of this century, the power of science to predict and effect successful mating in humans and animals was considerably enhanced. In short, reproductive biology progressed in its understanding of sex and was not merely an "immature" enterprise that served competing social interests.

But my point here is that new knowledge about sex did not in any way entail the claims about sexual difference made in its name. No discovery or group of discoveries dictated the rise of a two-sex model, for precisely the same reasons that the anatomical discoveries of the Renaissance did not unseat the one-sex model: the nature of sexual difference is not susceptible to empirical testing. It is logically independent of biological facts because already embedded in the language of science, at least when applied to any culturally resonant construal of sexual difference, is the language of gender. In other words, all but the most circumscribed statements about sex are, from their inception, burdened with the cultural work done by these propositions. Despite the new epistemological status of nature as the bedrock of distinctions, and despite the accumulation of facts about sex, sexual difference in the centuries after the scientific revolution was no more stable than it had been before. Two incommensurable sexes were, and are, as much the products of culture as was, and is, the one-sex model.

In this chapter and the next I will primarily be making the negative case that new scientific discoveries did not bring down the old model and enshrine the new. One sex, I want to emphasize again, did not die. But it met a powerful alternative: a biology of incommensurability in which the relationship between men and women was not inherently one of equality or inequality but rather of difference that required interpretation. Sex, in other words, replaced what we might call gender as a primary foundational category. Indeed, the framework in which the natural and the social could be clearly distinguished came into being.

Biological sex

In the late seventeenth and eighteenth centuries, science fleshed out, in terms acceptable to the new epistemology, the categories "male" and "female" as opposite and incommensurable biological sexes. One can sense this in subtle turns of phrase. Buffon, the encyclopedic Enlightenment naturalist, translates back and forth as if he senses that he is on the cusp of a momentous transformation: the peculiar correspondence between the parts of generation and the rest of the body might be called (with the ancients) "sympathy" or (with the moderns) "an unknown relation in the action of nerves."[11] A notion of order and coherence is replaced by corporeal wiring.

More generally, by the end of the seventeenth century the various intellectual currents that made up the transformation of human understanding known as the scientific revolution—Baconianism, Cartesian mechanism, empiricist epistemology, Newtonian synthesis—had radically undermined the whole Galenic mode of comprehending the body in relation to the cosmos.[12] This meant the abandonment, among other things, of the anatomical isomorphisms between man and woman and also the purging from scientific language of the old metaphors that had linked reproduction to other bodily functions, to the natural world, and to the great chain of being itself. Generation could now less plausibly be seen in terms of rennin and cheese; iron and loadstone lost their resonance as metaphors for semen and womb. The penis as plowshare and the womb as field did not quite capture Enlightenment views of fruitful intercourse. Hoary images drawn from agriculture—the vagina as an organ "inwardly wrinkled, like the inner skin of the upper jaw of a cow's mouth"—disappeared from works intended for a self-consciously sophisticated audience.[13] Indeed the term "generation" itself, which suggested the quotidian repetition of God's act of creation with all its attendant heat and light, gave way to the term "reproduction," which had less miraculous, more mechanistic connotations even if it did not quite capture the virtuosity of nature. As Fontanelle said, "Put a Dog Machine and a Bitch Machine side by side, and eventually a third little Machine will be the result, whereas two Watches will lie side by side all of their lives without ever producing a third Watch."[14] The importance in the eighteenth century of new theories of knowledge generally, and with respect to the body particularly, is a commonplace. Scientific race, for example—the notion that either by demonstrating the separate creation of various races (polygenesis) or by simply documenting difference, biology could account for differential status in the face of "natural equality"—developed at the same time and in response to the same sorts of pressures as scientific sex.[15] Claims of the sort that Negroes have stronger, coarser nerves than Europeans because they have smaller brains, and that these facts explain the inferiority of their culture, are parallel to those which held that the uterus naturally disposes women toward domesticity.[16] I want here simply to acknowledge that my particular story is part of what would be a more comprehensive history of exclusive biological categories in relation to culture.

Poullain de la Barre, one of the earliest writers in the new vein, illustrates the turn to biology when an old ordering of man and woman collapses. In his case the impetus to biology is twofold. In the first place de la Barre is committed to the Cartesian premise that the self is the thinking subject, the mind, and that it is radically not body. From this it follows that the mind, this decorporealized self, has no sex and indeed can have no sex. Gender, the social division between men and women, must therefore have its foundation in biology if it is to have any foundation at all. His version of Descartes' radical skepticism leads him to the same conclusion. He lists a number of views that the ignorant hold as unquestionable: that the sun moves around the earth; that traditional religion is true; that the inequality of man generally is evident in the "disparity of Estates and Conditions." And, "amongst these odd opinions," he writes, "there is not any mistake more Ancient, or Universal" than "the common Judgment which men make of the Difference of the two Sexes, and all that depends thereon"; ignorant and learned alike seem to think it "a paradox and piece of singularity" that woman might not be inferior to man in "capacity and worth."[17]

In other words, the usual views on sexual difference might simply be a mistake, like seeing a square tower as if it were round. It is not a Cartesian "clear and distinct" idea, as it would have been for Aristotle, but rather a question that can be decided on the same grounds as one judges whether the sun is the center of the solar system.[18] Given then that sexual difference is an empirical matter, even the most firmly held and seemingly secure views about women might turn out, upon further scrutiny, to be false. Moreover, de la Barre goes on, one can even demonstrate the precise, historically explicable causes of erroneous views: because the subject has been "but very lightly discoursed of"; because of "partiality"; because of the lack of "trial or examination." Once bias and superficiality have been dealt with, sexual difference is a question of biology that solely constitutes the category "sex." Specifically for de la Barre, the task is to demonstrate that the organic differences corresponding to the social categories of man and woman do not, or ought not to, matter in the public sphere. For others the project was quite the opposite. But whatever the political agenda, the strategy is the same: indeed, sex is everywhere precisely because the authority of gender has collapsed.[19]

Political theorists beginning with Hobbes had argued that there is no basis in nature, in divine law, or in a transcendent cosmic order for any specific sort of authority—of king over subject, of slaveholder over slave, or, it followed, of man over woman. For Hobbes, as for Locke, a person is essentially a sentient being, a sexless creature whose body is of no political relevance. Still, for both, males do end up being the head of households and nations. Men, not women, make the social contract. The reason for subordination, they want to hold, is not built into the world order; it does not arise from old-fashioned reasons like the superiority of spirit over matter or the historical dominance God granted Adam. Nor do they seem to want to attribute it to "mere nature," where a child would be more likely to obey its mother than its father. Instead it seems to have arisen in historical time as a consequence of a series of struggles that left women in the inferior position. Locke says simply that since "the last Determination, the Rule, should be placed somewhere, it naturally falls to the Man's share, *as the abler and the stronger.*"[20] In Hobbes it is much less clear, and one can only surmise that a woman's having a child puts her in a vulnerable situation, which allows the man to conquer her and her children and thereby create paternal rights by contract, by conquest in Hobbesian terms.[21] In any case he is adamant that paternal rights do not, as in the old model, arise from generation. However problematic, the tendency of early contract theory is to make the subordination of women to men a result of the operation of the *facts* of sexual difference, of their utilitarian implications. What matters is the superior strength of men or, more important, the frequent incapacity of women because of their reproductive functions.[22] Bodies in these accounts are not the sign of but the foundation for civil society.

Rousseau, arguing against Hobbes, takes a similarly biological tack. Hobbes, he says, erred in using the struggle of male animals for access to females as evidence for the natural combativeness of the primitive human state. True, he concedes, there is bitter competition among beasts for the opportunity to mate, but this is because for much of the year females refuse the male advances. Suppose they were to make themselves available only two months out of every twelve: "it is as if the population of females had been reduced by five-sixths." But women have no such periods of abstinence—love is "never seasonal" among the human species—and they are thus not in short supply; even among savages there are no "fixed periods of heat and exclusion" that produce in animals such "terrible moment[s] of universal passion."[23] Reproductive physiology and the nature of the menstrual cycle bear an enormous weight here, as the state of nature is conceptualized in terms of the supposed differences in the sexual receptivity of women and beasts.

And, to give a final example, Tocqueville argued that in the United States democracy had destroyed the old basis for patriarchal authority and that it was necessary to trace anew and with great precision "two clearly distinct lines of action for the two sexes."[24] In short, wherever boundaries were threatened or new ones erected, newly discovered fundamental sexual differences provided the material.

Their provenance was science. In the late eighteenth century, anatomists for the first time produced detailed illustrations of an explicitly female skeleton to document the fact that sexual difference was more than skin deep. Where before there had been only one basic structure, now there were two.[25] The nervous system assured, in still another realm, that the body "would be an observable and internally consistent field of signs," that female sympathy would be the result of female fibers.[26]

Gradually the genitals whose position had marked a body's place on a teleologically male ladder came to be rendered so as to display incommensurable difference. We can, already by the late seventeenth century, trace

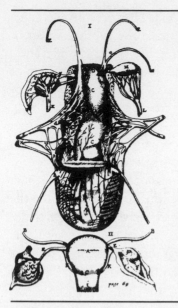

Fig. 51. The top drawing (I) shows a womb opened in relation to the "stones" and bladder. The lower drawing (II) shows the body of the uterus and the stones but, unlike earlier drawings, no vagina. From Bartholin, *Anatomy*.

the collapse of the old representations. Bartholin, who on occasion explicitly opposed the Galenic isomorphisms, produced in 1668 three separate drawings of the female genitalia: one that showed the whole generative system and pointedly left out the vagina and external pudenda; another that showed the womb open in relation to the "stones" (ovaries), again without a vagina; and finally one that showed the clitoris as a penis but rendered the vagina open so that it looked as little as possible like a penis (compare figs. 37 and 51). Even though these images belie the ancient construction of woman as an inferior, internalized man, their labels are still very much those of the old order: the "stones of woman" for the ovaries, the "deferent vessels" for the Fallopian tubes, the curiously metaphoric "sheath or scabbard of the womb" for what had been the neck of the womb and would become the vagina. Though the old representations were clearly no longer viable, genitals here were not yet doing the work of signification they would perform in the illustrations of the next century.

Just how shaky the new images still were is clear in the work of Regnier de Graaf (1641–1673). His discovery of the ovarian follicle provided the basis for much future discussion of sexual difference, but his illustrations of the female genitalia were more old-fashioned than Bartholin's. The

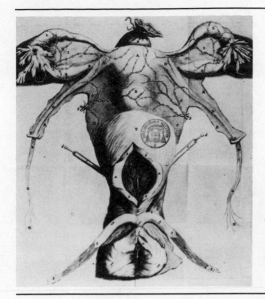

Fig. 52. The uterus, vagina, and ovaries—still labeled female testicles—from Regnier de Graaf, *De mulierum organis generationi inservientibus* (1672). If the vagina were not sectioned open, the picture would resemble earlier drawings produced to show the male and female organs as isomorphic.

entire vagina is still shown attached to the cervix, as in Renaissance texts, but de Graaf's depiction of the vagina opened just below the cervix and of the ovaries firmly attached to their ligaments tends to make the ensemble look considerably less penislike than its sixteenth- or early seventeenth-century counterparts (fig. 52).

By the late seventeenth century, the English anatomist William Cowper, like Bartholin, had separate drawings for the clitoris, for the pudendum and "fore part of the *vagina uteri,*" and for the uterus, ovaries, and Fallopian tubes. The only hints of the old formula are that he includes part of the vagina, albeit "divided so as to show its rouge," in his image of the uterus (thereby detracting from the penis effect) and that he has not quite adopted what would become modern nomenclature (figs. 53–54).

Indeed, "vagina" or equivalent words (*schiede, vagin*) standing alone to designate the sheath or hollow organ into which its *opposite,* the penis, fits during coition and through which the young are delivered only entered the European vernaculars around 1700. Other genital nomenclature also became more specific and laden with meaning. In a pornographic fantasy-travel book published in 1683, for example, the author describes a female-shaped island that had power over its male inhabitants through its

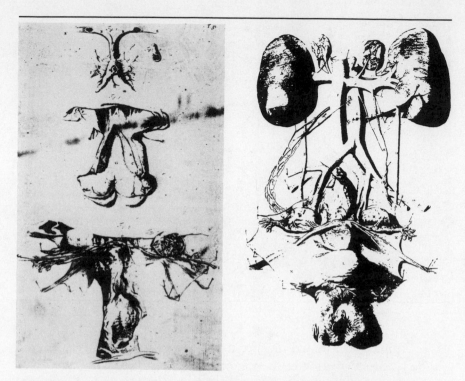

Figs. 53–54. The various parts of the female reproductive system and external genitalia are disaggregated. The vagina is opened so that it does not have the penislike effect of the closed organ shown in Renaissance illustrations. The clitoris, left top, is shown separately, and no effort is made to render the external pudenda as a female foreskin as before. On the right the uterus is shown in relation to the kidneys and their vasculature; the vagina is not shown. From William Cowper, *The Anatomy of Humane Bodies* (1697).

"soyl" and "mould" but definitely not through its sexual parts. Only the pregnant belly and what must be the urethra—it is never named—get specific references. But by the 1740s this erotic island is replete with the obvious modern genital landmarks: "the two forts called Lba"; "a metropolis called Cltrs."[27] Precisely during the intervening period, the hoary linguistic web in which words for womb and scrotum, penis and vagina, prepuce and vulva were entangled came unraveled. Whatever was there before, our forebears felt no need to name. Whatever came later is inseparable from the languages, largely scientific, through which it entered our subjectivity.

Organs that had been common to both sexes—the testicles—came as a result of the discovery of sperm and egg to have each its own name and to stand in synecdochal relationship to its respective sex. Sometime in the eighteenth century "testicle" could stand alone to designate unambiguously the male gonad; it no longer carries the modifiers "masculine" or "feminine." "Ovary," not "female stones" or "testicle feminine," came to designate its female equivalent. Moreover, the overtly political language of some earlier anatomical descriptions—Zacchia's description of a *beneficium* of the clitoris as leading to a false diagnosis of hermaphrodism, for example—gave way to the more clinical, organ-centered language of nineteenth-century medicine: "spurious" hermaphrodism due to "abnormal development or magnitude of the clitoris" reads a heading in one early nineteenth-century encyclopedia.[28]

The new relationship between generation and sexual pleasure, and hence the biological possibility of a passionless female, also had its origins in the late eighteenth century. In the 1770s the famous experimentalist Lazzaro Spallanzani succeeded in artificially inseminating a water spaniel, which suggested that in a dog, at least, orgasm was not necessary for conception.[29] Syringes could not "communicate or meet with joy," as a Scottish doctor observed.[30] (The surgeon John Hunter had earlier used a similar instrument to introduce the semen of a patient who suffered from a urethral defect into the vagina of the man's wife. But since the procedure took place after intercourse and with semen that had been ejaculated at the usual time, if not place, the experiment proved little about the role of female orgasm in conception.[31])

Pregnancy from rape provides the limiting case for a woman's conceiving without pleasure or desire. Samuel Farr, in the first legal-medicine text to be written in English (1785), argued that "without an excitation of lust, or enjoyment in the venereal act, no conception can probably take place."[32] Whatever a woman might claim to have felt or whatever resistance she might have put up, conception in itself betrayed desire or at least a sufficient measure of acquiescence for her to enjoy the venereal act. This is a very old argument. Soranus had said in second-century Rome that "if some women who were forced to have intercourse have conceived . . . the emotion of sexual appetite existed in them too, but was obscured by mental resolve," and no one before the second half of the eighteenth or early nineteenth century questioned the physiological basis of this judgment.[33] The 1756 edition of Burn's *Justice of the Peace,* the standard guide for English magistrates, cites authorities back to the *Institutes* of Justinian to the effect that "a woman can not conceive unless she doth consent." It does, however, go on to point out that as a matter of law, if not of biology, this doctrine is dubious.[34] Another writer argued that pregnancy ought to be taken as proof of acquiescence since the fear, terror, and aversion that accompany a true rape would prevent an orgasm from occurring and thus make conception unlikely.[35]

In practice it is doubtful whether these views had much effect on courts of law.[36] To begin with, some legal authorities held that the maxim "it can be no rape, if woman conceive with child" seemed not to form a law.[37] Then, because of the difficulty in proving rape, and more generally the common law's leniency in matters of personal assault, only the most egregious and repugnant rapes ever came to trial: attacks on young girls or pregnant women, violations of mistresses by servants, cases in which venereal disease was transmitted or the victim mutilated.[38] In such instances the niceties of whether orgasm occurred were probably not relevant. Finally, the pregnancy defense was known not to be entirely reliable. One doctor argued in 1823 that conception was possible even when intercourse had been involuntary or with a man for whom the woman felt repugnance because both states may lead to "so high a tone of constitutional orgasm" as to make ovulation possible. The orgasm in question here—a turgescence of the reproductive organs—need not have been felt or desired for it to do its work.[39]

But by the 1820s the medical doctrines upon which legal definitions of rape were based had changed dramatically. The view that rape was incompatible with pregnancy was proclaimed in a much-cited text as "an extraordinary dictum of the ancient lawyers," a "vulgar idea, from which some ignorant persons might still infer that a woman had consented, because she was proven pregnant," thus adding unmerited stigma to the other burdens of the unfortunate victim of crime.[40] While the eighteenth-century edition of Burn quoted above was vague on the scientific question of whether conception ruled out rape, its nineteenth-century version stated unequivocally that the notion was absurd, that it would be surprising if "any whose education and intellect were superior to those of an old nurse" still believed it. Whatever the vulgar might have believed—and, as suggested earlier, ordinary people might very well have continued to subscribe in a deep, inarticulate way to old notions still widely circulating in books and gossip—the learned world firmly rejected the connection of female pleasure and conception. This does not mean that experts embraced the hypothesis, which remained controversial for another century, that women could ovulate independently of intercourse. The point is rather that women could experience the tension of sexual intercourse and even orgasm, in the nineteenth-century sense of the word as a turgescence or pressure, without any concomitant sensation. The ovarian system, in other words, could work not only without the influence of the conscious self but without any phenomenal sign. "Physical constraint . . . sufficient to induce the required state" was all the ovaries needed.[41]

Even in the late eighteenth century, some writers had said that there was no relationship between the erogenous qualities of the external female genitalia and the serious work that went on within. One argued that the "lascivious susceptibility" of the external organs was materially useless to generation; another noted the "organization of the vagina for the purpose of exciting titillation and pleasure" only to follow this observation with the non sequitur that "it can and does accommodate itself to whatever size is necessary closely to embrace the penis in the act of copulation."[42] A major obstetrics textbook remarked casually that it would not dwell on the clitoris and other external organs because they were irrelevant to midwifery.[43] So, even if doctors in these and many similar texts did not directly address the question of whether women had sexual feelings or experienced orgasm, they considered these sensations as contingent to the order of things. No longer necessary for conception, they became something that women might or might not have, something to be doggedly and inconclusively debated rather than, as had been the case for so long, taken for granted.

And we must not take for granted the terms in which science defined the new sexes. It claimed that the body provided a solid foundation, a causal locus, of the meaning of male or female. The trouble here lies not with the empirical truth or falsity of specific biological views but with the interpretive strategy itself. Sexual difference no more followed from anatomy after the scientific revolution than it did in the world of one sex.

The aporia of biology

The aesthetics of anatomical difference. Anatomy, and nature as we know it more generally, is obviously not pure fact, unadulterated by thought or convention, but rather a richly complicated construction based not only on observation, and on a variety of social and cultural constraints on the practice of science, but on an aesthetics of representation as well. Far from being the foundations for gender, the male and female bodies in eighteenth- and nineteenth-century anatomy books are themselves artifacts whose production is part of the history of their epoch.

This is not to say, as we have seen in Chapter 3, that an anatomy text or illustration cannot be judged more or less accurate. There is progress in anatomy. There are bounds to the scientific imagination. Vesalius *was* wrong in depicting the *rete mirabele* in humans, although his eagerness to see it is understandable within the context of Galenic physiology. There are normally no holes in the septum of the heart as Renaissance anatomists thought, although again it is not difficult to see how a patent *foramen ovales,* present in a quarter of cases, and the myriad spaces between the *trabeculae carneae* that anchor the valves might not be mistaken for vents between the right and left sides. The ovaries *are* structurally dissimilar from the testicles, although not so much in their gross surface appearance as the early texts would have it.

But all anatomical illustrations, historical and contemporary, are abstractions; they are maps to a bewildering and infinitely varied reality. Representations of features that pertain especially to male or female, because of the enormous social consequences of these distinctions, are most obviously dictated by art and culture. Like maps, anatomical illustrations focus attention on a particular feature or on a particular set of spatial relationships. To fulfill their function they assume a point of view—they include some structures and exclude others; they strip away the plenum of sheer stuff that fills up the body—fat, connective tissue, and "insignificant variations" that are not dignified with names or individual identities. They situate the body in relation to death, or to this world, or to an identifiable face—or, as in most modern texts, they do not. As figs. 10–16 suggest, the social situation of cadavers was once far richer and more varied than it became in the nineteenth century. The compilers of anatomical texts use or eschew various techniques of the engraver or painter to gain specific effects. Anatomical illustrations, in short, are representations of historically specific understandings of the human body and its place in creation and not only of a particular state of knowledge about its structures.

Thus, for example, figs. 20–26, which make the vagina look like the penis, are not incorrect because they emphasize a relationship between the female reproductive organs that anatomists since the late seventeenth

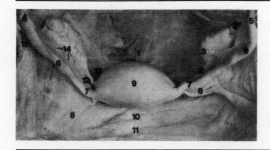

Fig. 55. Photograph of the uterus and ovaries from above, using embalmed material.

century have chosen to deemphasize; nor conversely are eighteenth-century illustrations (figs. 51–54) more correct because they do not emphasize this relationship. One could (figs. 28–29) produce a Renaissance look-alike from modern plates.

But the extent of interpretation inherent in any anatomical illustration is evident in less controversial contexts. Consider, for example, fig. 55, a photograph of the uterus and ovaries from above and in front. It is in no sense "ideological," but it is enormously selective. There is no blood or other fluid in the picture; most of the fat and connective tissue has been stripped away; the body in which the organ resided is scarcely in evidence; the tone is cool and neutral. Contrast this to two drawings of the same subject. The first (fig. 56), prepared to illustrate what was wrongly believed to be a human egg, looks almost like a Caspar David Friedrich landscape. Shaded valleys furrow the broad ligaments of the uterus; the trumpets of the Fallopian tubes look like exotic flowers growing out of a bank of billowing clouds. The second (fig. 57) is from a modern text and is in the tradition of schematic, almost architectural drawing introduced by the great German anatomist Jacob Henle, to show only particular features of an organ, salient for the occasion. There is almost no shading or sense of texture; the tone, as in the photograph, is detached and scientific; no affect mars its supposed objectivity; there is no sense of its being the organ of an individual. The final illustration of the same organ (fig. 58) operates at an even greater level of abstraction. Here is a blueprint, drawn to show a specific feature of the structure in question with no effort to situate it further, as if the organ were a machine. I do not want to maintain that these pictures are ideological in that they overtly distort observation in the interest of one political position or another. I simply want

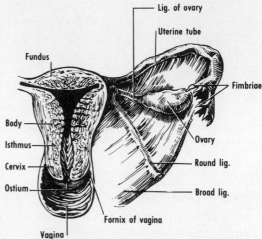

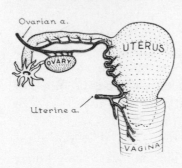

Fig. 56. (above, left) A richly textured drawing of the uterus, Fallopian tubes, and ovaries from an 1817 issue of *Philosophical Transactions* (no. 107). Note the way structures seem to flap in the wind and how shading creates a dramatic effect.

Fig. 57. (above, right) A modern, considerably less elaborated, and more abstract drawing of the structures seen in fig. 56.

Fig. 58. (left) A modern schematic drawing of the uterus, ovaries, and Fallopian tubes.

to point out what is already well established in the criticism of high art: pictures are the product of the social activity of picture making and bear the complex marks of their origins.

Still, anatomical illustrations that claim canonical status, that announce themselves to represent *the* human eye or *the* female skeleton, are more directly implicated in the culture producing them. Idealist anatomy, like idealism generally, must postulate a transcendent norm. But there is obviously no canonical eye, muscle, or skeleton, and therefore any representation making this claim does so on the basis of certain culturally and historically specific notions of what is ideal, what best illustrates the true nature of the object in question. Some texts, like the enormously successful Gray's *Anatomy*, blithely and unselfconsciously represent the general case of every feature as male. All the surface anatomy is demonstrated by male, though curiously unmuscular, subjects and thereby belies whatever objective claim one might want to make for the advantages of the male body in illustrating surface articulations. Even the schematically drawn cleavage lines that divide thorax from abdomen and the markings to show the course of blood vessels are shown on a male model; the hands in various stages of dissection are all male hands; the distribution of cutaneous nerves are shown on the schematic drawing of a man. It is simply assumed that the human body is male. The female body is presented only to show how it differs from the male.[44]

Samuel Thomas von Soemmerring, who produced one of two competing canonical illustrations of the female skeleton in the nineteenth century, was more straightforward in articulating his principles of selection. The anatomically normal was for him, as for much anatomy in the idealist tradition, the most beautiful. An anatomist was thus engaged in the same deeply serious task as a painter: to render the human form, and nature generally, in accord with the canons of art. In his comment on his illustration of the eye, Soemmerring argues:

> Just as, on the one hand, we assume that all works of art representing the human body and claiming ideal beauty for themselves must needs be correct from an anatomic point of view, so, on the other hand, should we as readily expect that everything that the dissector describes anatomically as a normal structure must needs be exceptionally beautiful.[45]

Like the distinguished anatomist Bernard Albinus, who counseled his colleagues to be like artists who "draw a handsome face, and if there happens to be a blemish in it, they mend it in the picture," Soemmerring promised to avoid anything in his representations that was "distorted, dried, shriveled, torn or dislocated."[46] Anything that failed to meet the highest aesthetic standards was banished from his representations of the body; the grand tradition of Sir Joshua Reynolds' prescriptions to painters in his *Discourses* was mirrored in the seemingly alien world of scientific illustration.

Soemmerring was dissatisfied with the d'Arconville/Sue female skeleton, the only alternative available in the 1790s, and set to construct an alternative based on the highest standards of observation and aesthetic judgment. Finding no skeleton in his collection suitable, he acquired one of a twenty-year-old girl of proven femininity (she had given birth); to this skeleton he apparently appended the well-known skull, from Johann Friedrich Blumenbach's collection, of a Georgian woman. He then went to great lengths to determine the appropriate pose, seeking the advice of artists and connoisseurs; he posed live models; and eventually he compared his product with the Venus de Medici and the Venus of Dresden. The canonical skeleton had to seem plausible as the foundation of the canonical female form.

All of this bears an uncanny resemblance to Alberti's account of the Athenian painter Xeuxis (fifth century B.C.):

> He thought that he would not be able to find so much beauty as he was looking for in a single body, since it was not given to a single one by nature. He chose, therefore, the five most beautiful young girls from the youth of the land in order to draw from them whatever beauty is praised in women. He was a wise painter.[47]

Thus the making of *the* female skeleton, or indeed of any ideal representation, is an exercise in a culturally bound aesthetic. And, as it happened, Soemmerring's beauty failed to meet the political standards of its day; the d'Arconville/Sue skeleton triumphed. Why? According to the Scots anatomist John Barclay, "although it is more graceful and elegant and suggested by men of eminence in modelling, sculpture and painting, it contributes nothing to the comparison which is intended."[48] The missed comparison of course was between men and women, and the specific mistake of which Soemmerring stood accused was his failure to represent with sufficient specificity the female pelvis, the most significant sign in the bones of sexual difference. To be sure that his readers fully comprehended the point, Barclay reproduced Albinus' male skeleton with George Stubbs's rendering of the musculature of a horse in the background and the Sue skeleton of the female with a skeletal ostrich looking on.[49] The iconography of the horse was transparent in a world in which the beast was bred for its speed, power, and endurance, in which a man on horseback still represented authority. The ostrich was a less usual sign, but it too must have been readable. Its enormous pelvis in proportion to its body directs the viewer's attention to the analogous feature in the accompanying human female, and its long neck must have been an allusion to the claim of phrenology that the characteristically long neck of women bore witness to their low "amativeness," their lack of passion.

Anatomical science was thus itself the arena in which representation of sexual difference fought for ascendancy. The manifest anatomical differences between the sexes, the body outside of culture, is known only through highly developed, culturally and historically bound paradigms, both scientific and aesthetic. The notion that scientific advance alone, pure anatomical discovery, could account for the extraordinary late eighteenth- and nineteenth-century interest in sexual dimorphism is not simply empirically wrong—it is philosophically misguided.

Embryogenesis and the Galenic homologies. A stranger surveying the landscape of mid-nineteenth-century science might well suspect that incommensurable sexual difference was created despite, not because of, new discoveries. Careful studies of fetal development would give credence not to new differences but to old androgynies, grounded this time not in myth or metaphysics but in nature. It had been known since the eighteenth century, for example, that the clitoris and the penis were of similar embryological origin. An early nineteenth-century textbook on forensic medicine, in a section on hermaphrodism and the difficulties of telling the sex of newborns, points out that at birth the clitoris "is often larger than the penis, and has frequently given rise to mistakes." The writer cites the *Memoirs de l'Academy Royal des Sciences de Paris* for 1767 to the effect that the seemingly disproportionate number of male miscarriages in the third and fourth months is due to the size of the clitoris in female embryos and the resulting confusion of sexual identification. (The error is understandable, as fig. 59 suggests.) More generally the triumph in embryology, during the first thirty years of the nineteenth century, of epigenesis (the view that complicated organic structures arise from simpler undifferentiated ones rather than from preformed entities inherent in the sperm or the egg) would seem to undermine root and branch difference. Science revealed an embryo in which the Wolffian duct, named after Kaspar Friedrich Wolff, was destined to become the male genital tract, and the Mullerian ducts, after Johannes Müller, would become the Fallopian tubes and the ovaries. Until about the eighth week, the two structures coexist. Furthermore, it was known by the middle of the nineteenth century that the penis and the clitoris, the labia and the scrotum, the ovary and the testes, begin from one and the same embryonic structure. The scrotal sac, for example, is a modification of the labia majora, a version of the embryonic labiscrotal swelling in which the lips grow longer, fold over, and join along the scrotal raphe.[50] Here, even more powerfully than

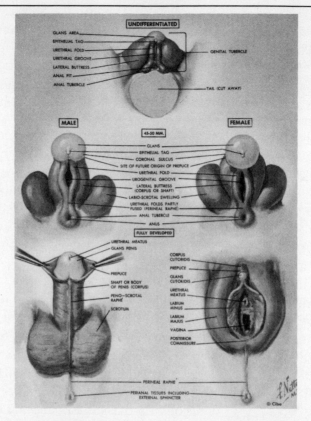

Fig. 59. At 40–55 mm in length, around two and a half months into gestation, the male and female genitalia are almost indistinguishable. Gradually, after the third or fourth month, it become easier to tell the sexes apart. Drawing by Frank Netter, *CIBA Collection of Medical Illustrations*.

in the early coexisting two ducts, the old Galenic homologies seem to find new resonance. Modern representations of the development of the external genitalia bear a remarkable resemblance to Vesalius' or Leonardo's illustrations, and modern charts of genital embryology seem faithfully to reproduce Galen's lecture on woman as inverted male.

Moreover, the idea of common embryological origins of various male and female organs, in the very different political climate of the 1980s, has engendered a modern version of ancient thought. One psychoanalyst in an effort to rehabilitate the vagina for its erotic and indeed erectile functions, after two decades of what he calls "clitorocentricity," marshals considerable evidence for the homology of male and female ejaculation. There are, he says, immunohistochemical homologies between the secretions of the male prostate and the female paraurethral glands, structures whose common roots in the embryonic urogenital sinus have been known since the nineteenth century. In fact, as he points out, the secretory glands that empty into the female urethra were known as prostates in both sexes until in 1880 they took the name of A. J. C. Skene, who extensively investigated them.[51] Thus a vast scientific literature—indeed, embryological investigation was the glory of nineteenth-century descriptive biology—provided a great repertory of new discoveries, which, far from destroying old homologies, could well have strengthened them. My point, however, is not to argue that scientific advances did somehow give greater credence to the ancient model. New cultural imperatives of interpretation simply had a larger field out of which to construct, or not construct, a biology of sexual difference.

Sperm and egg. The claim by Harvey in 1651 that all life comes from an egg; the subsequent announcement by de Graaf in 1672 that he had discovered the ovarian follicle that was thought to be, or to contain, that egg; and the revelation by Leuwenhoek and Hartsoeker, also in the 1670s, that semen contained millions of little animalcules: all this seemed to provide, in the microscopic generative products, an imaginatively convincing synecdoche for two sexes. The vaginal secretions that had for millennia been taken to be a thin, cooler, less perfect version of the male ejaculate turned out to be something entirely different: "since the discovery of the egg . . . that Liquor which has been taken by all preceding Ages for the Seed in [women], is found to be only a mucous Matter, Secreted from the Glands of the *Vagina.*" For a time it seemed, in fact, that the newly discovered egg would detract "much from the dignity of the Male sex" since it "furnish'd the matter of the Fetus," while the male only "actuated it." But then Anton van Leuwenhoek discovered that the male ejaculate was not just a thick liquid seed: "by the help of his Exquisite microscope . . . [he] detected Innumerable small Animals in the Masculine sperm, and by this Noble Discovery, at once removed that Difficulty."[52] Sperm and egg could now stand for man and woman; male dignity was restored.

Social sex thus projected downward into biological sex at the level of the microscopic generative products themselves. Very quickly the egg came to be seen as a merely passive nest or trough where the boy or girl person, compressed in each animalcule, was fattened up before birth. Fertilization became a miniaturized version of monogamous marriage, where the animalcule/husband managed to get through the single opening of the egg/wife, which then closed and "did not allow another worm to enter."[53] In other words, old distinctions of gender now found their basis in the supposed facts of life.

Moreover, the discoveries of egg and sperm marked the beginning of a long research program to find sexual reproduction everywhere.[54] For a time it succeeded in doing just that. Whether one believed that the egg or the sperm contained the new life already preformed, or that each contributed elements toward the epigenetic development of succeeding generations, sexual reproduction and the nature of sexual difference dominated thinking about generation.[55]

Very quickly sex also filtered down from animals to plants. The pistil, a word from the Latin *pistillium* (pestle), became an unlikely name for the seed-bearing ovary. The stamen—actually the anther at its end—from which the pollen emanates, became the botanical penis. Instantly plants were gendered, and sex was assimilated to culture: "hence it seems rational to denote these apices by a more noble name and attribute to them the importance of masculine sexual organs; it is there that the semen, the powder that constitutes the subtlest part of the plant, accumulates, and it is from there that it later flows forth."[56] The sexual nature of plants became the basis for Linnaeus' famous classificatory system. Further investigation found sexual products up and down the living world; beginning in the 1830s spermatozoa, for example, were located in every invertebrate group except Infusoria. The *Naturphilosophen* thus seemed to be right in viewing sexual difference as one of the fundamental dichotomies of nature, an unbridgeable chasm born not of the Pythagorean opposites but of the reproductive germs themselves and the organs that produced them.

As it turned out, however, the new discoveries were of only fitful utility. In the first place, the immediate, promiscuous projection of gender onto sex in Linnaeus' sexual system made even contemporaries blush. The group of plants classed as Monoecia, meaning "one house," took its name and character from the fact that "Husbands live with their wives in the same house, but have different beds [leaves]." The class Polygamia aequalis meaning "equal polygamy," was seen to "consist of many marriages with promiscuous intercourse."[57] Plant sex was so extremely gendered at its core that in his own day Linnaeus' taxonomy seemed quite indecent.

Furthermore, even in humans and other creatures in which egg and sperm were understood to be the distinct products of different sexes, the meanings of the terms were in constant flux. There was, in other words, no consensus as to what sperm and egg actually were or did, until the turn of the nineteenth century.[58] The synecdochic imagination was thus unfettered by the supposed discovery of distinctive generative products; the incommensurability of the sexes rested uneasily on microscopic bodies whose significance was much debated. Preformationists were unevenly divided between a majority who were ovists and a minority of animalculists. The choice between them was often ideological: among the main arguments against the animalculists was that God would never have devised so profligate a system that millions of preformed humans had to die in each ejaculation so that one might, on occasion, find food for growth in the egg. Insofar as observation had anything to do with theory—Haller, for example, was in part converted to preformationism and particularly to ovism because he thought that he could trace the continuity of the membranes of a chick embryo's intestines from the membranes of the yolk sac—gender played little role.[59]

So, even if some contemporaries spoke of the respective dignities of male and female being reflected in the two respective preformationist theories, the debate was really on different grounds. And in fact neither ovism nor animalculism suggested a world of two sexes but rather a world of no sex at all. Both bespoke parthenogenic reproduction: either the egg contained the new life and the sperm was just a living version of the glass rod that could make frog eggs develop on their own, or the sperm contained the new life and the egg was just a food basket. Technical developments in the explosively developing study of generation also undermined the supposed ubiquity of sexual reproduction. Charles Bonnet's proof in 1745 that aphids reproduced by parthenogenesis—a term coined by the great comparative anatomist Richard Owen in 1849—was the first step in finding that the development of unfertilized eggs from sexually mature females was far more widespread than had been thought possible. Abraham Trembley's demonstration, at about the same time, of the regenerative powers of hydra had general repercussions in discussions not only of sexuality but of generation at the theoretical level. Other developments and tendencies—the discovery of alternation of generations in 1842 and the increasing interest in hermaphroditic reproduction—also tended to push eighteenth-century models of universal sexual reproduction, insofar as such models existed, to the sidelines.[60]

I do not want to rehearse the long history of sperm-or-egg but only to point out that the gender claims made on their behalf were constantly being undermined by these sorts of controversies.[61] Until the 1850s it was unclear whether sperm merely stirred the semen—a wormlike mix-master—stimulated ovulation, touched the egg, or actually penetrated it. The conceptual triumph of cell theory and advances in microscopy and staining finally allowed Oskar Hertwig, in 1876, to demonstrate that the sperm did indeed penetrate the egg and that the actual joining of the egg and sperm nuclei *was* fertilization. (As I said, this seemed to provide an unassailable microscopic model for incommensurable sexual difference, until a move to the molecular, DNA level made it all less clear again.) Well into the twentieth century, the debate continued on whether all or only some of the nuclear material blended.

For much of the period under discussion here, the role and nature of the sperm remained obscure. Spallanzani had proven in the late eighteenth century that no amount of vapor from semen would fertilize frog eggs, that Harvey's *aura seminalis* was insufficient to cause the female mold to produce tadpoles, and that increasing filtrations of semen eventually rendered it impotent. He showed that naked male frogs mounting a female fertilized her eggs but that frogs wearing little taffeta trousers did not; he went on to demonstrate, furthermore, that the residue on their ludicrous garb was potent. (He had previously shown—by killing a female frog in the act of copulation and noting that the eggs still inside her did not develop while those that had been in contact with the sperm were fertile—that the eggs were fertilized outside the body.) Despite all of this, he continued to think that the little creatures in semen were mere parasites and that semen worked by stimulating the heart of a preformed fetus released from the ovary after fertilization.[62]

The debate between preformationists—ovists or animalculists—on the one hand and epigenesists on the other provides further evidence for just how irrelevant research on germ substances was to thinking about two sexes. The choice between preformation and epigenesis was made on philosophical rather than empirical grounds, but quarrels about gender played no part. Albrecht von Haller differed from Christian Woolf not on the interpretation of this or that piece of data—indeed they generally talked right by each other—but on basic issues in the philosophy of science: a mechanistic, Newtonian preformationism in which embryological development works out God's plan as against a rationalist, somewhat more vitalist epigenesis in which matter was not merely inert substance to be worked upon by God's laws.

Among epigenesists, a major figure like Buffon could still write in the cadences of the old biology of generation, as if nothing had happened, almost a century after the discovery of sperm and egg: "the female has a seminal liquor which commences to be formed in the testicles" and that "the seminal liquors are both [male and female] extracts from all parts of the body, and in the mixture of them there is everything necessary to form a certain number of males and females." The point is not that Buffon was wrong in his theories of pangenesis or right, for the wrong reasons, that there is a "moule intérieur" in the particles of male and female "semen" which organize matter into organic structures.[63] Rather I want to suggest that in the eighteenth and nineteenth centuries, and indeed today, at any given point of scientific knowledge a wide variety of contradictory cultural claims about sexual difference are possible. Pierre de Maupertuis, one of the major opponents of preformationism—he believed that atoms arranged one another according to some plan—in 1756 was still writing, as had Democritus in ancient Greece, about orgasm: "it is that moment, so rich in delight, which brings to life a new being."[64] Neither the level of scientific knowledge nor its "correctness" restrains the poetry written in its name.

But even if Maupertuis or other eighteenth- and nineteenth-century scientists had arrived at what we consider to be the correct interpretation of the data at hand, observation and experiment would still not have created a metaphor for maleness or femaleness. Translating facts about reproduction into "facts" about sexual difference is precisely the cultural sleight of hand I want to expose.

The ovary and the nature of woman. The most egregious instance of anatomical aporia, and the clearest case in which cultural assumptions fueled a research tradition whose results in turn confirmed those views, involved the ovary. "Propter solum ovarium mulier est id quod est" (it is only because of the ovary that woman is what she is), wrote the French physician Achille Chereau in 1844, forty years before there would be any evidence for the real importance of the organ in a woman's life. Here is a synecdochic leap to incommensurability that would in any circumstances be unsupportable.[65] But it is particularly ironic because the large role of the ovary in the biological lives of women—though certainly not making woman "what she is"—was finally established in the late nineteenth century by assuming that which was yet to be proven and using it as justification for the surgical removal of histologically normal ovaries. Bilateral ovariotomy—the removal of healthy ovaries—made its appearance in the early 1870s and became an instant success to cure a wide variety of "behavioral pathologies": hysteria,[66] excessive sexual desires, and more mundane aches and pains whose origins could not be shown to lie elsewhere. (The procedure was also called in German "die castration der Frauen," in French "castration chez la femme," or eponymously "Battey's or Hegar's operation" after Robert Battey and Alfred Hegar, the American and German surgeons who popularized it. It should be distinguished from what were usually called ovariotomies, the removal of cancerous or cystic ovaries for therapeutic reasons that would be regarded as medically sound today. The number of these operations also grew dramatically, as indeed did the number of all operations in the late nineteenth century, especially after the acceptance of Lister's aseptic techniques.[67])

Removing healthy ovaries in the hope of curing so-called failures of femininity went a long way toward producing the data from which the organ's functions could be understood. The dependence of menstruation on the ovary, for example, was shown by assuming that the swelling of the ovarian follicle produced heatlike, estrous symptoms in some women and that removal of the organ would therefore halt such sexual excesses.

There is a further irony in all of this because the operation both assumes and does not assume incommensurable sexual difference; it purports to create women who both are and are not more like men than they were before the procedure. The name itself, female castration, suggests the old view that the ovaries are female testicles, much like the male's. But doctors were quick to deny that ovariotomy was anything like castration in its psychological and social effects. There are no pictures comparable to fig. 60 in which roles are switched, in which instead of men, scalpel in hand, seen poised over the prostrate body of a woman, men (or more inconceivably yet, women) surgeons are preparing to castrate a man. There was no male castration, no removal of healthy testes, except in a few rare and quite specific instances for criminal insanity or to treat cancer of the prostate. While the female gonad was assumed, like its male coun-

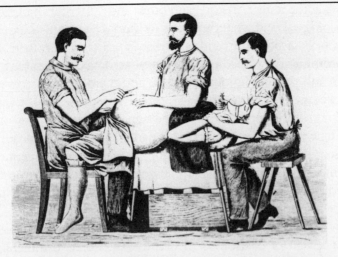

Fig. 60. Three male surgeons, c. 1880, performing an ovariotomy on a patient with a large cyst.

terpart, to have profound effects on various parts of the body, ovaries were not testicles in any cultural or metaphorical sense in the minds of the overwhelmingly male medical profession. They, unlike testicles, were not sacrosanct.

Yet the theoretical justification for "female castration" was that the ovaries, a woman's "stones" (once understood as a cooler version of the testes), were in fact the master organs of the female body so that if she lost them she would become more malelike, just as castrated males would become more femalelike. Ovariotomy did cause women to stop menstruating and did effect other changes in secondary sexual characteristics that made them more like men. On the other hand, removing the ovaries also made a woman more womanly, or at least more like what the operation's proponents thought women ought to be. Extirpating the female organs exorcised the organic demons of unladylike behavior.

All of this speculation about the synecdochic relationship between an organ and a person—a woman is her ovaries—or even between the ovary and some observable physiological or anatomical change was ideological hot air. Up to the late nineteenth century no one knew what removing the ovaries would do. (Even today the effects of postmenopausal ovariotomy are not well understood.) Far more was known about the effects of removing the testes. Aristotle and other ancient writers had recognized the physiological, and what they took to be psychological and behavioral, consequences of both pre- and postpubertal castration in men. Eunuchs figure prominently in medical and moral writings, in a variety of both Christian and pagan religious practices, and there are many observations on the effects of castration in male domestic animals.[68] But there are, as far as I know, no commentaries on the removal of ovaries in women and only a single reference to the procedure in animals: "The ovaries of sows are excised with the view of quenching in them sexual appetites and of stimulating fatness," wrote Aristotle; female camels, he continues, are mutilated to make them more aggressive for "war purposes" and to prevent their bearing young.[69]

Nothing was written on the relevance of such observations to humans until the advent of ovariotomy in the 1870s. For two millennia, from ancient Greece to late eighteenth-century London, there was no human case reported in medical or popular literature. Then Percival Pott, a distinguished surgeon at St. Bartholomew's Hospital in London, announced that he had examined a woman, age twenty-three, with two small soft masses, "unequal in their surface," one in each groin. She appeared healthy, menstruated regularly, and suffered no pain except when she stooped. Eventually she became "incapacitated from earning her bread" and, when nothing else alleviated her distress, agreed to have the growths removed. To Pott's apparent surprise, they were her ovaries. He notes that his patient returned to good health but that she appeared thinner and more muscular; "her breasts, which were large, are gone; nor has she menstruated since the operation, which is now some years." He offers no account of why all this happened.[70]

When in 1843 Theodor von Bischoff, the discoverer of spontaneous ovulation in dogs, wrote that the ovaries govern the human female reproductive cycle, he had but one further piece of evidence: the account of one Dr. G. Roberts, a medically trained traveler who claimed to have seen "castrated" women in India, aged about twenty-five, whose breasts were undeveloped, whose external pudenda lacked the usual fat deposits and covering hair, whose pelvises were deformed and buttocks male-like, who showed no signs of menstruation or any compensatory process, and who had absolutely no sexual drive.[71] Even if one credits this report and adds to it a series of casual clinical observations correlating malformation of the ovaries with absence of menstruation, the evidence available by the middle of the nineteenth century for the function of the ovary in the reproductive physiology of women remains slight.

The rise of "justifiable" ovariotomy after 1865—mostly for cysts, tumors, or other obvious pathologies—began to provide some quasi-experimental evidence for the ovary's functions, but since the workings of a healthy organ could not in many cases be reliably deduced from the effects of excising its diseased counterpart, such material was less than conclusive. Though an authoritative German handbook argues that there were so many cases on record attesting to the connection between the ovary and menstruation that further cases were scarcely worth noting, it still refers to Bischoff's by now forty-year-old citations of Roberts and Pott (whose report itself had by then been around for a century). Moreover, it proceeds to note that considerable weight was currently being placed on instances of menstruation continuing after removal of the ovaries and that, should a recent attack on such evidence prove inconclusive, one might have to reconsider whether the intimate relationship postulated between the uterus and the ovary had not been exaggerated.[72] In 1882 a French handbook cites both new material and much older evidence which suggested that the role of the ovary in menstruation and indeed in the whole reproductive cycle might well be as passive as that of the uterus.[73]

No one bothered to adduce age-old practical experience with oophorectomy in animals before 1873 when, a year after Battey began to advocate removal of the ovaries for various neurotic ills, a French physician remarked that in cows and pigs in which the operation was "commonly done during the first two months of life, the uterus ceases to grow and its volume remains stationary."[74] In short, when Battey and Hegar began removing healthy ovaries, and at the height of popular belief in the life-determining role of the organ, almost nothing was known of its function in women and no effort had been made to exploit what little veterinary experience existed. Here is a question not of the indeterminacy of anatomical and physiological knowledge but of willful ignorance.

Twenty years and the removal of thousands of healthy ovaries later, some of the assumptions on which the operation had been predicated finally rested on experimental evidence. It was Alfred Hegar, the distinguished professor of gynecology at Freiburg and the main European advocate of female castration, who brought the wisdom of generations of farmers together with his own clinical practice. Curious to know the

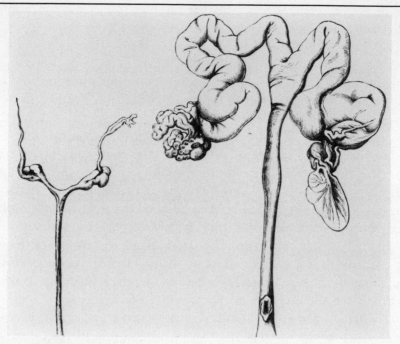

Fig. 61. Alfred Hegar's "first illustration of castrate atrophy of the uterus ever published."

long-term effects of the operations he was already performing, he searched the literature and found that female castration in animals was an ancient practice. He discovered that the castration of cows was popular in France in the 1830s but that the practice had fallen out of favor because the cows got too fat and stopped lactating. Veterinarians in his own day still removed ovaries but only when medically indicated: for "desire for the bull, a sort of nymphomania" (*Steiersucht, eine Art Nymphomanie*), which afflicted some 10 percent of the cows in certain regions![75]

Not to be deterred in his quest for knowledge, Hegar went back to the classics and to Aristotle's account of cutting out a sow's ovaries. He then sought out a *Schweine-Schneider*, "a cutter of pigs," whose basic technique, it turned out, was indistinguishable from that of his Greek predecessor, though from a nineteenth-century bourgeois perspective much more disgusting. The man took out a dirty knife, made a two-centimeter incision, put his dirty fingers around the ovaries, tubes, and ligaments, and cut them out. He then sewed up the incision with a needle and thread drawn from his "evil-smelling" trousers. (It has never been clear to me why, with such an exquisite sense of dirt and propriety, the idea of aseptic surgery did not occur to Hegar and his contemporaries in the decade before Lister. Hegar, by his own account, lost a third of his patients to sepsis.)

Having watched the pig cutter at work, Hegar tried the operation himself. He bought two female piglets and proceeded to remove both ovaries from one and only one from the other. When they had grown to maturity, he had them butchered and found that the completely spayed pig showed dramatic aplasia of the uterus, a uterus of infant size. He made a drawing of this specimen, had it engraved, and proudly published it as the "first illustration of castrate atrophy of the uterus ever published."[76] One need not deride the genuine contribution to knowledge that Hegar's experiments represents in order to condemn him, Battey, or other doctors for the mutilations they practiced in the name of therapy. The important point, however, is not simply that they were driven by a particular vision of woman to regard the ovary as the source of illnesses whose origins lay more in culture than in the body, but rather that they subscribed to an epistemology that regarded anatomy as the foundation for a stable world of two incommensurable sexes. Ovaries were removed not because they made women what they were, nor even just because of physicians' antifeminism, but because some doctors took literally the synecdoches they had invented. Ironically their practices did yield new knowledge about the ovaries' physiological functions. But their symbolic role, their function as a sign of difference, was untouched by progress.

Orgasm and sexual difference

On May 15, 1879, Mabel Loomis Todd—later the lover of Emily Dickinson's brother—carried out an extraordinarily precise experiment. Her hypothesis was that she would be fecund only at the moment of climax because afterwards her womb would close off, and "no fluid could reach the fruitful point." To test this proposition she allowed herself, she says, "to receive the precious fluid at least six or eight moments after my highest point of enjoyment had passed and when I was perfectly cool and satisfied." She got up and, since all of her husband's semen had apparently escaped, considered herself vindicated; their daughter Millicent, born nine months later, proved her wrong.[77]

Mabel Todd was very wrong. Unlike questions of anatomy and sexual difference, the question of whether women can conceive without orgasm—however culturally desirable "passionlessness" might be—can be definitively answered. So can the question of whether female orgasm closes off the womb. Empirical evidence can address even more complicated and problematic matters: whether women generally have orgasms during intercourse, or whether they have strong sexual—I mean here heterosexual—drives at all.[78] But, though science certainly articulated new views about female passionlessness as part of the making of two sexes, it provided only inconclusive and fragmentary evidence on orgasm until the early twentieth century, more than a century after the abandonment of the universally held view linking orgasm to generation and women to passion. New information, much less a coherent new paradigm in reproductive biology, did not render ancient wisdom out of date. (I will show, in some technical detail, that nothing about the discovery of the ovaries or their functions required major revisions in the physiology of pleasure and conception. Readers willing to accept this without elaborate documentation might want only to skim this section, especially the pages on the corpus luteum.)

De Graaf's careful dissections, which established that "female testicles should rather be called ovaries," inadvertently strengthened the link between intercourse and female "emission" because they showed that in rabbits the follicles, which de Graaf took to be eggs, "do not exist at all times in the testicles of females; on the contrary, they are only detected in them after coitus." Like other observers for at least the next century and a half, he was sure that ovulation occurred *only* as a result of intercourse, which simply by the nature of things had to be pleasurable: "if those parts of the pudendum [the clitoris and labia] had not been supplied with such delightful sensations of pleasure and of such great love, no woman would be willing to undertake for herself such a troublesome pregnancy of nine months." De Graaf's was the standard Renaissance account, except for his views on the female ejaculate: instead of being understood as weaker, more watery semen, it was construed as an egg in its surrounding liquid.[79]

There were actually very little new data on reproductive physiology. "The modus of conception," as the obstetrician William Smellie noted in 1779, "is altogether uncertain, especially in the human species, because opportunities of opening pregnant women so seldom occur."[80] One had to take the cases when they came along and make up a narrative as best one could.

Albrecht von Haller, for example, one of the giants of eighteenth-century biological science, simply projected male sexual experience onto women. He did this not because he had any particular interest in maintaining the skewed symmetry of the Galenic model, but because the analogy of the sexually aroused woman to the sexually aroused man seemed so commonsensical:

> When a woman, invited either by moral love, or a lustful desire of pleasure, admits the embraces of the male, it excites a convulsive constriction and attrition of the very sensible and tender parts, which lie within the contiguity of the external opening of the vagina, after the same manner as we observed before of the male.

The clitoris grows erect, the nymphae swell, venous blood flow is constricted, and the external genitalia become turgid; the system works "to raise the pleasure to the highest pitch." A small quantity of lubricating mucus is expelled in this process but, more important, "by increasing the heights of pleasure, [it] causes a greater conflux of blood to the whole genital system of the female," resulting in an "important alteration in the interior parts." Female erection, inside and out. The uterus becomes hard with inflowing blood; the Fallopian tubes engorge and grow "so as to apply the ruffle or fingered opening of the tube to the ovary." Then, at the moment of mutual orgasm, the "hot male semen" acting on this already excited system causes the extremity of the tube to stretch still further until, "surrounding and compressing the ovarium in fervent congress, [it] presses out and swallows a mature ovum." The extrusion of the egg, Haller points out finally to his learned readers, who would probably have read this torrid account in the original Latin, "is not performed without great pleasure to the mother, nor without an exquisite unrelatable sensation of the internal parts of the tube, threatening a swoon or fainting fit to the future mother."[81] The evidence for this scenario was scanty, but there is some in the literature. An English anatomist in 1716, for example, dissected a woman who had just been executed and purportedly found one tube "clasped around the ovarium"; upon investigating how this might have come about, he learned that "she had enjoyed a man in prison, not long before execution."[82]

Intercourse continued to be linked to ovulation and to an inner drama that, as in Haller's account, could be plausibly marked by pleasure. W. C. Cruickshank, searching for rabbit ova in 1797, found the corpus luteum only after coition, from which he concluded that "the ovum is formed in and comes out of the ovarium after conception." (The corpus luteum, the "yellow body," is formed after an ovarian follicle releases the egg. It is now known to secrete progesterone, which maintains the uterine lining in a state suitable for implantation. In most mammals it forms "spontaneously," independent of intercourse or conception, because ovulation occurs spontaneously; but in rabbits, which are generally coitally induced ovulators, it would not be present except in the circumstances Cruickshank describes.) But, more important, there seemed to be evidence for a real battle in wresting the egg from the ovary. The Fallopian tubes, he thought, "twisted like wreathing worms . . . [which] embraced the ovaria (like the fingers laying hold of an object) so closely, and so firmly, as to require some force, and even some laceration, to disengage them." Of course rabbits are not women, but Cruickshank clearly thought that his findings were applicable to humans, and so it would be surprising if so stormy a scene had no sensory correlative. The evidence would thus suggest that ovulation, like male ejaculation, would occasion some pleasurable feeling.[83]

C. E. von Baer (1792–1876), the German-Estonian biologist who was the first actually to see the mammalian ova, was still convinced when he reported on his extraordinary series of observations in 1828 that only a bitch who had recently mated could produce the egg he was seeking.[84] Indeed up to the early 1840s almost all authorities believed that coitally induced ovulation in humans as well as in other mammals was the norm. Thus in the two-sex model, as before, the generative substances in *both* men and women were believed to be produced only during intercourse; only now it was thought by some that these events could routinely occur, in women, without sensation.

This does not mean that no one advocated the view that ovulation occurred spontaneously. (If it did take place without intercourse, then a sort of mechanical, passionless conception would seem likely.) But what were later taken to be critical data against coitally induced ovulation in humans were, until the second half of the nineteenth century, interpreted as anomalous. There was nothing decisive in the existence of scars or "cicatrices," that is, the remains of the corpus luteum in the ovaries of virgins; burst follicles in the ovaries of women who died during or just after menstruation; or simply more scars in the ovary than could be accounted for by fruitful coition. Biologists seemed unwilling to let go of the idea that somehow the excitement of intercourse and sexual arousal was relevant to conception even if, miraculously, women did not feel any. Anesthetic conception, in other words, in no way followed from observation.

Thus John Pulley, an obscure eighteenth-century Bedfordshire doctor, found corpora lutea in virgins but argued that these scars were the result of uterine excitation induced through the unnatural "gratification" of desires, one presumes masturbation. Evidence from the dissection of "hysterical women" whose ovaries showed the signs of ovulation provided further proof, according to Pulley, for the role of sexual excitement in causing the extrusion of the egg.[85] Though forensic texts during the first half of the nineteenth century were generally skeptical of the notion that heightened pleasure signaled either conception or ovulation, and made much of the possibility of conception from nonconsensual intercourse, it remained perfectly plausible that ovulation did require the Sturm und Drang of coition or a reasonable facsimile. J. G. Smith wrote in a standard 1827 textbook that he could not deny that "there may be a sensible impulse conveyed by the excitement into which the uterine system appears to be thrown," when conception takes place. But, he said, many women are apt to imagine, out of hope or fear, that they have conceived—their reports on this matter are not to be trusted and can be of no practical concern.[86]

On the other hand, the question of whether a corpus luteum is evidence of past pregnancy or of intercourse *was* of considerable significance to forensic physicians: "it is a celebrated question, of great importance both in physiology and forensic medicine, and much agitated in recent years."[87] The answer was a qualified and complicated no. Women did show signs of ovulation without pregnancy or even intercourse, the majority view held, but only because the female reproductive system could be coaxed into action by lesser stimuli, strong desire for example. So, while generally speaking the presence of a corpus luteum could be taken as evidence for a woman's having had intercourse or a pregnancy, it was far from conclusive proof. Since "all those causes which excite greatly the sexual organs" can cause ovulation, the presence of corpus luteum is not "taken alone . . . a certain sign of sexual union having occurred"; but taken together with other signs it must be regarded as good presumptive evidence.[88] "A jury ought to be cautious," said one authority in jumping to the conclusion, based on signs of ovulation, that a woman had not been a virgin despite the "fact" that ovulation was generally occasioned only by fertile intercourse.[89] "Upon certain occasions," advised another, "excessive salacity may detach the ovum" and leave the scars in question.[90] (There is added confusion here because nineteenth-century doctors could not distinguish between the larger and more visible scars of the *corpus luteum verum*—the much enlarged corpus luteum that remains until the fifth or sixth month of pregnancy—and the smaller remains of the *corpus luteum spurium*, which fades rapidly after two weeks if pregnancy does not occur.[91])

A great deal rests on these controversies over the corpus luteum because they suggest that, as late as the early 1850s, no one had a clear idea of the circumstances governing the production of the egg. The evidence pointed to an even larger role for venereal excitement than in the old model of bodies and pleasures. Thus Johann Friedrich Blumenbach (1752–1840), professor of medicine at Göttingen and one of the most distinguished physicians of Europe, noted that ovarian follicles could burst without the effects of semen or even "without any commerce with the male," but concluded from this simply that on occasion "venereal ardor alone . . . could produce, among the other great changes in the sexual organs, the enlargement of the vesicles" and even cause their rupture. Far from undermining the old orgasm-conception link, Blumenbach's observations strengthened it; desire alone was enough to excite ovulation in certain sensitive systems. His English translator added supplementary anecdotal evidence: Valisneri's report of finding vesicles protruding from the ovaries of an eighteen-year-old woman who had been brought up in a convent and gave every appearance of being a virgin, a situation "frequently observed in brutes during heat"; Bonnet's report of a young woman who died "furiously in love with a man of low rank, and whose ovaria were turgid with vesicles of great size." Though not too confident of his position, Blumenbach ended up even more committed to the importance of sexual excitement than Galen was:

> On this point I find it difficult in the present state of knowledge to make up my mind; but I think it pretty evident that, although semen has no share in bursting the ovarium, the high excitement which occurs during the heat of brutes and the lascivious states of the human virgin is sufficient frequently to effect the discharge of ova. It is perhaps impossible otherwise to explain the fact that ova are so commonly expelled from the ovaria, and impregnated whenever a connection is arbitrarily or casually brought about.[92]

Johannes Müller (1801–1858), a brilliant teacher and a leading proponent of physiological reductionism, also downplayed the evidence that might have suggested spontaneous ovulation in women. He argued that the presence of scars in the ovaries of virgins were merely signs of anomalous ovulation and not of normal ovulation independent of coition and conception. Though the exact forces that caused the thrusting of the egg into the Fallopian tube remained obscure, most of the evidence suggested that the egg itself was generated only as an immediate part of the process of fertilization itself. Humans worked like that ubiquitous experimental creature of the nineteenth century, the rabbit. Something spectacular was still thought to happen in coition, and medicine lent little technical support for the rise of passionlessness.[93]

Nineteenth-century accounts of the mechanics of conception also offered no technical support for the notion of anesthetic intercourse and conception. What emerges is a new and vastly inflated role for semen, which somehow pushes, squeezes, or otherwise excites a woman's insides and which, judging from the silence on the matter, is able to do so without her feeling anything. The distinguished Edinburgh physician John Bostock argued that in women "certain causes and especially the excitement of the seminal fluid" produced "an unusual flow of blood to the ovaria"; amid all the "excitement" a vesicle bursts and discharges a drop of albuminous fluid (the egg was still only imprecisely imagined), which is picked up by the erect Fallopian tubes embracing the ovary and carried down to the uterus.[94] Once again, we have a projection of male physiology inward. Another eminent obstetrician thought that the male sperm worked like an electric current coursing through the Fallopian tubes and causing the expression of the ovum; a major English medical handbook in 1836 postulated the swelling of the follicle as a consequence of sexual excitement and its bursting as the result of "an action which begins usually during sexual union, but which may also occur without any venereal orgasm."[95]

The remarkable thing about all these accounts is not that they are wrong by modern standards—humans ovulate, and the corpus luteum is formed, independent of intercourse, orgasm, or conception—or even that they are so rich in what today seem like improbable metaphors, but rather that they grant so large a role to female sexual excitement and genital arousal. More remarkable still is that they say so little about the accompanying sensations. Orgasm continues to play a critical part in conception but now those who suffer it need feel nothing.

In part this has nothing specifically to do with women or with intercourse. Sexual pleasure was not the only subjective quality to lose its place in the new medical science. The power of the anatomical-pathological model, as it emerged from Paris hospitals in the late eighteenth century, lay in its capacity to strip away individual differences, affective and material, so as to perceive the essence of health or disease in organ tissues. The autopsy, not the interview, was the moment of truth; corpses and isolated organs could not speak of pleasures.

The nineteenth century was the great age of the post-mortem, of pathology's ascendancy. During his career as pathological anatomist, Karl von Rokitansky, one of the founders of the discipline, is said personally to have made some 25,000 diagnoses. His department at the Vienna General Hospital performed some 2,000 autopsies a year during his tenure—over 80,000 by this estimate—probably more than had been performed in the entire previous history of medicine.[96] Because of the advent of large teaching hospitals with an almost endless supply of poor patients in most of the major cities of Europe, and because of increasing state interest in the causes of death, the number of bodies and organs available to the medical profession for research was almost unlimited. A new kind of medicine, and the new institutions in which it was practiced, made subjectively reportable states, such as pleasure, of relatively little scientific interest. The state of organs was what mattered, and indeed almost all of the evidence for the reproductive physiology of women prior to the end of the nineteenth century came from the ovaries, uteruses, and tubes removed from the dead or from surgical patients: "I now send for your inspection the ovaries of a young unmarried woman who died a few days ago," wrote the surgeon Mr. Girdwood to his colleague Robert Grant; on July 2, 1832, Sir Astley Cooper sent Robert Lee the ovary of a woman who died from cholera while menstruating; Emma Bull, who had only one period and who died of dropsy on May 23, 1835, was opened in the morning to reveal one smooth ovary and one with a single scar; a twenty-year-old virgin's ovaries showed all the stages of ovulation, thus providing still more evidence, a French doctor thought, for the independence of the process from sexual feeling.[97]

The erasure of women's orgasm from accounts of generation is also not the simple result of male ignorance of, or willful blindness to, female genital anatomy. One of the obstetricians quoted above notes that the clitoris is "strictly analogous" to parts of the penis and that it contributes "a large share, and perhaps the greater part, of the gratification which the female derives from sexual intercourse."[98] The 1836 handbook cited says straightforwardly that the "lower part of the vagina and the clitoris are possessed of a high degree of sensibility" but then claims, with no supporting evidence, that in *"some* women, but not in all" they are "the seat of venereal feelings from excitement" and that "in many women such feelings are altogether absent." Feelings were considered irrelevant to both the "fecundating power" of the male and the "liability of conception" of the female, but our author makes no similar claim about the absence of male pleasure. The argument seems to be that only women have an orgasm—how else does the egg get out?—but do not feel it. They have this capacity, as I reconstruct the argument, because human sexual feelings are under "the intellectual and moral powers of the mind." Civilization in all its political, economic, and religious manifestations mercifully leads mankind from "scenes and habits of disgusting obscenity among those barbarous people whose propensities are unrestrained by mental cultivation" to a state in which "the bodily appetites or passions, subject to reason, assume a milder, less selfish, and more elevated character."[99] In the literature I have examined, women's bodies in particular bear the marks of this civilizing process. The physiology of their bodies—in this instance, in many like it, and most powerfully in Freud—adapts to the demands of culture. Although women, like men, were held to experience erection (both of the clitoris and of the internal organs) excitement, and ejaculation, "many" could somehow do so without feeling anything. Again, the point is not to sort out what is, by modern standards, right or wrong about these propositions, but rather to note that culture and not biology was the basis for claims bearing on the role and even the existence of female sexual pleasure. As in the one-sex model, the body shifted easily in the nineteenth century from its supposedly foundational role to become not the cause but the sign of gender.

If one regards the question of female passionlessness as an essentially epidemiological question, about the correlation between orgasm and ovulation or conception, there was equally little known on either side of the issue. No one before the twentieth century had inquired into the incidence of women's pleasure during heterosexual intercourse and, as Havelock Ellis pointed out in 1903, "it seems to have been reserved for the nineteenth century to state that women are apt to be congenitally incapable of experiencing complete sexual satisfaction, and peculiarly liable to sexual anesthesia." He proceeds to cite scores of studies that purport, on the basis of almost no evidence, to speak to this novel issue.[100] Adam Raciborski, the French physician who claimed to have discovered spontaneous ovulation in women, simply declares that three quarters of all women merely endure the embrace of their husbands, just as William Acton in the midst of his book about men thought that he need do no more to make his case than pronounce, "the majority of women are not much troubled by sexual feeling of any kind."[101]

No one knew the answer. One English writer pointed out in his chapter on "the relative amorousness of males and females" that in a field "so characterized by delicacy and silence," most people "judge others in the light of their own limited experiences." Or, as he might more accurately have observed, according to what they would have liked to believe. His own answer, with no supporting data, is that there are three, roughly equal classes of women: (1) those as passionate and responsive as the average man; (2) those less passionate but still taking pleasure "in sexual congress—*especially just preceding menstruation and immediately following its periodical cessation*"; and (3) those who experience no physical passion or pleasurable sensation and who endure sex out of duty. He concludes, disagreeing with his initial hypothesis, that category two is probably the largest after all, category one the smallest.[102] Otto Adler, a late nineteenth-century German expert on these matters, presents an even less ingenuous case of passing off personal or social prejudice for scientific fact. He concludes that as many as 40 percent of women suffered "sexual anesthesia," among whom he included ten who reported that they either masturbated to orgasm or were subject to unconsummated but nevertheless powerful sexual appetites, and one who actually had an orgasm on the examining table as the good doctor examined her genitalia.[103]

The peculiar problems of research in relating sexual pleasure to reproduction were due not only to biases but to professional politics and to the very doctrines of female passionlessness and delicacy that science was called upon to support. The comparative anatomist and birth-control advocate Richard Owen lamented that all theories of generation were "mere speculation": "Would more time have been spent on collecting the actual experiences of human beings." But such work was too difficult for the ignorant and beneath the dignity, or so they thought, of the learned.[104] A German physician, puzzled over how the ovaries became involved in reproduction, surmised that perhaps "libido" was after all the primary agent. In animals, he reasoned, the ovaries changed in time of heat; from a fellow physician he learned that a colleague's wife had long been barren and "bore the masculine embrace without pleasure" but that "she felt libido once and immediately became pregnant." On the other hand, he also knew from his own practice that women became pregnant without feeling anything. There must be "many supremely interesting confidences" told to doctors by their patients, which if correlated would provide the answer. But, alas, politics and prudery stood in the way of epidemiology.[105] A Sicilian physician reported that patients spoke of nothing so much as sex, but that reporting to the profession on such matters was out of the question.[106]

If the respectable physician had no direct access to information about the sexual experiences of women, they could sometimes report on what the husbands of these women had to say. An English writer with a determined empirical streak did just this. Forty out of fifty-two men said that the sexual feelings of their wives had indeed been dormant prior to marriage. This is no surprising result, given each man's presumed pride in his own awakening powers; more surprising is that fourteen out of the fifty-two husbands reported that their wives continued to feel no sexual desire.[107] Clearly the data are flawed by a less than satisfactory survey technique.

The first systematic modern survey of normal women's sexual feelings was one conducted by Clelia Duel Mosher starting in 1892. Based on the answers of some fifty-two respondents, it was inconclusive. True, 80 percent reported having orgasms, leading one historian to argue against the stereotype of the sexually frigid Victorian woman.[108] But as Rosalind Rosenberg points out, most of the women also reported considerable reluctance to have sex and that they would be happier left alone.[109] In short, almost nothing was known about sexual responsiveness among women in general, much less about its relation to ovulation or conception. (There was perhaps even less known about the sexual responsiveness and habits of men, but that is another story.)

Similarly, the epidemiology of infertility in relation to orgasm remained a cipher. In the old model, an ungendered absence of heat as suggested by lack of sexual desire or orgasm was regarded as a common and remediable cause of barrenness. In the new model, which questioned the very existence of female sexual desire, such matters ought to have been irrelevant. They were not. The first systematic survey on the subject, published in 1884, accepts the ancient account as its initial hypothesis. Matthews Duncan, a well-known London gynecological surgeon, was convinced that the absence of sexual pleasure was a major cause of infertility. Yet he found that 152 out of 191 sterile women who consulted him (79 percent) said that they desired sex and that 134 out of 196 (68 percent) reported sexual pleasure, if not orgasm, in coition. Without comparable statistics for fertile women, these numbers mean little, but they seem to suggest quite the opposite of his initial hypothesis and also, incidentally, that English women did not merely lie back and think of Empire.[110]

Other than Duncan's survey, there is little except for a few impressionistic reports, all of which support not the new view of passionlessness but the old link between desire and conception. E. H. Kisch, a German specialist and spa doctor, was convinced that sexual excitement in women was "a necessary link in the chain that leads to impregnation." This conviction derived from his research into 556 cases of first pregnancy, which he found occurred seldom after first coition and most often between ten to fifteen months after marriage (a dubious claim) and from his personal experience that an unfaithful wife was more likely to conceive with her lover than with her husband. The inference from date of first pregnancy to the role of passion depended on the more fundamental observation that most women were sexually unawakened until marriage and that their capacity for erotic pleasure flowered slowly. Presumably, pregnancy coincided with full bloom.[111] B. C. Hirst, in a leading American obstetrics text from 1901, repeated the sort of impromptu clinical lore that had been around for centuries: the ideal condition for conception was mutual synchronous orgasm; conversely, in one of his cases a married woman had endured six years of frigid, infertile intercourse but had become pregnant when coitus and orgasm finally coincided.[112] But how this was to be interpreted remained problematic. Commenting on female pleasure, the *Reference Handbook of Medical Sciences* (New York, 1900–1908) casually states: "Conception is probably more likely to occur when full venereal excitement is experienced."

In short, there was almost no specific new epidemiological information available during the nineteenth century on the incidence of female sexual desire or on its relation to conception. Indeed, as the next chapter will show, "moral" causes of infertility and other repercussions in the body of "good order" gone awry make their way into the world of scientific sex.

SIX

___

Sex Socialized

The form of representation cannot be divorced from its purpose and the requirements of the society in which the given language gains currency.

E. H. GOMBRICH
ART AND ILLUSION

In this chapter I will offer a series of narratives drawn from the middle of the eighteenth to the early twentieth centuries in western Europe. The first two—about politics and political theory and about the fairly technical question of when ovulation occurs during the menstrual cycle—are intended to show how, in specific contexts, incommensurable, opposite sexes came into being. The second two—an account of why masturbation and prostitution are not so much sexual as they are social pathologies with sexual consequences and a reading of Freud's argument about the transition from clitoral to vaginal sexuality as a case of near universal hysteria—are intended to show the contrary tendency: how the one-sex model with its interpenetration of the body and culture flourished at the same time in other, quite specific contexts. Having argued in Chapter 5 that the two-sex model *was not* manifest in new knowledge about the body and its functions, I will argue here that it *was* produced through endless micro-confrontations over power in the public and private spheres. These confrontations occurred in the vast new spaces opened up by the intellectual, economic, and political revolutions of the eighteenth and nineteenth centuries. They were fought in terms of sex-determinant characteristics of male and female bodies because the truths of biology had replaced divinely ordained hierarchies or immemorial custom as the basis for the creation and distribution of power in relations between men and women. But not all confrontations of sex and gender were fought on this ground, and one-sex thinking flourished still. The play of difference never came to rest.

Politics and the biology of two sexes

The universalistic claims made for human liberty and equality during the Enlightenment did not inherently exclude the female half of humanity. Nature had to be searched if men were to justify their dominance of the public sphere, whose distinction from the private would increasingly come to be figured in terms of sexual difference. The Encyclopedists' argument that marriage is a voluntary association between equal parties—a relationship in which neither partner has an intrinsic claim to power—is immediately met by the counterargument that someone has to be in charge in the family and that someone is the male, because of his "greater force of mind and body" (essentially Locke's position). Biology thus assures marital order, but it also sets the terms for still another counterclaim: "man does not invariably have more strength of body," from which it follows that the exceptional circumstances in which women do control families and kingdoms are not unnatural.[1]

Sex was also a major battleground of the French Revolution: "a contestation between male and female, in which the middle-class revolutionary creation of political culture was to validate the political culture of men and culpabilize that of women." However much class lines might be blurred, "that between men and women had at all costs to be made visible."[2] The promises of the French Revolution—that mankind in all its social and cultural relations could be regenerated, that women could achieve not only civil but personal liberties, that family, morality, and personal relations could all be made afresh—gave birth not only to a genuine new feminism but also to a new kind of antifeminism, a new fear of women, and to political boundaries that engendered sexual boundaries to match. The creation of a bourgeois public sphere, in other words, raised with a vengeance the question of which sex(es) ought legitimately to occupy it. And everywhere biology entered the discourse. Obviously those who opposed increased civil and private power for women—the vast majority of articulate men—generated evidence for women's physical and mental unsuitability for such advances: their bodies unfit them for the chimerical spaces that the revolution had inadvertently opened. But revolutionary feminists also spoke in the language of two sexes. It is ridiculous, Condorcet argues, to exclude women from the political franchise because of biology: "Why should individuals exposed to pregnancies and other passing indispositions be unable to exercise rights which no one has dreamed of withholding from persons who have the gout or catch cold quickly." On the other hand, he is sure that women—and here he is speaking of fundamental sexual characteristics—"are superior to men in gentle and domestic virtues."[3] Olympe de Gouges, in her famous declaration of the rights of women, says that "social distinctions can be founded only on general utility," but already in the previous paragraph she announced that she speaks in the name of "the sex that is superior in beauty and in courage of maternal suffering."[4] For both, a woman's place is determined by her body; revolution and not simply male bourgeois desire for a "haven in a heartless world" generated "separate spheres."

Whatever other ideological work the doctrine of separate spheres did in the nineteenth century—and it will turn up both to justify and to condemn woman's political action—it explicitly shattered the notion of a hierarchy of the sexes and served as the cornerstone of a powerfully multi-valent alternative model. Women as beings who are "little affected by sensuality," "a species of angel," "a purer race . . . destined to inspire in the rest of the human race the sentiments of all which is noble, generous and devoted" (this is from a French feminist of the revolutionary era), were the cultural creation of the middle classes, men and women, with a variety of political agendas.[5] But woman so construed is *not* a lesser man, measured on a male scale of virtue, reason, or sensuality.

Various doctors also wrote for diverse political and cultural purposes and consequently produced a variety of accounts of sexual difference. But their professional prestige and right to speak on such matters rested on the conviction that these differences resided fundamentally in the body. Thus Auguste Debay, author of the leading nineteenth-century marriage manual in France, seems anxious to encompass a wide range of human physiology, especially of male and female sexual experience, to stake as broad a claim as possible against the clergy who traditionally spoke on such matters. His vision and sympathies are clearly male; he counsels women to fake orgasms if necessary and never to refuse their husbands. (He counsels husbands never to demand sex from unwilling wives, though how they are to know this in the midst of so much dissimulating is not clear.) But Debay has no interest in a biology of passionlessness: he goes into great detail regarding clitoral orgasm, notes that a woman's pleasure during intercourse arises from the rubbing (*frottement*) of the pubis of the male on the clitoris and not from rubbing in the vagina.[6] The urologist William Acton, on the other hand, famous for his claim that "the majority of women (happily for them) are not very much troubled by sexual feelings of any kind," was obsessed with masturbation and various defects of the seminal economy. He wrote for men, about men's problems, and was interested in women primarily as a healthy place for his patients economically to deposit their sperm.[7] Hence his shrill, even by nineteenth-century standards, condemnation of masturbation, to which women are connected through passive exchange.

This sort of list is endless. Supposed biological differences between male and female bodies were generated in a variety of contexts. Roussel and Moreau and Cabinis, the most prominent moral anthropologists of the French Revolution, wrote as part of the Napoleonic retrenchment in matters of family and gender, arguing that corporeal differences demanded the social and legal differences of the new Code. Differences were propounded in conflict. Susanna Barrows has shown how fears born of the Paris Commune and of the political possibilities opened up by the Third Republic engendered an extraordinarily elaborated physical anthropology of sexual difference, to justify resistance to change.[8] In Britain the rise of the women's suffrage movement in the 1870s elicited similar responses: women were construed as creatures who for various reasons, in many respects like those that disadvantaged the darker races, were incapable of assuming civic responsibility.[9]

But reinterpretations of the body had roots in less worldly circumstances as well. Social-contract theory at its most abstract postulated a body that, if not sexless, is nevertheless undifferentiated in its desires, interests, or capacity to reason. In striking contrast to the old teleology of the body as male, liberal theory begins with a neuter individual body: sexed but without gender, in principle of no consequence to culture, merely the location of the rational subject that constitutes the person. The problem for this theory is how to legitimate as "natural" the real world of male dominion of women, of sexual passion and jealousy, of the sexual division of labor and of cultural practices generally from such an original state of no-gender. The answer to making their "natural beings recognizable," as Carole Pateman puts it, was for social-contract theorists to "smuggle social characteristics into the natural condition."[10] However the argument works in detail, the end result is that women are absent from the new civil society for reasons based in "nature." A biology of sexual incommensurability offered these theorists a way of explaining—without resorting to the natural hierarchies of the one-sex model—how in the state of nature and prior to the existence of social relations, women were already subordinated to men. Therefore the social contract could then be created between men only, an exclusively fraternal bond. Ironically, the genderless rational subject engendered opposite, highly gendered sexes.

The ostensibly neutral language of liberalism also left women themselves without a voice of their own and initiated a feminist discourse of difference in search of one.[11] If women were simply lesser versions of men, as the old one-sex model had it, then there would be no need for them to write or take public action or make any other claims for themselves as women; men could represent them far better than they could represent themselves. But the same unacceptable consequences arise if they are in all respects the same: if women have no special interests or legitimate grounds for their social being, men could speak for them as they had in the past. (This is the "difference dilemma," as Martha Minnow calls it.) Hence feminism too, or at least versions of feminism, turned to a biology of incommensurability to replace both the teleologically male interpretation of bodies, on the basis of which a feminist stance is impossible, and the view that all bodies in public discourse are sexless, in which case it is irrelevant. "We do not advocate the representation of women because there is no difference between men and women; *but rather because of the difference between them*," argued the nineteenth-century feminist Millicent Fawcett. "We want women's special experiences as women . . . to be brought to bear on legislation," she says, and offers the hope that "by giving women greater freedom . . . the truly womanly qualities in them will grow in strength and power."[12] (This need not be a claim about biology, but in the context of nineteenth-century debates on the "woman question" it almost invariably was.)

Though I will illustrate the political generation of two sexes in the works of various thinkers and activists, I do not want to suggest that this process is somehow abstracted from day-to-day reality. Two incommensurable sexes *are* the result of discursive practices, but they become possible only within the social realities to which these practices give meaning. Thus Rousseau was enraged by the cultural influence of women not only for idiosyncratic reasons or because relations with women represented the prototypical case of man's slavish dependency: his obsessions on the matter developed in the great age of the salon, where women had in fact created a genuine new public space within the old regime. This historical development is part of what I take to be the discursive creation of difference. More generally, as Joan Landes puts it, "an ideologically sanctioned order of gender differences and public-private spheres . . . grounds the institutional and cultural geography of the new public sphere."[13]

And now some contexts for the making of incommensurable difference. Rousseau's complicated antifeminist account is perhaps the most theoretically elaborated of the liberal theories of bodies and pleasures, and the most concretely concerned with the relation of sexual difference to the origins of society, but it is only one among a great many examples of how deeply a new biology was implicated in cultural reconstruction.[14] In the state of nature, as he imagines it in the first part of *A Discourse on Inequality*, there is no social intercourse between the sexes, no division of labor in the rearing of young, and, in a strict sense, no desire. There is of course brute physical attraction between sexes, but it is devoid of what he calls "moral love," which "shapes this desire and fixes it exclusively on one particular object, or at least gives the desire for this chosen object a greater degree of energy." In this world of innocence there is no jealousy or rivalry, no marriage, no taste for this or that woman; to men in the state of nature "every woman is good." Rousseau is remarkably precise in specifying the reproductive physiology of women that must underlie this condition. Against Hobbes, he argues that there is no violent competition among human males for females in the state of nature because women, unlike other female animals, do not have alternating periods of heat and abstinence and are thus always sexually available. Humans, moreover, are spared the "terrible moment of universal passion" that occurs in some animals when "the whole species goes on heat at the same time."[15] Reproductive physiology and the nature of the menstrual cycle bear enormous weight here; the state of nature is conceptualized as dependent on the biological differences between women and beasts. (The jurist Samuel von Pufendorf, incidentally, derives quite the opposite conclusion from the same "facts." The state of nature is violent, not pacific, and in desperate need of law precisely because of the absence of seasonality in human passion. Animals, he argued, feel the "stings of love" only in order to propagate and once "they have arrived at their end," the passions cease; but in humans the passions "are aroused more frequently than seems necessary for the propagation of the species" and they are in need of civilization to control them. Again, much depends on the physiology of passion.[16])

But what happened to Rousseau's primitive and supposedly pacific state of desire? He gives an account of the geographical spread of the human race, of the rise of the division of labor, of how in developing dominion over animals man "asserted the priority of his species, and so prepared himself from afar to claim priority for himself as an individual." But the individuation of desire, the creation of what he calls the moral part of love ("an artificial sentiment"), and the birth of imagination ("which causes such havoc amongst us") are construed as the creation of women, specifically as the product of female modesty. The *Discourse* presents this modesty as volitional, instrumental, clearly postlapsarian: "[It is] cultivated by women with such skill and care in order to establish their empire over men, and so make dominant the sex that ought to obey." But in *Emile* modesty is naturalized and definitely not the product of culture: "While abandoning women to unlimited desires, He [the Supreme Being] joins modesty to these desires in order to constrain them." Somewhat later, in a note, Rousseau adds: "The timidity of women is another instinct of nature against the double risk they run during their pregnancy." Indeed, throughout *Emile* he argues that natural differences between the sexes are represented and amplified in the form of moral differences that society erases only at its peril.[17]

Book 5 begins with the famous account of sexual difference and sameness. "In everything not connected with sex, woman is man . . . In everything connected with sex, woman and man are in every respect related but in every respect different." But of course a great deal about women *is* connected with sex: "The male is male only at certain moments. The female is female her whole life . . . Everything constantly recalls her sex to her." "Everything," it turns out, is everything about reproductive biology: bearing young, suckling, nurturing, and so on. Indeed the chapter becomes a catalogue of physical and consequently moral differences between the sexes; the former, as Rousseau says, "lead us unawares to the latter." Thus "a perfect woman and a perfect man ought not to resemble each other in mind any more than in looks." From the differences in each sex's contribution to their union it follows that "one ought to be active and strong, the other passive and weak." After announcing that the problem with Plato is that he excludes "families from his regime and no longer knowing what to do with women, he found himself *forced to make them men*," Rousseau concludes that "once it is demonstrated that man and woman are not and ought not to be constituted in the same way in either their character or temperament, it follows that they ought not to have the same education."[18]

For Rousseau a great deal depends on the natural (biological), modesty of women and on their radically distinct role in reproducing the species. Indeed, all of civilization seems to have arisen in consequence of the secular fall from innocence when the first woman made herself temporarily unavailable to the first man. But Rousseau is simply pushing harder on a set of connections that are commonplace in the Enlightenment. In his article on *jouissance,* Diderot locates the creation of desire, marriage, and the family, if not love itself, at the moment when women first came to withhold themselves:

> when woman began to discriminate, when she appeared to take care in choosing between several men upon whom passion cast her glances . . . Then, when the veils that modesty cast over the charms of women allowed an inflamed imagination the power to dispose of them at will, the most delicate illusions competed with the most exquisite of senses to exaggerate the happiness of the moment . . . two hearts lost in love vowed themselves to each other forever, and heaven heard the first indiscreet oaths.[19]

Diderot, like Rousseau, seems to believe there was a time before female modesty, a time "when woman began to discriminate." But modesty and the possibility of sexual restraint, however cagey and conniving, are nevertheless natural qualities of women. Diderot's and Rousseau's stories had to go that way; to be a woman in civil society is to be modest, to create but not to have desire. To be otherwise is to be "unnatural."

The special qualities of female sexual desires become in the eighteenth century a key element in understanding the meaning of human history. Most prominently among figures of the Scottish Enlightenment, for example, John Millar argues for the crucial role of women and their virtues in the progress of civilization. Far from being lesser men, they are treated in his *Origin of the Distinctions of Ranks* as both a moral barometer and an active agent in the improvement of society.[20] Millar's case begins with the claim that sexual relations, being most susceptible "to the peculiar circumstances in which they are placed and most liable to be influenced by the power of habit and education," are the most reliable guide to the character of a society. In barbarous societies women accompanied men to war and were scarcely different from them; in peaceful societies that had progressed in the arts, a woman's rank and station were dictated by her special talents for rearing children and by her "peculiar delicacy and sensibility," whether these derived from her "original constitution" or her role in life. (These sentiments will of course be echoed in the far more explicitly biological context of Darwinism a hundred years later.) Thus civilization in Millar's account leads to an increasing differentiation of male and female social roles; conversely, a greater differentiation of roles and specifically greater female "delicacy and sensibility" are signs of moral progress. But women themselves in more civilized societies are also the engines of further advance: "In such a state, the pleasures which nature has grafted upon love between the sexes, become the source of an elegant correspondence, and are likely to have a general influence upon the commerce of society." In this, the highest state, he is thinking of French salon society and the *femme savante*. Women are "led to cultivate those talents which are adapted to the intercourse of the world, and to distinguish themselves by polite accomplishments that tend to heighten their personal attractions, and to excite those peculiar sentiments and passions of which they are the natural objects." Thus desire among civilized men is inextricably bound up in Millar's moral history with the history of specifically female accomplishment.[21]

It is hardly surprising in the context of Enlightenment thought and postrevolutionary politics that the moral and physical differentiation of women from men should also be critical to the political theories of women writers, from the early socialism of Anna Wheeler, at one end of the political spectrum, through the radical liberalism of Mary Wollstonecraft, to the domestic ideology of the conservative Hannah More and the progressive Sarah Ellis. For Wheeler and others, the denial or devaluation of specifically female passion is part of a general devaluation of passion.[22] Reason, they dared to hope, would be triumphant over the flesh. Wheeler and the utopian socialists were, after all, writing out of the tradition that produced William Godwin's argument that civilization would ultimately eliminate destructive passions, that the body would be curbed by the enlightened mind. Women could be in the vanguard of this triumph. (It is against this view, as Catherine Gallagher argues, that Thomas Malthus rehabilitates the body and insists upon the absolute irreducibility of its demands, especially its sexual demands.[23])

But the new claims by women to heightened moral sensitivity were claims not only *against* the flesh but *for* new political space. Wheeler makes this quite clear, though she is ambivalent about the purported passionlessness of women—a version of the difference dilemma—which adjusts itself to the rhetorical demands of the moment. Her book, jointly written with William Thompson, is a sustained attack on James Mill's argument that the interests of women and children are subsumed in the interests of husbands and fathers. They hold to the contrary that women must speak for themselves and that they have something distinctive to say. But the important aspect of their battle with Mill, for my purposes, is that it was fought specifically over the nature of female passion and the bargaining power it supposedly bestows. Mill's "moral miracle" would be credible, they admit for argument's sake, if he were justified in holding that women are protected against abuse because, themselves free from sexual desire, they are in an excellent negotiating position: men, who are decidedly not liberated from their bodies, "will act in a kind way toward woman in order to procure from her those gratifications, the zest of which depends on the kindly inclinations of one party yielding them." But if women are not "like the Greek Asphasia," cold and sexless, then Mill's argument is nonsense. Not only are women sexed and desirous; in the current state of affairs, "woman is more the slave of man for gratification of her desires than man is to woman." The double standard allows men to seek gratification outside of marriage but forbids it to women.[24]

On the other hand, Wheeler and Thompson's analysis of the sorry state of the male world and their need to claim some political ground for women leads them, in other rhetorical circumstances, to change their emphasis dramatically. In a chapter tellingly entitled "Moral Aptitude for Legislation More Probable in Women than Men," women are represented *not* as men's equal in passion but as superior in morality and empathy, generally better able than men to act in accord with the common interest. Whether women had these traits in some hypothetical state of nature or acquired them through a kind of moral Lamarckianism is unclear, but in the modern world women demonstrate a greater susceptibility to pain and pleasure, a more powerful desire to promote the happiness of others, and a more developed "moral aptitude" than men. These, Wheeler and Thompson argue, are the most important qualities in a legislator. It is, moreover, precisely women's inferior strength and their inability to oppress others through force that would make them fair and just rulers.

Women as mothers and as the weaker sex need a world at peace far more than men, and they would be constitutionally more likely to legislate ways to obtain it. Wheeler and Thompson's arguments are more poignantly put than this summary suggests, but they contribute to a construction of woman not very different from that of far more conservative domestic ideologists. Whether through inherent nature—because they have more sensitive nervous systems, as many eighteenth- and nineteenth-century doctors held—or through centuries of suffering, women are construed, in and through their bodies, as being less in the thralls of passion and unreason and hence morally more adept than men.[25]

Mary Wollstonecraft is caught in much the same dilemma. Liberal theory pushes her to declare that the rational subject has, in essence, no sex; yet she was only too aware of the power—in her own life, the destructive violence—of sexual passion. She also believed, with Rousseau, that civilization increased desire and that "people of sense and reflection are most apt to have violent and constant passions and to be preyed on by them." Finally, for Wollstonecraft to subscribe to the notion of the subject as genderless was to deny the manifestly particular qualities of women's experiences. Her *Vindication of the Rights of Women,* as Mary Poovey points out—and this is even more the case in her other works— thus takes up a peculiarly defensive posture toward female sexuality and its control. "Men are certainly more under the influence of their appetites than women," she says straightforwardly; women have the capacity to lead almost bodiless existences. But she is compelled to warn of the "nasty" and "immodest habits" girls acquire at boarding school (masturbation, I suppose) and consistently denies the existence of the very desire whose presence she senses so acutely and finds so threatening and distasteful.[26]

Wollstonecraft's contemporary, the German liberal Theodor Gottlieb von Hippel, whose *On Improving the Status of Women* has much in common with the *Vindication,* reveals similar tensions. First he argues that "nature does not appear to have intended to establish a noteworthy difference or to have favored one sex at the expense of another." But he, like the domestic ideologists, also wants to create a separate, equal, unhierarchical but nevertheless radically different sphere for women which also is grounded in nature: perhaps, he postulates, women are more attracted to, and potentially better able to create beauty in poetry and painting, because "men are more alienated from nature than women"; women's potential influence on morality arises from the fact that "a soft and moderate character is peculiar to the opposite sex."[27]

Wollstonecraft's tentative and always tensile solution was, like Hippel's, for women to take the moral high ground. Because they are blessed with a unique susceptibility "of the attached affections," it was women's special role in the world to civilize men and raise children in virtue. In the *Female Reader* Wollstonecraft lays on a heavy dose of religion, which she says will be the solace of her readers when they find themselves, as they often will, "amidst the scenes of silent unobserved distress," which it is their duty to alleviate. And Wollstonecraft shares with early socialist feminists a commitment to passionlessness, whether out of personal distaste, some sense of its political possibilities, an acute awareness of passion's dangers, or a belief in the special undesiring qualities of the female body.[28]

Wollstonecraft's arguments for the differences between the sexes begin to sound very much like those of Sarah Ellis, however profound the political chasm that divided the two women. Indeed domestic ideology, in England at any rate, united not only Anglicans and Dissenters but, as Davidoff and Hall show, Radicals, Liberals, and Tories, men and women.[29] Ellis wrote from a progressive, explicitly antiaristocratic position, which sought to lift women out of an ornamental role and to give them a base of real influence. There is always a tension in her work—and in domestic ideology generally—between woman as "a relative creature," a version of the older view that she is a lesser man who exists in relation to him, and woman as an independent being who wields potentially enormous power in her own sphere. It is the latter position that is most powerfully articulated and that comes to be grounded in *sexual* difference. In *The Wives of England,* one of the canonical works of domestic ideology, Ellis argues that wife and mother are "at the center of a circle of influence, which will widen and extend itself to other circles, until it mixes with the great ocean of eternity." This influence is born of the heightened moral sensibilities of the female organism. Though women are to have no role in the world of mundane politics, they are to confront issues "such as extinction of slavery, the abolition of war in general, cruelty to animals, the punishment of death, temperance, and many more, on which, neither to know, nor to feel, is almost equally disgraceful. In short, women's politics must be the politics of morality."[30] Women, in short, are creatures less plagued by passion, a selfish and destructive tendency, and more fully endowed with fellow feeling and the sort of corporeal tranquility required to be the radiant centers of a new morality. Passionlessness is thus born of a particular political moment and of a strategy for staking out a public arena of action grounded in the virtues of the female private sphere.

The immediate, political demands for the creation of biologically distinct sexes and the specific role of science in this enterprise are especially clear in one late nineteenth-century instance. The physician Elizabeth Blackwell, in her construction of mentally dominated sexuality in women, wrote as a professional: "in guarding the human faculties" and in furthering "the gradual growth of thought, which leads to ever higher forms of society," the physiologist and the physician have "very important aid to render." Physiology was important because she believed that cultural progress, increasingly moral behavior, was imprinted on the flesh of succeeding generations just as individual habits became second nature to the body. Beasts, she argued, have no mental component in their sexual relations; primitive people and the working classes have relatively little and are thus unchaste; civilized people have a dominant mental component and thus value chastity highly.[31]

Progress is marked, in other words, by the subordination of the brutishly physical in sex; chaste sexual relations, a cultural triumph for the race, become "inseparably interwoven with the essential structure of our physical organization." Progress in this fashion leaves its mark on the race. Men of course can practice chastity, but the real job of "interweaving," Blackwell argued, belonged to women. Although she was almost Galenic in regarding certain fluids and functions as common to both sexes—the organs that produce egg and sperm are strictly analogous; "sperm emission" is a version of menstruation, and both are mechanisms of natural balance; each part in the female corresponds to a part in the male—she reversed the valences. Men's functions are lesser versions of women's. More important, she thought that men and women differ in two crucial respects, which makes her vision fit into the two-sex model: women have a uterus and men do not; woman's sexual urges are primarily mental and men's are not. (Here again is a reversal of the usual formula. Not having a uterus defines man, as opposed to not having a penis defining woman; woman is associated with mind and man with nature, rather than the other way around.) Blackwell did not deny physical desire in women but argued that their sexual feelings arose primarily from the depths of the mind:

This mental element of human sex, exists in major proportion in the vital force of women . . . Those who deny sexual feeling to a woman, or consider it so light a thing as hardly to be taken into account in social arrangements, quite lose sight of this immense spiritual force of attraction, which is distinctly human sexual power, and which exists in so very large a proportion in their nature.

Unlike men's, women's sexual pleasure is thus not attached "chiefly to the act of coition" but to higher things. But what drives woman's mental sexuality to be in civilization's moral vanguard? "The pure sentiment of maternity . . . the special aptitude given to women by the power of maternity . . . the enlarged intelligence of mothers [which] will be welcomed as the brightest harbinger of sexual regeneration."[32] And with sexual regeneration will come social regeneration. Women for Blackwell, no less than for Millar and the giants of the Scottish Enlightenment, both caused and reflected cultural progress.

Yet there is obviously a more critical edge to Blackwell's account. Implicit is a deep hostility to what she perceived as the aggression, the brutishness and insensitivity, of men toward women. She campaigned against the physician's speculum as an assault on women's purity; and in another tract she explicitly developed her notion of the dominant mental qualities of woman's sexuality as an attack on male sexuality. Consider, Blackwell urges, a neo-Malthusian pronouncement against using coitus interruptus and the sheath: "Any preventive means, to be satisfactory, must be used by the *woman* [Blackwell's italics], as it spoils the passion and impulsiveness of the venereal act, if the man have to think of them." Here the "cloven foot is fully revealed"; women are meant somehow to manage male passion. Ideal marital sexuality, on the other hand, would be based on "positive physical facts," which meant that "the wife must determine the times of union." This was in part because a woman's intimate knowledge of her reproductive cycle—like all her contemporaries, Blackwell wrongly thought that the middle of the menstrual cycle was the least likely time for conception to take place—allowed for "a natural method of regulating the size of families," and because the powerful mental component in woman's sexual makeup made her a more promising moderator of desire. These arrangements would further world historical progress: "the regulation of sexual intercourse in the best interests of womanhood, is the unrecognized truth of Christianity, toward which we are slowly groping."[33] Passionlessness in this account allows women not only a major role in the advancement of humanity but a defense against and a justification for control of their bodies.[34]

I do not want to suggest that all of these writers and causes, from Rousseau's reconstruction of the state of nature to Blackwell's attack on male sexuality, were part of the same theoretical or political undertaking. Rather I have sought to give examples of the wide range of apparently unrelated political agendas in which a new differentiation of the biological sexes occupied a central place. Desire was given a history and the female body was distinguished from the male's, as the shattering transformations of European society between the seventeenth and the nineteenth centuries put unbearable pressure on old views of the body and its pleasures. A biology of cosmic hierarchy gave way to a biology of incommensurability, anchored in the body, in which the relationship of men to women, like that of apples to oranges, was not given as one of equality or inequality but rather of difference. This required interpretation and became the weapon of cultural and political struggle.

The cultural politics of cyclical fertility

"The sciences of life can confirm the intuitions of the artist, can deepen his insights and extend the range of his vision."[35] In discussing menstruation, ovulation, and cyclical desire during the nineteenth century, I want to describe how facts, or what were taken to be facts, became the building blocks of social visions: the dry and seemingly objective findings of the laboratory, the clinic, or the "field" became, within the disciplines practiced there, the stuff of art, of new representations of the female as a creature profoundly different from the male. This "art," clothed in the prestige of natural science, in turn became the supposed foundation of social discourse.

I am not so much interested here in the gynecologist's or physiologist's overtly polemical pronouncements on women, though there is an abundance of them. From little-known doctors to the giants of nineteenth-century medicine—Charcot, Virchow, Bischoff—came the cry that claims for equality between the sexes were based on profound ignorance of the immutable physical and mental differences between the sexes and that these, not legislative whim, determined the social division of labor and rights. The certain and impartial methods of science proved, most doctors thought, that women are not capable of doing what men do, and vice versa (including studying medicine). Instead I am more concerned in this chapter to show how, in the poetically unpromising domains of histology and physiology, observations were turned into the materials for art—for the artifices of sex—which were then claimed to have a prior natural existence. Reproduction and its relationship to pleasure has been one of the axes of this book, but I want to make clear that this was by no means the only arena for the construction of sexual difference. So I will begin with two unclinical examples.

The Darwinian theory of natural selection provided and still provides seemingly limitless material for imagining the process of sexual differentiation.[36] Sexual selection: among animals a passive female selects as mates the most aggressive males or the most attractive, the most gorgeously plumed, the most melodious. Having given numerous examples, Darwin concludes: "Thus it is, I believe, that when the males and females of any animal have the same general habits of life, but differ in structure, color or ornament, such differences have been mainly caused by sexual selection."[37] The process works in humans as well; modesty is selected for among women and prowess among males—despite the fact that in our species only males do the choosing—because the males with a choice will pick the most beautiful, and by implication the most modest, of the women available.[38] Sexual selection "apparently has acted on man, both on the male and female side, causing the two sexes to differ in body and mind." (The same process, Darwin said, creates racial divergence and the differentiation of species generally.) In each generation men and women are thus a bit more different from each other than in the previous one, Darwin suggests, quoting approvingly the German materialist thinker Karl Vogt: "The difference between the sexes as regards the cranial cavity increases with the development of the race, so that the European male excels much more the female, than the negro the negress."[39] If one believes this, then the divergence of all manner of gender characteristics could be imagined as a special case of the general process of divergence through which species are born. This seems to be what Vogt had in mind. In fig. 62 "a^{14}" and "z^{14}" become ciphers to be filled in as needed, and the differences between the sexes become the product of the grand and inevitable process of selection that governs all life.

Difference could also be generated and imagined in what are today disreputable "sciences." Phrenology, for example, was the nineteenth century's equivalent of modern biological determinism. The untutored (who insisted on a phrenological analysis of the condemned in any complete

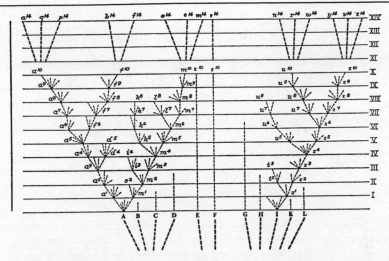

Fig. 62. Darwin's schema for how species are produced seems to provide a model as well for how relatively small differences between men and women in more primitive societies have been exaggerated through the process of civilization.

execution account) as well as the learned (those at least with a taste for materialist explanation) subscribed to its tenets. By a careful analysis of head shape, of the strengths of certain key features, some thirty-seven components of human character could, it was thought, be assessed for each individual. Though education played some part in creating personality, a person was fundamentally the product of an admixture of inborn traits: combativeness, sublimity, benevolence, and so forth. Different parts of the brain were thought responsible for specific characteristics and the shape of the head and neck reflected the nature of the brain beneath. The cerebellum, for example, was regarded as the seat of sexual instinct, of what popular phrenologists called "amativeness," and women, as might be expected, were said to have smaller cerebella than men, "moderate" on a scale ranging from very large in highly sexed males to very small in children. This "just right" quantity of passion, combined with the other qualities that women enjoyed, resulted in a creature who

> will exercise more of pure love and virtuous affection towards the opposite sex, than of mere amative passion—of chaste Platonic affection, than of sexual love—of pure and sentimental friendship, than mere animal feeling . . . This is the kind of attachment generally exercised by females, in whom adhes[iveness] is commonly altogether larger than amat[iveness].[40]

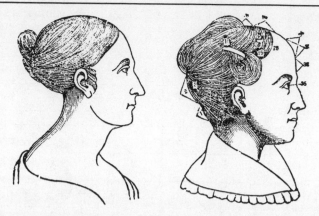

Fig. 63. The woman on the left has an abnormally small cerebellum; the other woman's cerebellum and hence her "amativeness" are just right. From Fowler's *Practical Phrenology*.

One could look at the small, delicate female neck, manifestly incapable of hiding a large cerebellum, and tell that amativeness was moderate; the raised area above indicated that adhesiveness, on the other hand, was well developed (see fig. 63 for the ideal woman's head). This theory of a cerebellum locus for passion also solved one of the tensions in the doctrine of passive female sexuality—why women, with their exquisitely sensitive nervous systems, would not find the pleasures of sexual intercourse even more delightful than would men but would instead find it relatively easy to renounce passion. Answer: "Her smaller cerebellar organ of will renders her less determined in pleasure and enables her to yield to suspense and renunciation." (Why a less developed will would make renunciation easier is left unexplained.) But, as was so often the case with nineteenth-century efforts to circumscribe with words the sexuality of women, this effort foundered as soon as it got underway. While women could more easily renounce sex, they could not forgo it entirely; and, if renunciation was not absolutely voluntary, it would have far greater pathological consequences than it would have in men. Nevertheless, earnest attempts to derive the passive female from her anatomy were not to be deterred.

My concern, however, is less with such grand theories or with the endless pronouncements of doctors than with how real science—careful work within a research paradigm that draws reasonable conclusions from its results—contributes to the artifice of sexual difference.[41] I begin with the elegantly simple, critical experiment that established spontaneous ovulation in dogs and by extension in other mammals. In the novelistic style that characterizes so much early nineteenth-century scientific reporting, Theodor von Bischoff tells his reader that on December 18 and 19, 1843, he noted that a large female dog in his possession had begun to go into heat. On the 19th he allowed her contact with a male dog, but she refused its attentions. He kept her securely caged for two more days and then brought on the male dog again; this time she was interested, but he separated the animals before coition could take place. At ten o'clock two days later, he cut out her left ovary and Fallopian tubes and carefully closed the wound. The Graafian follicles in the excised ovary were swollen but had not yet burst. Five days later he killed the dog and found in the remaining ovary four developing corpus lutei filled with serum; careful opening of the tubes revealed four eggs. He concluded:

> I do not think it is possible to demonstrate with any more thoroughness the whole process of the ripening and expulsion of the eggs during heat, independently of coition, than through this dual observation on one and the same animal.[42]

And of course if ovulation occurred independently of coition, it must also occur independently of fecundation as well. The naturalist F. A. Pouchet considered the latter discovery so major that he formulated it as his fifth and critical law of reproductive biology, "le point capital" of his 476-page magnum opus.[43] The historian Jules Michelet was enraptured and hailed Pouchet for having formulated the entire science of reproductive biology in a daring, definitive work of genius.[44]

Granted that dogs go into heat and ovulate whether they mate or not, what evidence was there that women's bodies function in a similar manner? Almost none. No one before the twentieth century claimed to have seen a human egg outside the ovary. Bischoff admitted that there was no direct proof for the extension of his theory to women, but he was sure that an egg would be found soon enough.[45] In 1881 Victor Hensen, professor of physiology at Kiel, notes in a standard textbook that human eggs still eluded investigators, though with unwarranted optimism he adds in a footnote that it should not be too difficult to find one in the Fallopian tubes.[46] In fact it was. An unfertilized egg was not reported until 1930, and then in the context of an argument against the nineteenth-century view relating heat to menstruation.[47] Thus the crucial experimental link—the discovery of the egg—between menstruation, on the one hand, and ovulation and the morphology of the ovary, on the other, was lacking in humans.

Investigators could only note in the cases that came their way that women were menstruating, or that they were at some known point in their menstrual cycles, and then attempt to correlate these observations with the structural characteristics of the ovary removed in surgery or autopsy. From these characteristics, supposedly, the timing of ovulation could be deduced. But researchers lacked as a biological triangulation point the actual product of the ovary, and the results of their studies were unsatisfactory.

It is not surprising, in itself, that these investigators should have thought that ovulation occurred just after the menses because, in the learned and popular literature from antiquity to the late eighteenth century, it was a commonplace that the purgation of menstruation made women more likely to conceive in the days following its abatement. New evidence that ovulation accompanied heat in some animals, coupled with the assumption of a certain uniformity among all mammals, gave new life to ancient wisdom. Autopsy evidence gave further credence to the view that ovulation occurred at or just after menstruation.[48] (Such evidence was problematic. One could always come up with counterexamples, such as Sir James Paget's report of his autopsy on Mrs. M, executed twelve hours after her period began, whose ovaries showed "no appearance of recent rupture of a vesicle, or the discharge of an ovum."[49])

Evidence for the timing of ovulation based on pregnancy from a single coition, whose occurrence in the menstrual cycle was supposedly known, also tended to support the old views in their new guise. A vast number of observations—some 50,000 in all the studies—suggested that day 8 from the onset of menstruation was the most likely for conception and that by days 12–14 the chances were a good deal less.[50] (These studies were generally based on the reports of women, gathered considerably after the event, as to when during their menstrual cycle they had become pregnant. In the absence of diaries or other records to jog their memories, women tended to report what was generally held to be the most likely times for conception.) Popular tracts reinforced these beliefs. Dr. George Napheys recommended that "a time about midway between the monthly recurring periods is best fitted for consummation of marriage," because "this is the season of sterility."[51] The Roman Catholic physician

Carl Capellman gave the same sort of advice in one of the earliest, and from the perspective of more knowledge completely wrong, expositions of the rhythm method.[52] Mary Stopes likewise told her hundreds of thousands of readers that conception took place during or just after the period and that the middle days were relatively safe.[53]

The trouble was that these sorts of studies—based on efforts to correlate date of coition with pregnancy or ovarian morphology with menstruation—never yielded consistent results. The role of the ovaries in the reproductive cycle of mammals was very imperfectly understood until the publication of a series of papers beginning in 1900: Papanicolaou's studies of the cytology of the cervical mucosa around 1910 provided the first reliable marker of the ovarian cycle in humans; appropriate hormone assays appeared a little later; finally, by the 1930s, the hormonal control of ovulation by the ovary and the pituitary was generally understood.[54]

But far more was at stake in Bischoff's experiment than proof of ovulation in dogs and pigs, independently of coition or fecundation, at the time of heat or the extension of this insight to humans during the menses. The discovery of spontaneous ovulation in some mammals was of enormous historical importance in how women's bodies were represented. Menstruation, which had been a relatively benign purging of plethora, not unlike other forms of corporeal self-regulation common to men and women, became the precise equivalent of estrus in animals, marking the only period during which women are normally fertile. Widely cited as Pouchet's "eighth law," the view was that "the menstrual flow in women corresponds to the phenomenon of excitement which is manifested during the rut *l'epoque des amours* in a variety of creatures and especially in mammals."[55] The American physician Augustus Gardiner drew out the implications of the menstruation-rut analogy less delicately: "The bitch in heat has the genitals tumefied and reddened, and a bloody discharge. The human female has nearly the same."[56]

With these interpretations of spontaneous ovulation, the old physiology of pleasure and the old anatomy of sexual homologies were definitely dead. The ovary, whose distinction from the male testes had only been recognized a century earlier, became the driving force of the whole female economy, with menstruation the outward sign of its awesome power. The engorged and finally burst follicle set in motion uterine carnage, with the external bleeding characteristic of such wounds. As the distinguished British gynecologist Matthews Duncan put it in an excessively rich image: "Menstruation is like the red flag outside an auction sale; it shows that something is going on inside."[57] And that something was not a pretty notion: the social characteristics of women seemed writ in blood and gore and cyclic rages scarcely containable by culture. The silent workings of a tiny organ weighing on the average seven grams in humans, some two to four centimeters long, and the swelling and subsequent rupture of the follicles within it came to represent what it was to be a woman.

But why would anyone believe this story, this culturally explosive fiction that menstruation was in women what heat was in dogs, when all the behavioral signs suggested nothing of the sort? Bischoff's answer was simple: the equivalence of menstruation and heat is simply common sense. If one accepts spontaneous ovulation during periods of heat in mammals generally, it "suggests itself." In any case, he adds, there is much indirect evidence for the equation of heat and menstruation, as well as the authority of the "most insightful physicians and naturalists" from the earliest times on.[58]

In fact the analogy was far from evident, there was little indirect evidence, and most of those from antiquity to Bischoff's day who gave their views on the subject denied its existence. Aristotle did equate the bleeding in animals and menstruation in women, but only because he thought that all blooded animals, male *and* female, produced residues—"the greatest of all in human beings"—from which both semen and the catamenia were concocted.[59] Pliny asserted flatly that a woman is "the only animal that has monthly periods."[60] Nothing new was said on the subject for almost two millennia, and when Haller took up the question in the 1750s, he was quite explicit on the point that while there are "some animals, who, at the time of their venal copulation, distil blood from their genitals," menstruation is peculiar "to the fair sex [of] the human species." Moreover, in contrast to bleeding in animals, menstruation in Haller's view was entirely independent of sexual desire. Intercourse did not increase or decrease the menstrual flow, nor did menstruation excite intercourse: women denied a heightened "desire of venery" during their periods and reported rather being "affected by pain and languor." Finally, sexual pleasure was localized "in the entrance of the pudendum" and not in the uterus, from which the menses flow.[61] J. F. Blumenbach, among the most widely reprinted and translated textwriters of the next generation, joined Pliny in arguing that only women menstruated, though he cautioned his readers that the investigation of the "periodical nature of

this hemorrhage is so difficult that we can obtain nothing beyond proba-
bility" and should thus be careful not to offer conjecture as fact.[62]

What scant facts there were seemed more anthropological than biological, and these came under heavy attack. In a masterful review of the literature up to 1843, Robert Remak, professor of neurology at Posen,
argued that even if one granted that all mammals have regularly recurring
periods of bleeding and that it originated in the uterus and not from the
turgescent external genitalia—neither concession was warranted by the
evidence—there remained "one further circumstance on which to ground
the most radical difference between menstruation and the periodical flow
of blood from the genitals of animals": the absence of marked periodicity
in the sexual desire of women in contrast to the beasts:

> In female animals the bleeding accompanies heat (*Brunst*), the period of the
> most heightened sexual drive, the only time the female will allow the male
> access, and the only time she will conceive. Quite to the contrary, in women
> the menstrual period is scarcely at all connected to increased sexual desire
> nor is fecundity limited to its duration; indeed a kind of instinct keeps men
> away from women during the menses—some savage people like certain Af-
> rican and American tribes isolate menstruating women in special quarters—
> and experience shows that there is no time during the inter-menstrual pe-
> riod when women can not conceive. It follows therefore that the animal
> heat is totally missing in women . . . Indeed the absence of menstruation in
> animals is one of the features that distinguish man from the beasts.[63]

Johannes Müller, in his 1843 textbook, came to similar conclusions. He
modestly pointed out that neither the purposes nor the causes of the
periodical return of the menses were known. Quite probably, however, it
existed to "*prevent* in the human female the periodical return of sexual
excitation (*Brunst*)" that occurs in animals.[64] Common sense, in short,
did not explain why nineteenth-century investigators would want to view
the reproductive cycle of women as precisely equivalent to that of other
animals.

Professional politics and the imperatives of a particular philosophy of
science perhaps offer a better answer. As Jean Borie points out, Pouchet
in 1874 pursued "une gynaecologie militante"; the same could be said of
many of his colleagues.[65] Their mission was to free women's bodies from
the stigma of clerical prejudice and centuries of popular superstition and,
in the process, to substitute the physician for the priest as the moral preceptor of society. (One might also want to argue that the insistence on

woman as a species of beast would have particular cachet in the context of French attacks on a church that was increasingly appealing to female piety.) At the heart of the matter lay the faith that reproduction, like nature's other mysteries, was in essence susceptible to rational analysis. Pouchet explicitly calls his readers' attention to the pristinely scientific, experimentally grounded character of his work and its avoidance of metaphysical, social, and religious concerns. In the absence of specific evidence of human ovulation, "logic" would dictate that women functioned no differently from the bitch, sow, or female rabbit, who in turn followed the same fundamental laws as mollusks, insects, fishes, or reptiles.[66] Thus there were considerable professional and philosophical attractions to the position that menstruation was like heat and that a sovereign organ, the ovary, ruled over the reproductive processes that made women what they were.

This radical naturalization, the reduction of women to the organ that now, for the first time, marked an incommensurable difference between the sexes and allegedly produced behavior of a kind not found in men, did not itself logically entail any particular position on the social or cultural place of women. What mattered was the mode of argument itself, the move from sex to gender, from body to behavior, from menstruation to morality. The actual content of purported sexual differences varied with the exigencies of the moment. Thus the equation of heat and menstruation could be the basis for a case against women's participation in public activities, which required steady, day-to-day concentration. Women were too bound to their bodies to take part in such endeavors. But the supposed equivalence of heat and menstruation could also be construed as evidence for women's superior capacity to transcend the body. Women could be the engineers of civilization precisely *because* every month they had to overcome the debilities of a brutish condition.

Arguing against those who held that the lack of animal-like lust or behavioral disturbances in women belied the new theory of spontaneous ovulation, one noted authority, G. F. Girdwood, draws attention to "the influence exercised by moral culture on the feelings and passions of humanity." Observe "the marvellous power exercised by civilization on the mind of her who, from her social position, is rendered the charm of man's existence." It is no wonder that the creature who can subjugate her own feelings, simulate good cheer when her heart is rent in agony, and in general give herself up to the good of the community can exercise control "the more energetically, at a time [menstruation] when she is taught that a stray thought of desire would be impurity, and its fruition pollution." But then, as if to back off from this model of woman as simultaneously a time bomb of sexuality and a living testimony to the power of civilization that keeps it from exploding, Girdwood concludes that "to aid her in her duty, nature has wisely provided her with the sexual appetite slightly developed."[67]

The indigestion of this passage, its sheer turning in on itself, bears witness to the extraordinary cultural burden that the physical nature of women—the menstrual cycle and the functions of the ovaries—came to bear in the nineteenth century. Whatever one thought about women and their rightful place in the world could, it seemed, be understood in terms of bodies endlessly open to the interpretive demands of culture.

All in all, the theory of the menstrual cycle dominant from the 1840s to the early twentieth century rather neatly integrated a particular set of real discoveries into an imagined biology of incommensurability. Menstruation, with its attendant aberrations, became a uniquely and distinguishingly female process. Precisely those phenomena whose absence spoke against its analogy to heat in animals now provided, through their veiled but nevertheless real presence in humans, the most compelling evidence for a view of women as radically different from men, their bodies and souls enslaved to a uniquely female cycle, both awesome and compelling. Behavior hidden in women, just as ovulation is hidden, could be made manifest by associating it with the more transparent behavior of animals. But as this story was being elaborated, very different social constructions were also being made. The body could mean almost anything and hence almost nothing at all.

Nowhere is such an ingenuous line of argument—that menstruation is as dangerous as heat because it seems so little to resemble it—more fully developed than in the massively learned and comprehensive synthesis by Adam Raciborski, a man Michelet hailed as a Prometheus who almost miraculously shed light on the darkly mysterious nature of women. The work's full title shows the extent of its claims: *Traité de la menstruation, ses rapports avec l'ovulation, la fécondation, l'hygiene de la puberté et l'age critique, son rôle dans les différentes maladies, ses troubles and leur traitment.* Here is the moral physiology of the Enlightenment brought to fruition, the physician firmly ensconced as its prophet. Early in the work, in a section on "The Physiology and Symptoms of Heat (*epoques de rut*)," Raciborski writes—without apparent motivation since this is not a book on veterinary medicine—about the mad behavior of dogs and cats during heat. Dogs, who in normal circumstances never leave their master's side for an instant, race off during rut to satisfy the instinct "which dominates all else." When they return home they seem excessively affectionate toward their owners, "visibly humbled, as if they had done something requiring pardon." Cats in heat race around the apartment, leap from one piece of furniture to another, lunge at windows without regard to the danger. If their venereal desires go unsatisfied, these behavioral aberrations repeat themselves "so to speak, indefinitely."[68]

But how is all of this relevant to menstruation in women, the ostensible subject of Raciborski's 631-page tome? Because, he says, it furnishes crucial new evidence for the analogous relationship between the menses and heat. The overt breakdown of socialization in animals, the collapse of the master-pet relationship, is hidden only by the precarious veneer of civilization in humans. "We will see that the turgescence—the crisis—of menstruation (*l'orgasme menstruel*) is one of the most powerful causes of nervous over-excitement in women." Many nervous afflictions begin precisely at the moment when the whole system is preparing itself for the onset of menstruation; others visibly worsen with the approach of each successive period; and still others recur only at these moments and cease during the intermenstrual interval. One must concede, Raciborski concludes, that "the swelling of ovulation (*l'orgasme de l'ovulation*) must be intimately linked to the [human] nervous system since similar disturbances arise in it and in that of animals."[69] The supposed, and to the untrained eye entirely hidden, behavioral perturbations of menstruating women, only hinted at in our species, turn out to resemble closely the quite visible aberrations of animals in heat. Animal madness, in other words, acts as a sort of magnifying glass for what women experience during menstruation and thereby provides further evidence for the analogy of the two processes. The interpretive loop goes round and round.

Similarly, with a not so subtle linguistic sleight of hand, the emotional baggage of periodically recurring animal lust and unbridled passion was dumped, for the first time in the nineteenth century, onto the bodies of women. The German word *Brunst,* for example, the period of heat in animals said to be equivalent to menstruation, hitherto used especially for the rutting season of the buck, bears brazen testimony to the enormous shift of sexual meanings underway in nineteenth-century writing. The term derives from the Old High German *Brunst* (a glowing ember or fire) and is related to the Gothic *Brunst* (a sacrifice conducted by fire). It has an old association with fire in *Feuerbrunst* (a large conflagration) and with affective perturbations through *Inbrunst* (a sort of mystical exaltation) and the Middle High German *Inbrunstig* (intense desire).

Brunst thus combines, as does the English word "heat," the sense of being physiologically hot—in the old model being ready to procreate, to concoct the seed—with the sense of violent action, intensity as in "the heat of battle," and the elemental power of fire. Thus female animals during breeding season and women during their monthly periods are both in a state of "burning" passion. The hero in Wagner's *Siegfried* is in "Brunst" after making his way without protective armor through the magic fire that guards Brunhilde: "Es braust mein Blut in bluhender Brunst; ein zehrendes Feuer is mir entzundet" (My blood rages in radiant passion; a consuming fire is ignited within me). If this sort of superexcitation is dangerous to a hero, it must be rather incapacitating in the ordinary woman, however much its most overt symptoms might be hidden during her reproductive cycles.

The English word *estrus* (also *oestrus, estrum*), especially in the adjectival form *estrous cycle* used in referring to the female of all higher animals, has an equally curious pedigree. It derives from the Latin *oestrus,* meaning literally a gadfly and figuratively a frenzy. The linguistic connection to the menstrual cycle is at first not apparent. There is a German close relative: Dr. Carl Franz Nägele argued that both the precursors and the accompanying conditions of the "oestrus venerus" of female animals bore certain similarities to the prodromata of menstruation, though he was loath to commit himself to the analogy so widely accepted after 1840.[70]

The connection of estrus specifically with sexual excitement is, however, somewhat more firmly established. Elliotson, in his 1828 English translation of Blumenbach, notes that "during the venereal *oestrum,*" in the throes of sexual passion, the Fallopian tubes become turgid and embrace the ovaries.[71] Bartholomew Parr's *London Medical Dictionary* (1819) in its entry "clitoris" gives as a synonym "oestrum veneris"; Joseph Thomas' American *Medical Dictionary* (1886) defines "orgasm" as "eager desire or excitement, especially venereal," and urges the reader to "see *Oestrum.*" According to the OED, in Billings' *Medical Dictionary* (1890) "oestrus" is defined *tout court* as "rut, orgasm, clitoris."

The final linguistic link between estrus as a moment of sexual frenzy, heat in animals, and menstruation in women comes in the last quarter of the nineteenth century. "The rutting, heat, oestrum, or venereal oestrum of animals," declared the veterinarian George Fleming in 1876, "is analogous to 'menstruation' in women." Then in 1900 Walter Heape, a Cambridge don and an immensely influential researcher on reproductive biology, not to mention a rabid antifeminist, brought estrus into regular use as describing the reproductive cycle of mammals, including humans: "The sexual season of *all* mammals is evidenced by . . . one oestrous cycle . . . or a series of oestrous cycles."[72] Heape realized that the swelling ovary did not actually cause menstruation, or vice versa, and that some exogenous agent caused the sexual cycle in animals, a "generative ferment" that, he admits quite ingenuously, he had wanted to call an "oestrus toxin," changing his mind only when he realized that there seems to exist a substance stimulating sexual activity in *men* and that there is no reason to assume the presence of a poison in his own sex.[73]

In stories like these, from the 1840s on, menstrual bleeding became the sign of a periodically swelling and ultimately exploding ovarian follicle whose behavioral manifestations I have described. But matters were even worse. What one saw on the outside was only part of the story; the histology of the uterine mucosa and of the ovaries revealed far more. Described in seemingly neutral scientific language, the cells of the endometrium or corpus luteum became re-presentations, rediscriptions of the social theory of sexual incommensurability. The militant Heape, for example, is absolutely clear on what he thinks of the female body in relation to the male. Though some of the differences between men and women are "infinitely subtle, hidden" and others are "glaring and forceful," the truth of the matter is that "the reproductive system is not only structurally but functionally fundamentally different in the Male and the Female; and since all other organs and systems of organs are affected by this system, it is certain that the Male and Female are essentially different throughout." They are, he continues, "complementary, in no sense the same, in no sense equal to one another; the accurate adjustment of society depends on proper observation of this fact."[74]

A major set of these facts, for Heape and many others, pertained to the uterus in relation to menstruation. At the time Heape wrote, the basic histology of menstruation—let alone its causes—was little understood. Earlier descriptions, as the young Viennese gynecologists Adler and Hitschmann noted in their classic 1908 paper, were demonstrably inadequate.[75] But the point here is not that so little was known about menstruation but rather that what was known became, through extraordinary leaps of the synecdochic imagination, the cellular correlative to the socially distinguishing characteristics of women.

Today the uterus is described as passing through three stages, rather colorlessly designated proliferative, secretory, and menstrual, the first two defined by the operative hormones, the last by the sloughing off of cells. In the nineteenth and early twentieth centuries it was said to proceed through a series of at least four and as many as eight stages, all defined by histology. Its normal stage was construed as "quiescence"; this was followed by "constructive" and "destructive" stages and a stage of "repair." Menstruation, as one might surmise, was defined as occurring at the destructive stage, when the uterus gave up its lining. As Heape puts it, in an account redolent of war reportage, the uterus during the formation of the menstrual clot is subject to "a severe, devastating, periodic action." The entire epithelium is torn away at each period, "leaving behind a ragged wreck of tissue, torn glands, ruptured vessels, jagged edges of stroma, and masses of blood corpuscles, which it would seem hardly possible to heal satisfactorily without the aid of surgical treatment."[76] Mercifully, this is followed by the recuperative stage and a return to normality. Little wonder that Havelock Ellis, steeped in this rhetoric, would conclude that women live on something of a biological roller coaster. They are "periodically wounded in the most sensitive spot in their organism and subjected to a monthly loss of blood." The cells of the uterus are in constant dramatic flux and subject to soul-wrenching trauma. Ellis concludes, after ten pages of still more data on the physiological and psychological periodicity in women, that the establishment "of these facts of morbid psychology, are very significant; they emphasize the fact that even in the healthiest woman a worm however harmless and unperceived, gnaws periodically at the roots of life."[77]

The gnawing worm is by no means the only image of pain and disease employed to interpret uterine or ovarian histology. One could put together an extraordinary chamber of narrative horrors about the lives of cells from the writings of leading nineteenth-century scientists and intellectuals. The bursting of the follicle is likened by Rudolf Virchow, the father of modern pathology, to teething, "accompanied with the liveliest disturbance of nutrition and nerve force."[78] For the historian Michelet, woman is a creature "wounded each month," who suffers almost constantly from the trauma of ovulation, which in turn is at the center of a physiological and psychological phantasmagoria dominating her life.[79] Less imaginatively, a French encyclopedia likens follicular rupture to "what happens at the rupture of an acute abscess."[80] The eminent physiologist E. F. W. Pflüger likens menstruation to surgical debridement, the creation of a clean surface in a wound, or alternatively to the notch used in grafting a branch onto a tree, to the "innoculationschnitt."[81]

Elie Metchnikoff, who won a Nobel Prize in 1908 for the discovery of phagocytosis, the process in which white blood cells ingest threatening bacteria, views the purported prevalence of such cells in menstrual blood as an indication of the presence in the uterus of noxious, proto-inflammatory material. Like firemen arriving after the blaze has already been put out, the leucocytes have been called to the scene for nought: the shedding of the uterine mucosa and the foul substances it contains, followed by a healing of the inner wound, gets rid of the materials that the phagocytic cells had been summoned to clear.[82] Such descriptions are legion, but it should be abundantly clear that imperatives of culture or the unconscious dictated the language of sex, of how the female body was defined and differentiated from the male's. Sex and sexual difference are not simply there, any more than gender is.

Though all of the evidence presented so far is by men and produced in a more or less antifeminist context, image making and the construction of the body through science occurs in feminist writers as well. Mary Putnam Jacobi's *The Question of Rest for Women During Menstruation* (1886), for example, is a sustained counterattack against the view that "the peculiar changes supposed to take place in the Graafian vesicles at each period . . . involve a peculiar expenditure of nervous force, which was so much dead loss to the individual life of the woman," thereby rendering women unfit for higher education, a variety of jobs, and other activities that demand large expenditures of the mental and physical energy that was, because of the supposed ovarian drain, in short supply. Since the "nervous force" was commonly associated in higher animals and in women with sexual arousal, Jacobi's task became one of severing the sexual from the reproductive life of women, of breaking the ties between the two postulated in the ovarian theory of Bischoff, Pouchet, Raciborski, and others.[83]

Much of her book is taken up with a compilation of the real or supposed empirical failings of this view. Neither menstruation nor pregnancy, she argues, is tied to the time of ovulation; indeed, as several hundred cases of vicarious menstruation in women suggest, menstruation itself is only statistically, not fundamentally, bound to ovulation and thus to reproduction. The amount of blood that flows to the uterus even in women who feel particular pelvic heaviness is but a tiny proportion of the body's blood, far less than the proportion of blood transferred to the stomach and intestines during the daily processes of digestion. So there is no evidence, Jacobi continues, that the uterus, ovaries, or their appendages become turgid during the menstrual period, and the effort to link a sort of histological tension of the reproductive organs to sexual tension, to the excitement of heat, is pointless. But though many of her criticisms are well taken, she neither offers a more compelling new theory of the physiology of ovulation nor gives a clearer picture of cellular changes in the uterine mucosa than do those she is arguing against.

Jacobi does, however, offer a new metaphor: "All the processes concerned in menstruation converge, not toward the sexual sphere, but the *nutritive,* or one department of it—the reproductive." The acceleration of blood flow to the uterus "in obedience to a *nutritive* demand" is precisely analogous to the "afflux of blood to the muscular layer of the stomach and intestines after a meal." In this debate Jacobi, like her opponents, tended to reduce woman's nature to woman's reproductive biology. But for her the essence of female sexual difference lay not in periodically recurring nervous excitement or in episodes of engorgement, rupture, and release of tension, but rather in the quiet process of nutrition. Far from being periodical, ovulation in Jacobi's account is essentially random: "The successive growth of the Graafian vesicles strictly resembles the successive growth of buds on a bough." (Here she might well be borrowing metaphors from studies of asexual reproduction in lower animals.) Buds, slowly opening into delicate cherry or apple blossoms and, if fertilized, into fruit, are a far cry from the wrenching and sexually intense swellings of the ovary imagined by the opposing theorists.[84]

Indeed, Jacobi's woman is in many respects the inverse of that of Pouchet, Raciborski, or Bischoff. For these men the theory of spontaneous ovulation demanded a woman shackled to her body, woman as nature, as physical being, even if the tamed quality of her modern European avatar spoke eloquently of the power of civilization. For Jacobi, on the other hand, biology provides the basis for a radical split between woman's mind and body, between sexuality and reproduction. The female body carries on its reproductive functions with no mental involvement; conversely, the mind can remain placidly above the body, free from its constraints. Jacobi's first effort at a metaphorical construction of this position uses fish whose ova are extruded without "sexual congress, and in a manner analogous to the process of defecation and micturation." In higher animals sexual congress is necessary for conception, but ovulation remains spontaneous and independent of excitement. From this it follows that "the superior contribution of the nutritive element of reproduction made by the female is balanced by an inferior dependence upon the animal or sexual element: in other words, she is sexually inferior."[85]

Of course, Jacobi cannot deny that in lower animals the female sexual instinct is tied exclusively to reproduction and that ruptured follicles are invariably found during heat. She nevertheless maintains that there is no proof of anything but a coincidental relationship between the state of the ovaries and the congested state of the external and internal genitalia that seems to signal sexual readiness. In women, she adamantly maintains, "the sexual instinct and reproductive capacity remain distinct; there is no longer any necessary association between sexual impulse, menstruation, and the dehiscence of ova." Indeed, her entire research program is devoted to showing that the menstrual cycle may be read as the ebb and flow of female nutritive rather than sexual activity, that its metabolic contours are precisely analogous to those of nutrition and growth. And this brings us back to the metaphor of the ovary as fruit blossom: "The woman buds as surely and as incessantly as the plant, continually generating not only the reproductive cell, but the nutritive material without which this would be useless." But how, given that women generally eat less than men, do they obtain a nutritive surplus? Because "it is the possibility of making this reserve which constitutes the *essential peculiarity* of the female sex."[86]

The point here is not to belittle Jacobi's scientific work but rather to emphasize the power of cultural imperatives, of metaphor, in the interpretation of the rather limited body of data available to reproductive biology during the late nineteenth century. At issue is not whether Jacobi was right in pointing out the lack of coincidence between ovulation and menstruation and wrong in concluding that there is no systematic connection between the two. It is rather that both she and her opponents emphasized some findings and rejected others largely on ideological grounds, seeing woman either as civilized animal or as mind presiding over a passive, nutritive body. But even the accumulation of fact, even the coherent and powerful modern paradigm of reproductive physiology in contemporary medical texts, offers but slight restraint on the poetics of sexual difference. The very subject seems to inflame the imagination. Thus, when W. F. Ganong's 1977 *Review of Medical Physiology*, a standard reference work for physicians and medical students, allows itself one moment of fancy, it is on the subject of women and the menstrual cycle. Amid a review of reproductive hormones, of the process of ovulation and menstruation described in the cold language of science, one is quite unexpectedly hit by a rhetorical bombshell, the only lyrical moment linking the reductionism of modern biological science to the experiences of humanity in 599 pages of compact, emotionally subdued prose: "Thus, to quote an old saying, 'Menstruation is the uterus crying for lack of a baby.'"[87] Cultural concerns have free license here, however cradled they may be in hard science. As in nineteenth-century texts, woman is viewed as the uterus, which in turn is endowed, through the familiar turn of the pathetic fallacy, with feelings, with the capacity to cry.

The menstrual cycle, even if it did not for all times and places differentiate women from men, nevertheless was the prism through which modern sexual difference could be historically understood. Rousseau, as I pointed out in Chapter 5, argued against Hobbes that one could not draw any inferences about humans in the state of nature from the fights that take place among certain animals for the possession of the female. Among humans, there are always enough females to go around, since they are never physiologically unavailable and since sex ratios are roughly equal: a peaceable kingdom of sexual plenty. Pufendorf drew precisely the opposite conclusions from the human female's constant availability and argued that the condition required legal regulation.

This sort of thinking, from widely varying perspectives, has a continuous tradition up to the present. Edward Westermarck, a major late nineteenth-century anthropologist, used the vast new ethnographic literature, generated in part of course by political pressures for a natural history of sexual differences, to make menstruation and constant female desire a product and not a cause of civilization. His interest in the subject is generated by disputes with cultural anthropologists like Morgan or Bachofen, who regarded human marriage as a response to primitive promiscuity, and his strategy is to present vast quantities of "evidence" for the primitive seasonality of female desire: the Amazons, according to Strabo, lived ten months without the company of men and then every spring descended to breed with the males of a neighboring tribe; the Indians of California, belonging "to the lowest of races on earth," have, according to Westermarck's source, "their rutting seasons as regularly as have the deer, the elk, the antelope, or any other animal"; the Australian aborigine, "like the beasts in the field . . . has but one time for copulation in the year"; female domestic animals breed more often than those in the wild. From all this it followed, according to Westermarck, that the less civilized a female creature, the less sexually active it was. Therefore "it must be admitted that the continued excitement of the sexual instinct could not have played a part in the origin of human marriage."[88] Conversely, the sort of sustained desire permitted by a monthly menses, and thus female desire generally, is not natural but itself generated by culture.

Elizabeth Wolstenholme in 1893 gives an extraordinary and angry account of menstruation as a sign of male oppression, fixed in the female body by the inheritance of acquired characteristics:

> For carnal servitude left cruel stain,
> And galls that fester from the fleshly chain;
> Unhealed the scars of man's distempered greed,
> The wounds of blind injustice still they bleed . . .
> Her girlhood's helpless years through cycles long
> Had been a martyrdom of sexual wrong . . .
> Action repeated tends to rhythmic course,
> And thus the mischief, due at first to force,
> Brought cumulative sequences to the race,
> Till habit bred hereditary trace.

These traces, the monthly cycle, are then "misread by man, the sign of his misdeed . . . as symptom of her nubile need"; menstruation, "not more to woman natural than to brute," will end when women take control of their bodies.[89]

A tremendous amount seems to be at stake in the nature of woman's reproductive cycle and its relationship to desire. The problem is less the role of the sexual impulse in human life generally than it is in the life of women. Whereas the male sexual impulse, in Havelock Ellis' words, is open, aggressive, unproblematic, in a woman we encounter "elusiveness," a "mocking mystery." He and others have explored the subject now for two centuries; various just-so stories were and are still told about the "peculiarities" of women. The nature and even the existence of cycles of sexual interest in relation to the menstrual cycle continue to be explored.[90] The puzzle, posed once it was definitively established that menstruation was not heat and that ovulation in women was indeed hidden, generated a new batch of tales that are close relatives of the nineteenth-century narratives, though based on a different set of biological beliefs. (I mean here stories like the one that says ovulation is hidden so as to keep females from knowing when they are fertile. Were they to know, far from desiring motherhood, they would shun intercourse to avoid its dangers.[91])

The solitary vice, the social evil, and pouring tea

Wolstenholme and Westermarck wrote as if the body were only the sign of social practices, not their foundation: menstruation was not the cause of a peculiar female way of being in the world, as it had been for the doctors cited earlier; it was the consequence. Already the epistemological sands of the two-sex model have shifted so that culture and the body are no longer distinct and isolated categories. But still the focus there is on a peculiarly female function. Here I want to see how two human activities, masturbation and prostitution, may be regarded as social perversions visited upon the body rather than as sexual perversions with social effects.

It is often thought that the eighteenth- and nineteenth-century obsession with masturbation and prostitution are part of a new literature "dominated by a tone of total and repressive sexual intolerance."[92] I want to argue instead that the "*solitary* vice" and the "*social* evil" were believed to be, as their new names imply, social pathologies that visited destruction on the body in the same way that in ages past blasphemy or lechery produced monsters. The insane, pale, quivering masturbator and the coarse, barren prostitute were the miscreants of the modern age produced, as had been their deformed predecessors, by a moral sickness.

As a very one-sex sort of activity, masturbation was also a one-sex vice. Although nineteenth-century worries about masturbatory perturbations have been given special attention by historians, the underlying pathogenesis of masturbatory disease in both sexes was thought to be the same: excessive and socially perverted nervous stimulation. Hence the supposed connection of tuberculosis and masturbation: "Let it be known that pulmonary consumption, whose horrible ravages in Europe ought to give alarm to all governments, has drawn from this very source [masturbation] its fatal activity."[93] "Girls hide most of the ravages of the vice under 'general nervous excitement'; boys have not this convenient refuge."[94]

One need also only read the various editions and translations of *Onania, or the Heinous Sin of Self-Pollution* or the Swiss Dr. Tissot's *L'Onanisme* or their many imitators—R. L. Perry's *The Silent: A Medical Work on the Dangerous Effects of Onanism,* for example—to know that Foucault was right: here is a literature that generates erotic desire in order to control it.[95] Story after story of young men and women discovering in their genitals the pleasures of solitary sex form a vast corpus of incendiary porn whose erotogenic power is not diminished by the obligatory horrifying, cautionary end.

Rousseau, who thought deeply about sexual desire and the making of the social order, condemned masturbation severely, as a social wrong. In *Emile* he cautions against it because it might substitute for marriage; in the *Confessions* he says he permits himself the practice because his involvement with Thérèse represented the undifferentiated desire of the state of nature (it was not "moral") while masturbation was the product of his own "lively imagination," a sort of moral self-love.[96]

Although in traditional church teaching fornication was thought far worse than onanism, in the post-eighteenth-century world the "crime of solitude" was thought to "undermine the constitution and poison the mind ten times more than illicit commerce with a woman."[97] An advertising booklet that must have circulated in the tens if not hundreds of thousands in the nineteenth century cautions that indulging in the passions during youth "in a manner which is contrary to nature" is the road to ruin, and then goes on to lament that these practices arise only because of the "rigid custom" that allows unmarried females to indulge "in the natural gratification of the master-passion" only at the cost of total loss of reputation.[98] (A less commercially interested authority, R. D. Owen, son of the utopian socialist Robert Owen, makes the same point when he argues that the origins of onanism were probably in the convents of Europe while its growing popularity in the nineteenth century was the result of the continued "un-natural separation of the sexes."[99]) "Immoderate use of enjoyment, even in a natural way," is debilitating, warns a nineteenth-century doctor, echoing ancient lore. But speaking as a modern man he asks: "What must then be the consequences when nature is forced [through self-abuse] against her will?"[100] The real trouble with masturbation in these dire warnings is not that it robs the body of precious fluids but that it violates Aristotle's dictum, given new life during the industrial revolution by fears that it might not be true, that man is a social animal.

The political and sexual radical Richard Carlile (1790–1843) makes the best argument for how masturbation must be construed as a threat to "the nature of human solidarity," and how little it appears to be a problem of excess or wicked sexual desire. Sociability, not repression, is at stake. Carlile's *Every Woman's Book* is a sustained attack on conventional sexual morality, a plea for freeing the passions, and a practical guide to birth control. Love is natural, only its fruits should be controlled, marriage laws constrain a passion that should not be shackled, and so on. Carlile advocates Temples of Venus for the controlled, healthy, extramarital satisfaction of female desire—five sixths of the deaths from consumption among young women resulted from want of sexual commerce, he thought, and perhaps as much as nine tenths of all other illness as well. But on the subject of masturbation, Carlile the sexual radical is as shrill as the most evangelically inspired moralist or alarmist physician. Born of the cloister or its modern equivalents, where diseased religion turns love into sin, "the appeasing of lascivious excitement in females by artificial means" or the "accomplishment of seminal excretion in the male" is not only wicked but physically destructive. Masturbation leads to disease of mind and body. Indeed, the "natural and healthy commerce between the sexes" for which he offers the technology is explicitly linked to the abolition of prostitution, masturbation, pederasty, and other unnatural practices.[101]

The contrast could not be clearer between a fundamentally asocial or socially degenerative practice—the pathogenic, solitary sex of the cloister—and the vital, socially constructive act of heterosexual intercourse. But the supposed physical effects of masturbation seem almost a secondary reaction to its underlying social pathology. The emphasis in the solitary vice should perhaps be less on "vice," understood as the fulfillment of illegitimate desire, than on "solitary," the channeling of healthy desire back into itself. The debate over masturbation that raged from the eighteenth century on might therefore be understood as part of the more general debate about the unleashing of desire in a commercial economy and about the possibility of human community in these circumstances— a sexual version of the classic "Adam Smith Problem."[102] And, as in the one-sex model, violating the social norm had horrible physical consequences as well. The monster born to colonial Anne Hutchinson's follower lives on in the suicidal masturbator whose faculties are greatly impaired, whose thinking is impractical, memory weak, and body reduced to skin and bones. But even if not a complete wreck, the masturbator will never find comfort in married love and thus contributes to the social monstrosity of sterility.[103]

Prostitution is the other great arena in which the battle against unsocialized sex was fought. Here too society and the body are intertwined. Whoring, of course, had long been regarded as wicked and detrimental to the commonweal, but so had drunkenness, blasphemy, and other disturbances of the peace. Not until the nineteenth century did it come to be *the* social evil, a particularly disruptive, singularly threatening vice. How this happened is a long story, and I will tell only part of it.

Prostitutes were generally regarded as an unproductive commodity. Because they were public women; because their reproductive organs bore such heavy traffic; because in them the semen of so many different men was mixed, pell-mell, together; because the ovaries of prostitutes, through overstimulation, were seldom without morbid lesions; because their Fallopian tubes were closed by excessive intercourse; or, most tellingly, because they did not feel affection for the men with whom they had sex, they were thought to be barren, or in any case unlikely to have children. One writer went so far as to argue that when prostitutes did become pregnant it was by men they especially liked; and when prostitutes who had been transported to Van Dieman's Land reformed and set up domestic situations, they suddenly found themselves fertile.[104]

Of course not every expert would agree. Indeed, Jean-Baptiste Parent-Duchâtelet, a genuinely gifted nineteenth-century specialist in public health, insisted that there was nothing physically unusual about prostitutes. They did not have unusually large clitorises—only three out of six thousand did—and were therefore not attracted to prostitution by excessive sexual desire; if they had fewer children, it was because they practiced abortion or birth control. Prostitution, he argued, is not inherent in bodies; in its modern form it is purely a pathology of commercial urban society. But in disagreeing with the general wisdom, Parent-Duchâtelet is allying himself with what I take to be the main interpretive thrust of the idea of the barren prostitute: a confusion between the dangerously asocial world of commercial exchange and the healthy social world of married love.[105]

To get at this, let me go back to the high Middle Ages when the observation that prostitutes are barren first appears. Aristotle, among others, had pointed out that the womb of a woman who was too hot—and the lascivious nature of prostitutes suggested this excess of *calor genitalis*—might well be inhospitable to conception: it might burn up the conjoined seeds. But Aristotle did not actually equate prostitution with excess heat. Lucretius points out that prostitutes use lascivious movements that inhibit conception by diverting "the furrow from the straight course of the plowshare and make[ing] the seed fall wide of the plot." But this observation is in the course of a discussion of why "obviously our wives can have no use" for such twists and turns.[106]

The reasons given in late medieval and Renaissance literature for the barrenness of prostitutes are several: excess heat, a womb too moist and slippery to retain the seed, and the mingling of various seeds, reasons very much like those given by nineteenth-century doctors. But I want to draw attention to a less explicitly physiological explanation, which links the problem of barrenness with a more general derangement of the body politic. A twelfth-century encyclopedist, William of Conches, explains why prostitutes rarely conceive. Two seeds are necessary for conception, he reminds his readers, and prostitutes "who only perform coition for money and who because of this fact feel no pleasure, emit nothing and therefore engender nothing." A sixteenth-century German doctor makes a similar argument. Among the causes of barrenness, Lorenz Fries notes, is "a woman's lack of passion for a man as, for example, the common women (*gemeynen Frawlin*) who work only for their sustenance." One might construe "common women" to mean not prostitutes but peasants who work *only* to earn their keep rather than, as Luther would have preached, for the greater glory of God. This would fit in with the analogies made by Calvin and others between sexual heat or passion and the ardor the heart ought to feel for God. It also fits in with the fact that Fries was a teacher at the new Protestant university in Strasbourg.[107]

Here is yet another version of the old saw that orgasm is necessary for conception. But why do prostitutes not experience pleasure, and why are "common women" chosen to illustrate the point that an absence of passion ensures sterility? The friction of intercourse must be as warming in harlots as in other women, but their bodies respond differently. In the examples I have cited, money, or more precisely an illegitimate exchange of money, provides the missing middle term. Prostitution is sterile because the mode of exchange it represents is sterile. Nothing is produced because, like usury, it is pure exchange. As R. Howard Bloch argues, it was precisely in the twelfth century and in response to a nascent market economy that usury became of urgent concern to the church. And the particular wickedness of charging interest, it was held, is that nothing real is gained by it. Indeed, as Aristotle argues, usury is "the most hated sort" of exchange and is to be particularly censured because it represents the antithesis of the natural, the productive, household economy. A perverted economic practice, like perverted sex, breeds abominations or nothing: "Interest, which means the birth of money from money, is applied to the breeding of money because the offspring resembles the parent. That is why of all modes of getting wealth this is the most unnatural" (*Politics*, 1.10.1258b5–7). It is as if usury were incestuous intercourse. In Catherine Gallagher's terms, "what multiplies through her [the prostitute] is not a substance but a sign: money." (I have in a sense been arguing that this distinction between sign and substance is untenable in dealing with the history of the body.) Prostitution becomes, like usury, a metaphor for the unnatural multiplication not of things but of signs without referents.[108]

A deep cultural unease about money and the market economy is couched in the metaphors of reproductive biology; this is in Aristotle's formulation. But, more to the point here, fear of an asocial market takes on a new guise in the claim that sex for money, coition with prostitutes, bears no fruit. This sort of sex is set in sharp contrast—one senses this especially in the German example—to the household economy of sex, which is quintessentially social and productive. Fries elsewhere in the text cited develops the metaphor of the womb protecting the fetus just as the crust of bread protects the crumbs. The image of baking, warmth, and kitchen contrasts with the cool barrenness of those who work, have intercourse, *only* for pay, outside the bounds of the household.

By the nineteenth century, the trope of the barren prostitute had a respectable seven-century pedigree. But the boundaries it guarded—between home and economy, private and public, self and society—were both more sharply drawn and more problematic in the urban-class society of Europe after the industrial revolution. Or at least so thought contemporary observers. Society seemed to be in unprecedented danger from the market place; the sexual body reflected all the anxieties of this danger; and, in this new version of the one-sex model, cultural meaning caused the flesh to bend to its dictates.

The problem with both masturbation and prostitution was essentially quantitative: doing it alone and doing it with lots of people rather than doing it in pairs. Such sex is thus in the same category as other misdeeds of number, the withdrawal of the protagonist of Florence Nightingale's *Cassandra,* for example, who refuses to pour tea for the household and withdraws to her solitary couch. The social context, not the act, determines acceptability. The paradoxes of commercial society that had already plagued Adam Smith and his colleagues, the nagging doubts that a free economy might not sustain the social body, haunt the sexual body. Or, the other way around, the perverted sexual body haunts society and reminds it of its fragility, as it had done in other ways for millennia.

Freud's problem

Freud's account of how the clitoral sexuality of young girls gives way to the vaginal sexuality of mature women powerfully focuses on the issues of my book. On the one hand, Freud is very much a man of the Enlightenment, inheritor of its model of sexual difference. Anatomy is destiny, as he said in a phrase he did not really mean; the vagina is the opposite of the penis, an anatomical marker of woman's lack of what a man has. Heterosexuality is the natural state of the architecture of two incommensurable opposite sexes. But Freud, more than any other thinker, also collapses the model. Libido knows no sex. The clitoris is a version of the male organ—why not the other way around?—and only by postulating a sort of generalized female hysteria, a disease in which culture takes the causative role of organs, does Freud account for how it supposedly gives up its role in women's sexual lives in favor of the "opposite organ," the vagina. Here, in other words, is a version of the central modern narrative of one sex at war with two.

The story begins in 1905 when Freud rediscovered the clitoris, or in any case clitoral orgasm, by inventing its vaginal counterpart. (Recall Renaldus Columbus' prior sixteenth-century claim.) After four hundred, perhaps even two thousand, years there was all of a sudden a second place from which women derived sexual pleasure. In 1905, for the first time, a doctor claimed that there were two kinds of orgasm and that the vaginal sort was the expected norm among adult women. This generated an immense polemical and clinical literature. More words have been shed, I suspect, about the clitoris than about any other organ, or at least about any organ its size.[109]

I want to make two points in particular. In the first place, before 1905 no one thought that there was any other kind of female orgasm than the clitoral sort. It is well and accurately described in hundreds of learned and popular medical texts, as well as in a burgeoning pornographic literature. It simply is not true, as Robert Scholes has argued, that there has been "a semiotic coding that operates to purge both texts and language of things [the clitoris as the primary organ of woman's sexual pleasure] that are unwelcome to men." The clitoris, like the penis, was for two millennia both "precious jewel" and sexual organ, a connection not "lost or mislaid" through the ages, as Scholes would have it, but only (if then) since Freud.[110] To put it differently, the revelation by Masters and Johnson that female orgasm is almost entirely clitoral would have been a commonplace to every seventeenth-century midwife and had been documented in considerable detail by nineteenth-century investigators. A great wave of amnesia descended on scientific circles around 1900, and hoary truths were hailed as earth-shattering in the second half of the twentieth century.

My second point, more central to the concerns of this book, is that there is nothing in nature about how the clitoris is construed. It is not self-evidently a female penis, and it is not self-evidently in opposition to the vagina. Nor have men always regarded clitoral orgasm as absent, threatening, or unspeakable because of some primordial male fear of, or fascination with, female sexual pleasure. The history of the clitoris is part of the history of sexual difference generally and of the socialization of the body's pleasures. Like the history of masturbation, it is a story as much about sociability as about sex. And once again, for the last time in this book, it is the story of the aporia of anatomy.

"If we are to understand how a little girl turns into a woman," Freud says in the third of his epochal *Three Essays on the Theory of Sexuality*, "we must follow the further vicissitudes of [the] excitability of the clitoris." During puberty, the story goes, there occurs in boys "an accession of libido," while in girls there is "a fresh wave of repression in which it is precisely cliteroidal sexuality that is affected." The development of women as cultural beings is thus marked by what seems to be a physiological process: "what is overtaken by repression is a piece of masculine machinery."[111]

Like a Bahktiari tribesman in search of fresh pastures, female sexuality is said to migrate from one place to another, from the malelike clitoris to the unmistakably female vagina. The clitoris does not, however, entirely lose its function as a result of pleasure's short but significant journey.

Instead it becomes the organ *through which* excitement is transmitted to the "adjacent female sexual parts," to its permanent home, the true locus of a woman's erotic life, the vagina. The clitoris, in Freud's less than illuminating simile, becomes "like pine shavings" used "to set a log of harder wood on fire."

This strangely inappropriate identification of the cavity of the vagina with a burning log is not my concern here. Stranger still is what happens to biology in Freud's famous essay. A little girl's realization that she does not have a penis and that therefore her sexuality resides in its supposed opposite, in the cavity of the vagina, elevates a "biological fact" into a cultural desideratum. Freud writes as if he has discovered the basis in anatomy for the entire nineteenth-century world of gender. In an age obsessed with being able to justify and distinguish the social roles of women and men, science seems to have found in the radical difference of penis and vagina not just a sign of sexual difference but its very foundation. When erotogenic susceptibility to stimulation has been successfully transferred by a woman from the clitoris to the vaginal orifice, she has adopted a new leading zone for the purposes of her later sexual activity.

Freud goes even further by suggesting that the repression of female sexuality in puberty, marked by abandonment of the clitoris, heightens male desire and thus tightens the web of heterosexual union on which reproduction, the family, and indeed civilization itself appear to rest: "The intensification of the brake upon sexuality brought about by pubertal repression in women serves as a stimulus to the libido of men and causes an increase in its activity."[112] When everything has settled down, the "masculine machinery" of the clitoris is abandoned, the vagina is erotically charged, and the body is set for reproductive intercourse. Freud seems to be taking a stab at historical bio-anthropology, claiming that female modesty incites male desire while female acquiescence, in allowing it to be gratified, leads humanity out of the savage's cave.

Perhaps this is pushing one paragraph too hard, but Freud in these passages is very much in the imaginative footsteps of Diderot and Rousseau, who argued that civilization began when woman began to discriminate, to limit her availability. Freud in the *Three Essays* is not quite so explicit, but he does appear to be arguing that femininity, and thus the place of women in society, is grounded in the developmental neurology of the female genitals.

But could he really have meant this? In the first place, the long written history of the body would have shown that the vagina fails miserably as a "natural symbol" of interior sexuality, of passivity, of the private against the public, of a critical stage in the ontogeny of woman. In the one-sex model, dominant in anatomical thinking for two thousand years, woman was understood as man inverted: the uterus was the female scrotum, the ovaries were testicles, the vulva was a foreskin, and *the vagina was a penis*. This account of sexual difference, though as phallocentric as Freud's, offered no real female interior, only the displacement inward to a more sheltered space of the male organs, as if the scrotum and penis in the form of uterus and vagina had taken cover from the cold.

If Freud was not aware of this history, he surely must have known that there was absolutely no anatomical or physiological evidence for the claim that "erotogenic susceptibility to stimulation" is successfully transferred during the maturation of women "from the clitoris to the vaginal orifice." The abundance of specialized nerve endings in the clitoris and the relative impoverishment of the vagina had been demonstrated half a century before Freud wrote and had been known in outline for hundreds of years. Common medical knowledge available in any nineteenth-century handbook thus makes Freud's story a puzzle, if it is construed as a narrative of biology. Finally, if the advent of the vaginal orgasm were the consequence of neurological processes, then Freud's question of "how a woman develops out of a child with bisexual dispositions" could be resolved by physiology without any help from psychoanalysis.

Freud's answer, then, must be regarded as a narrative of culture in anatomical disguise. The tale of the clitoris is a parable of culture, of how the body is forged into a shape valuable to civilization despite, not because of, itself. The language of biology gives this tale its rhetorical authority but does not describe a deeper reality in nerves and flesh.

Freud, in short, must have known that he was inventing vaginal orgasm and that he was at the same time giving a radical new meaning to the clitoris. Richard von Krafft-Ebing may have anticipated him a bit when in the 1890s he wrote that "the erogenous zones in women are, while she is a virgin, the clitoris, and, after defloration, the vagina and cervix uteri." But this is in the context of a discussion of a variety of erogenous zones; immediately following is the observation that "the nipple particularly seems to possess this [erogenous] quality." Krafft-Ebing, like many of his contemporaries, believed that the "normally developed mentally and well bred" woman's sexual desires were small. He also regarded woman's supposed sexual passivity (a symbol for her passivity in public life) as imbedded in "her sexual organization."[113]

But neither he nor anyone else drew social consequences from the distinction between vaginal and clitoral eroticism. There was, in fact, no evidence at all in the contemporary literature for the sort of vaginal sexuality Freud postulates. Nor was there any special interest in denying it. The stark contrasts we shall see below are the result of a historical juxtaposition of texts. Authorities in French, German, and English during Freud's time, and stretching back to the early seventeenth century, were unanimous in holding that female sexual pleasure originated in the structures of the vulva generally and in the clitoris specifically. No alternative sites were proposed.

The major English-language medical encyclopedia of Freud's day begins the "clitoris" subheading of a lengthy and up-to-date entry on "Sexual Organs, Female" by citing the Viennese anatomist and philologist Joseph Hyrtl, who derived the word "clitoris" from a Greek verb meaning "to titillate" and observed that these etymological roots are reflected in the German colloquial term *Kitzler* (tickler).[114] Its anatomy is presented as the homologue of the penis, although the clitoris' nervous supply is "far greater, in proportion to its size." Indeed,

> its cutaneous investment is supplied with special nerve endings, which give it remarkable and special sensitivity . . . At the base of the papillae are the endings which Krause believes to be related to the peculiar sensibility of the organ and has named corpuscles of sexual pleasure (*Wollustkörperchen*). They are usually called genital corpuscles.[115]

On the other hand, the upper and middle portions of the vagina are enervated by "the same sources as the uterus." It is "not very sensitive," and indeed the anterior wall is so insensitive that it "can be operated on without much pain to the patient." This may be hyperbole, but it suggests that to nineteenth-century authorities the vagina was an unlikely candidate for the primary locus of sexual pleasure in women.

No one took it to be such. Freud's contemporary, the gynecologist E. H. Kisch, for example, cites Victor Hensen's article on the physiology of reproduction in the authoritative *Handbuch der Physiologie* (1881) to the effect that direct stimulation of sexual feeling is through the dorsal nerve of the penis and the clitoris. Kisch then notes that sexual pleasure in women is due chiefly to friction on the clitoris through the intromitted penis that stimulates the nerve fibers connected to Krause's genital ("voluptuary") corpuscles.[116] The major French medical reference work of the late nineteenth century describes the clitoris as an erectile organ situated at the upper end of the vulva which has the same structure as the corpus cavernosum of the penis, the same erotic functions, but lacks a urethra. The vagina, on the other hand, is defined simply as the passage from the vulva to the uterus which serves to evacuate the menses, contain the male organ during copulation, and expel the product of fecundation. Most of the article is devoted to its pathologies.[117]

As early as 1844, with the publication of Georg Ludwig Kobelt's massively documented *The Male and Female Organs of Sexual Arousal in Man and Some Other Mammals,*[118] the anatomy of genital pleasure was firmly established. Kobelt, first of all, devised a technique for injecting the vasculature of the clitoris so that an organ notoriously difficult to study in post-mortem material could be readily examined. He then proceeded to describe its structure and function in exquisite detail and concluded, based on the clitoris' erectile tissues and its blood and nerve supply, that the glans cliteroides was the primary locus of sexual arousal in both humans and other mammals; it was the precise homologue of the male organ, the glans penis. (Kobelt distinguished the passive male and female organs, or the glans of the penis and clitoris, from the active organs, or the shafts of these structures.) The function of all this machinery, according to Kobelt, is to provide sexual pleasure, which will make women want to have intercourse despite the dangers of pregnancy and the trials of motherhood.[119] Its physiology is described in clinical detail. When outside stimuli

> come into contact with the glans of the clitoris, then the blood which is causing the *bulbus* to swell, by way of the reflex spasms of the *musculus constrictor cunni,* is propelled through the exposed *pars intermedia* into the glans, now ready for the stimulus; and thereby the purpose of the entire passive apparatus (the sensation of sexual pleasure) is achieved. The sexually pleasurable titillation increases with continuing stimulation up to its final transformation into indifference [orgasm] and return to the usual quiescent state of the affected parts. The process is further supported by the same sort of auxiliary means as in the male.

The vagina, Kobelt thinks, is so well known that it warrants no extended description. But he nevertheless pauses to point out that it plays a minimal role in genital orgasm: "The small number of nerves which, singly, make their way down into the voluminous vaginal tube puts the vagina so far behind the glans—small but very rich in nerves—that we can grant the vagina no part in the creation of the specific pleasurable sex feelings in the female body."[120]

Kobelt's book was by far the most detailed account of the clitoris ever published, but it did not radically revise established views. An earlier French medical encyclopedia came to roughly the same conclusions. "Clitoris," it says, derives from the Greek verb *kleitoriazein,* meaning to touch or titillate lasciviously, to be inclined to pleasure. A synonym is "oestrus veneris," a frenzy of sexual passion. The clitoris is like the penis in form and structure and "enjoys an exquisite sensibility," which makes it highly susceptible to "abuse." The author of this entry disapproves strongly of titillating the clitoris, as some colleagues recommend, to cure certain nervous disorders like catalepsy. (Although unacknowledged, this was a therapy derived from a famous case of Galen's in which a widow, laboring under a purported backup of "semen," suffered from backaches and other pains until the pressure was relieved by a midwife who rubbed her genitals.) A subsequent entry on "clitorisme," the female equivalent of masturbation, discusses further abuses invited by this site of pleasure.[121]

In the "vagin" entry, on the other hand, the subject is defined as the "cylindrical and elastic passage from the uterus to the external parts." There follows a short discussion of nomenclature which warns against confusing the vagina with the cervix, the part that used to be called "the neck of the womb," but there is no discussion of its innervation or erotic functions.[122]

These articles from the nineteenth century refer back in turn to a seventeenth-century text by François Mauriceau, one of the luminaries of French obstetrics. He notes that the clitoris is "where the Author of Nature has placed the seat of voluptuousness—as He has in the glans penis—where the most exquisite sensibility is located, and where he placed the origins of lasciviousness in women." Indeed, the pudendum more generally has the capacity to engender delight because the nerves that supply the clitoris supply it too. Mauriceau, after describing for almost six pages the clitoris' muscles, nerves, and vasculature, concludes that it functions just like the penis.[123]

The vagina is a far duller organ. It is the tube leading from uterus to the outside, "a slack canal (*mol & lache*) which during coition embraces the penis." Only the glands near its outer end are relevant to sexual pleasure because they pour out great quantities of a saline liquor during coition, which increases the heat and enjoyment of women. These are the substances, Mauriceau suggests, to which Galen was referring when he spoke of needing to use other means to cause their release when the caresses of a man were not available. And this takes the history of the clitoris back to where I left it earlier. In 1612 Jacques Duval wrote: "In French it is called temptation, the spur to sensual pleasure, the female rod and the scorner of men: and women who will admit their lewdness call it their *gaude mihi* [great joy]."[124]

The French physician echoes the certainties and tensions of later as well as earlier accounts. On the one hand, the clitoris is the organ of sexual pleasure in women. On the other, its easy responsiveness to touch makes it difficult to domesticate for reproductive, heterosexual intercourse. This was Freud's problem, and I will now return to it.

Although Freud may not have been aware of all the detailed history of genital anatomy I have just recounted, it is impossible that he would not have been familiar with what was in the standard reference books of his day. He was, after all, especially interested in zoology during his medical-student days and was an expert neurologist. Furthermore, one did not have to be a scientist to know about clitoral sexuality. Walter, protagonist of the notorious *My Secret Life,* notes in his review of the copulative organs that the clitoris is an erectile organ which is "the chief seat of pleasure in a woman." Probably thousands of tracts about masturbation proclaimed its sensitivity. And of course Freud himself points out that biology has been "obliged to recognize the female clitoris as a true substitute for the penis," though it does not follow from this that children recognize that "all human beings have the same (male) form of genital" or that little girls therefore suffer penis envy because their genital is so small.[125]

Freud, in short, must have known that what he wrote in the language of biology regarding the shift of erotogenic sensibility from the clitoris to the vagina had no basis in the facts of anatomy or physiology. Both the migration of female sexuality and the opposition between the vagina and penis must therefore be understood as re-presentations of a social ideal in yet another form. On a formal level, the opposition of the vagina and penis represents an ideal of parity. The social thuggery that takes a polymorphously perverse infant and bullies it into a heterosexual man or woman finds an organic correlative in the body, in the opposition of the sexes and their organs. Perhaps because Freud is the great theorist of sexual ambiguity, he is also the inventor of a dramatic sexual antithesis: between the embarrassing clitoris that girls desert and the vagina whose erotogenic powers they embrace as mature women.[126]

More generally, what might loosely be called patriarchy may have appeared to Freud as the only possible way to organize the relations between the sexes, leading him to write as if its signs in the body, external active penis versus internal passive vagina, were "natural." But in Freud's question of how it is that "a woman develops out of a child with a bisexual disposition," the word "woman" clearly refers not to natural sex but to theatrical gender, to socially defined roles. The supposed opposition of men and women, "exclusive gender identity," in Gayle Rubin's terms, "far from being an expression of natural differences . . . is the suppression of natural similarities."[127] In *Civilization and Its Discontents* Freud seems poignantly aware of the painful processes through which body parts are sorted out and come to represent the most telling of differences. Civilization, like a conquering people, subjects others to its "exploitation," proscribes "manifestations of sexual life in children," makes "heterosexual genital love" the only permitted sort, and in so doing takes the infant, "an animal organism with (like others) an unmistakably bisexual disposition" and molds it into *either* a man or a woman.[128] The power of culture thus represents itself in bodies, forges them, as on an anvil, into the required shape. What Rosalind Coward has called in another context "ideologies of appropriate desires and orientations" must struggle—one hopes unsuccessfully—to find their signs in the flesh.[129] Freud's argument, flying as it does in the face of centuries of anatomical knowledge, is a testament to the freedom with which the authority of nature can be rhetorically appropriated to legitimize the creations of culture.

It is, however, an argument that works on its own terms and thereby illustrates just how powerfully culture operates on the body. In the first place, Freud remained a Lamarckian all his life. He believed in the inheritance of acquired characteristics, which he generalized to include traits of the psyche—aggressions and need, for example. Need, he wrote to his colleague Karl Abraham, is nothing other than the "power of unconscious ideas over one's own body, of which we see remnants in hysteria, in short, 'the omnipotence of thought.' "[130]

Hysteria is the model for mind over matter. The hysteric, like the patient who feels pain or itches in a missing limb, has physical symptoms that defy neurology. The hysteric's seizures, twitches, coughs, and squints are not the result of lesions but of neurotic cathexes, of the pathological attachment of libidinal energies to body parts. In other words, parts of the body in hysterics become occupied, taken possession of, filled with energies that manifest themselves organically. (Freud's term *Besetzung* is translated by the English neologism "cathexis." The verb *besetzen* also has the sense of "charge," as with a furnace, or "tamp," as with a blasting charge, or "set in place," as with a paving stone or a jewel.)

Freud knew that the natural locus of woman's erotic pleasure was the clitoris and that it competed with the culturally necessary locus of her pleasure, the vagina. Marie Bonaparte reports that her mentor gave her Felix Bryk's *Neger Eros* to read. The author argued that the Nandi tribes engaged in clitoral excision on nubile seventeen- and eighteen-year-old girls so as to encourage the transfer of orgiastic sensitivity from its "infantile" zone to the vagina, where it must necessarily come to rest. The Nandi were purportedly not interested in suppressing female pleasure but merely in facilitating its redirection to social ends. Freud drew Bonaparte's attention to the fact that Bryk must have been familiar with his views and that the hypothesis regarding Nandi orgasmic transfer was worth investigating.

Bonaparte's efforts to discover the fortunes of "clitoroidal" versus "vaginal" sexuality in women whose clitoris had been excised proved inconclusive, but she did offer a theoretical formulation of the transfer of erotic sensibility that fits my understanding of Freud's theory of female sexuality. "I believe," writes Bonaparte, "that the ritual sexual mutilations imposed on African women since time immemorial . . . constitute the exact physical counterpart of the psychical intimidations imposed in childhood on the sexuality of European little girls."[131] "Civilized" people no longer seek to destroy the old home of sensibility—an ironic observation for Bonaparte, since she collected cases of European excision and herself underwent painful and unsuccessful surgery to move her clitoris nearer her vaginal opening so that she might be "normally orgasmic"—but enforce the occupation, or cathexis, of a new organ by less violent means.

If we put all of this together, Freud's argument might work as follows. Whatever polymorphous perverse practices might have obtained in the distant past, or today among children and animals, the continuity of the species and the development of civilization depend on the adoption by women of their correct sexuality. For a woman to make the switch from clitoris to vagina is to accept the feminine social role that only she can fill. Each woman must adapt anew to the redistribution of sensibility that furthers this end, must reinscribe on her body the racial history of bisexuality. But neurology is no help. On the contrary. Thus the move is hysterical, a recathexis that works against the organic structures of the body. Like the missing-limb phenomenon, it involves feeling what is not there. Becoming a sexually mature woman is therefore living an oxymoron, becoming a lifelong "normal hysteric," for whom a conversion neurosis is termed "acceptive."

And this gets us back to Freud's concern, which like Shakespeare's at the end of *Twelfth Night* is somehow to assure that bodies whose anatomies do not guarantee the dominance of heterosexual procreative sex nevertheless dedicate themselves to their assigned roles. But Freud is at the same time a product of nineteenth-century biologism, which postulates two sexes with distinctive organs and physiologies, and of an evolutionism that guarantees the adaptation of genital parts to heterosexual intercourse. In the end, the cultural myth of vaginal orgasm is told in the language of science. And thus, not thanks to but in spite of neurology, a girl becomes the Viennese bourgeois ideal of a woman.

I end this book with Freud not because he comes at the end of the making of sexual difference but because he posed its problems so richly. I might have ended with the scientists, including my great-uncle Ernst Laqueur who in the 1930s worried about endocrinological androgeny when male hormones were found in the female and female hormones in the male. But that worry is only a chemical version of the sorts of issues already raised by nineteenth-century embryology. Freud, precisely because he shattered the old categories of man and woman, had to work hard and ingeniously to establish new ones. With all his passion for biology, this preeminent twentieth-century thinker showed how difficult it is for culture to make the body fit into the categories necessary for biological and thus cultural reproduction. Two sexes are not the necessary, natural consequence of corporeal difference. Nor, for that matter, is one sex. The ways in which sexual difference have been imagined in the past are largely unconstrained by what was actually known about this or that bit of anatomy, this or that physiological process, and derive instead from the rhetorical exigencies of the moment. Of course the specific language changes over time—Freud's version of the one-sex model is not articulated in the same vocabulary as Galen's—and so does the cultural setting. But basically the content of talk about sexual difference is unfettered by fact, and is as free as mind's play.

Notes

1. Of Language and the Flesh

1. Jacques-Jean Bruhier, *Dissertation sur l'incertitude des signes de la mort* (Paris, 1749, 2nd ed.), 1.74–79.
2. Antoine Louis, *Lettres sur la certitude des signes de la mort, où l'on rassure les citoyens de la crainte d'être enterrés vivans* (Paris, 1752), pp. 53–54. On the preceding pages he reproduces Bruhier's text verbatim.
3. John Maubray, *The Female Physician* (London, 1724), p. 49. See Philippe Ariès, *The Hour of Our Death* (New York: Knopf, 1981), pp. 377–381, for the connection between erotic literature and the medical literature of death in the eighteenth century.
4. Unconscious conception, however, was not regarded as impossible. There is a folklore tradition on this theme that would be worth exploring. Lot, it will be recalled, was so drunk when he begat children by his two daughters in turn that "he knew not when she lay down or when she arose" (Genesis 19.31–35). In the Italian tale "The Sleeping Queen," the youngest son of the King of Spain finds "a maiden of angelic beauty" who had clearly been "put under a spell while she slept." He undresses, gets into bed with her, and passes "a delightful night with her without her giving any sign she knew he was there." He leaves a note when he departs; she is delivered of a boy nine months later. See Italo Calvino, *Italian Folktales,* trans. George Martin (New York: Pantheon, 1980), pp. xxiv, 207–213.
5. Nicholas Venette, *Conjugal Love; or the Pleasures of the Marriage Bed Considered in Several Lectures on Human Generation* (London, 1750), p. 41; this English translation is designated the "twentieth edition." There were at least twenty-three French editions in the eighteenth century, eight before Venette's death in 1698. See Roy Porter, "Spreading Carnal Knowledge or Selling Dirt Cheap? Nicholas Venette's *Tableau de l'amour conjugal* in Eighteenth-Century England," *Journal of European Studies,* 14 (1984), 233–255.
6. *Aristotle's Master Piece* in *The Works of Aristotle the Famous Philosopher* (New York: Arno Press, 1974), p. 9; *Aristotle's Masterpiece or the Secrets of Generation Displayed* (London, 1684), p. 29. This work, loosely based on the pseudo-Aristotelian

Problemata, was continuously reprinted from the middle of the fifteenth century to the 1930s, if not to the present. See D'Arcy Power, *The Foundation of Medical History* (Baltimore: Williams and Williams, 1931), pp. 147–178; Roy Porter, "The Secrets of Generation Display'd: *Aristotle's Masterpiece* in Eighteenth Century England," special issue of *Eighteenth Century Life,* 11 (1985), 1–21; Janet Blackman, "Popular Theories of Generation: The Evolution of Aristotle's Works," in J. Woodward and D. Richards, eds., *Health Care and Popular Medicine in Nineteenth Century England* (London: Croom Helm, 1977), pp. 56–88. There were more than twenty-seven editions in America alone before 1820; see O. T. Beall, "*Aristotle's Masterpiece* in America: A Landmark in the Folklore of Medicine," *William and Mary Quarterly,* 20 (1963), 207–222.

7. Michael Ryan, *A Manual of Jurisprudence and State Medicine* (London, 1836, 2nd ed.), pp. 246, 488. Ryan gives Robert Gooch, *A Practical Compendium of Midwifery* (London, 1831), as the source of the ostler's story and for similar stories refers his readers to E. Kennedy, *Obstetric Medicine* (London, 1834), which is indeed a rich source. The ostler's story is a variant on that of the farmhand in Montaigne's essay "On Drunkenness": a "widow of chaste reputation" finds herself inexplicably pregnant; she promises to forgive and marry the child's father if he will only present himself. One of her farmhands confesses that he came upon her "so fast asleep by her fireplace, and in so indecent a posture, that he had been able to enjoy her without waking her." *The Complete Essays of Montaigne,* trans. Donald M. Frame (Stanford: Stanford University Press, 1965), p. 246. Stories of this sort did not become evidence for any general truths about the relationship of orgasm to conception until the nineteenth century. See also Heinrich von Kleist's "The Marquise of O . . ." in which the protagonist finds herself inexplicably pregnant. Mary Jacobus gives a rich account of this story in "In Parenthesis: Immaculate Conception and Feminine Desire," *Body/Politics: Women and the Discourses of Science,* ed. Mary Jacobus, Evelyn Fox Keller, and Sally Shuttleworth (London: Routledge, 1990), pp. 11–28.

8. Philo, *Legum allegoriae,* 2.7, cited in Peter Brown, "Sexuality and Society in the Fifth Century A.D.: Augustine and Julian of Eclanum," *Tria corda: Scritti in onore di Arnaldo Momigliano,* ed. E. Gabba (Como: New Press, 1983), p. 56.

9. I take the term "passionlessness" and an understanding of its political meaning in the early nineteenth century from Nancy Cott's pioneering article, "Passionlessness: An Interpretation of Victorian Sexual Ideology, 1790–1850," *Signs,* 4.2 (1978), 219–236.

10. Nemesius of Emesa, *On the Nature of Man,* ed. William Tefler (Philadelphia: Westminster Press, 1955), p. 369.

11. *Aristotle's Master Piece,* Arno Press ed., p. 3.

12. Galen, *De semine,* 2.1, in *Opera omnia,* ed. C. G. Kuhn, 20 vols. (Leipzig, 1821–1833), 4.596.

13. Heinrich von Staden, *Herophilus: The Art of Medicine in Early Alexandria* (Cambridge: University Press, 1989), pp. 168, 185–186, 234.

14. Michel Foucault, *The History of Sexuality,* trans. Robert Hurley, vol. 1 (New York: Pantheon, 1978); Lawrence Stone, *Family, Sex and Marriage in England, 1500–1800* (New York: Harper and Row, 1977); Ivan Illich, *Gender* (New York: Pantheon, 1982).

15. Jacques-Louis Moreau, *Histoire naturelle de la femme,* vol. 1 (Paris, 1803), p. 15, which sounds the theme of the entire volume.

16. J. L. Brachet, *Traité de l'hysterie* (Paris, 1847), pp. 65–66, cited in Janet Beizer, "The Doctor's Tale: Nineteenth Century Medical Narratives of Hysteria," manuscript.

17. Patrick Geddes and J. Arthur Thompson, *The Evolution of Sex* (London, 1889), p. 266. Geddes and his colleague develop further their view that the "sexes differ fundamentally in the life-ratio of anabolic to katabolic changes," in *Sex* (London: Williams and Norgate, 1914), pp. 77–80.

18. John J. Winkler, "Laying Down the Law: The Oversight of Men's Sexual Behavior in Classical Athens," in David Halperin, John J. Winkler, and Froma Zeitlin, eds., *Sex Before Sexuality* (Princeton: Princeton University Press, 1990), pp. 171–209.

19. Peter Brown, *The Body and Society: Men, Women, and Sexual Renunciation in Early Christianity* (New York: Columbia University Press, 1988), pp. 167–168, 294–295.

20. Barbara Metcalf, *Perfecting Women: Maulana Ashraf ʿAli Thanawi's Bihisti Zewar* (Berkeley: University of California Press, 1990).

21. Caroline Bynum, *Holy Feast and Holy Famine: The Religious Significance of Food to Medieval Women* (Berkeley: University of California Press, 1987).

22. This genetic disorder is common in three Dominican Republic villages where it is known as the "penis at twelve" condition. See Julianne Imperato-McGuinley et al., "Steroid 5-Alpha-Reductase Deficiency in Man: An Inherited Form of Male Pseudo-Hermaphroditism," *Science,* 186 (1974), 1213–15.

23. Angus McLaren, "The Pleasures of Procreation: Traditional and Bio-Medical Theories of Conception," in W. F. Bynum and Roy Porter, eds., *William Hunter and the Eighteenth-Century Medical World* (Cambridge: University Press, 1985), p. 340.

24. Esther Fischer-Homberger, "Herr und Weib," *Krankheit Frau und andere Arbeiten zur Medizingeschichte der Frau* (Bern: Huber, 1979). This account of the decline in the social status of procreation is part of a sophisticated argument for a decline in the importance of sexual potency and a rise in the significance of "mental" potency in men, which in turn the author regards as an indicator of the shift from family to public functions as marks of status. Doctors increasingly viewed the nervous system and the brain as the organizing structure of the human body; reproduction, now regarded as a female process, was demoted as a sign of status.

25. George W. Corner, "The Events of the Primate Ovarian Cycle," *British Medical Journal,* 4781 (August 23, 1952), 403.

26. Anne Fausto-Sterling, *Myths of Gender* (New York: Basic Books, 1985). This book is not so much concerned with debunking studies on biological difference as in showing that so-called sex differences in behavior are actually gender differences.

27. See Londa Schiebinger, *The Mind Has No Sex? Women in the Origins of Modern Science* (Cambridge: Harvard University Press, 1989), pp. 191–200.

28. Michel Foucault, *The Order of Things* (New York: Vintage Books, 1973), pp. 30–31.

29. Maurice Godelier, "The Origins of Male Domination," *New Left Review*, 127 (May–June 1981), 17.

30. For three recent and varying formulations of this question, see Evelyn Fox Keller, "The Gender/Science System: or, Is Sex to Gender as Nature Is to Science?", *Hypathia*, 2 (Fall 1987), 37–49; Donna Haraway, "Situated Knowledges: The Science Question in Feminism and the Privilege of Partial Perspective," *Feminist Studies*, 14 (Fall 1988), 575–599; Linda Alcoff, "Cultural Feminism versus Post-Structuralism: The Identity Crisis in Feminist Theory," *Signs*, 13 (Spring 1988), 405–436.

31. Gayle Rubin, "The Traffic in Women: Notes on the 'Political Economy' of Sex," in Rayna R. Reiter, ed., *Toward an Anthropology of Women* (New York: Monthly Review Press, 1975), pp. 158–159. In Nancy F. Cott's words, "Feminism is nothing if not paradoxical. It aims for individual freedoms by mobilizing sex solidarity. It acknowledges diversity among women while positing that women recognize their unity." See her "Feminist Theory and Feminist Movements: The Past Before Us," in Juliet Mitchell and Ann Oakley, eds., *What Is Feminism: A Re-Examination* (New York: Pantheon, 1986), p. 49.

32. Sherry B. Ortner and Harriet Whitehead, "Introduction: Accounting for Sexual Meanings," in Ortner and Whitehead, eds., *Sexual Meanings: The Cultural Construction of Gender and Sexuality* (Cambridge: University Press, 1981), p. 1.

33. "Variations on Common Themes," in Elaine Marks and Isabelle de Courtivron, eds., *New French Feminisms* (New York: Schocken, 1981), p. 218.

34. In addition to Alcoff, note 30 above, see Joan W. Scott, "Deconstructing Equality versus Difference: Or, the Uses of Post-Structuralist Theory for Feminism," and Mary Poovey, "Feminism and Deconstruction," in *Feminist Studies*, 14 (Spring 1988), 33–50, 50–66.

35. Julia Kristeva, "Women's Time," trans. Alice Jardine and Harry Blake, *Signs*, 6 (Fall 1981), 33–34.

36. Joan Scott, "Gender: A Useful Category of Historical Analysis," *American Historical Review*, 91 (December 1986), 1065, 1067; italics mine.

37. Catharine A. MacKinnon, in *Toward a Feminist Theory of the State* (Cambridge: Harvard University Press, 1989), p. xiii, states that she will use sex and gender "relatively interchangeably"; the definition of gender is from her "Feminism, Marxism, Method and the State: An Agenda for Theory," *Signs*, 7 (Spring 1982), 533, cited in a useful editorial on various meanings of gender in *Signs*, 13 (Spring 1988), 399–402. For MacKinnon on "gender difference," see *Feminism Unmodified* (Cambridge: Harvard University Press, 1987), pp. 3, 46–62.

38. Ruth Bleier, *Science and Gender: A Critique of Biology and Its Theories on Women* (New York: Pergamon Press, 1984), p. 80. When she speaks of sexual differences, Bleier is by and large, but not always, referring to behavioral and not morphological or biochemical differences. I understand her claim to be that not only are so-called gender differences not natural but that prior politically salient understandings of sex as a biological category lead to the search for behavioral correlatives.

39. Foucault, as feminists have pointed out, restricts himself almost entirely to the making of the male self. His use of the masculine pronoun is thus more than

conventional. Still there is no reason why his method is not applicable to the making of the self, gendered or—if such a thing is possible—ungendered. For Nietzsche's notion of the world as a work of art and its relevance to Foucault's antiessentialism, I have drawn heavily on Alexander Nehamas, *Nietzsche: Life as Literature* (Cambridge: Harvard University Press, 1985); quote from p. 3. I am sympathetic with Foucault, and by extension Nietzsche, but I agree with Nehamas that some interpretations of the world are better than others.

40. Jeffrey Weeks, *Sexuality and Its Discontents* (London: Routledge, 1985), p. 122. This is an immensely useful, learned, and insightful guide to "the subject of sex."

41. Foucault, *History of Sexuality,* 1.157.

42. Ernst Laqueur was one of the discoverers of estrogen. He isolated the "female" hormone from the urine of stallions, thereby raising the uncomfortable possibility of endocrinological androgyny at the very moment when science seemed to have finally discovered the chemical basis of sexual difference.

43. Werner Laqueur's article was published in *Acta Brevia Neelandica*, 6 (1936), 1–5. The "uterus masculinus," now called the prostatic utricle, is a small hollow sac that extends into the body of the prostate. It is the "remains of that part of the Mullerian duct [the urogenital sinus] out of which, in the female, the vagina forms." The uterus masculinus, in other words, is the vestigial vagina, so named because it was once thought to represent the remains of a structure from which the uterus and upper vagina derive. See also Keith L. Moore, *The Developing Human* (Philadelphia: Saunders, 1977, 2nd ed.), pp. 235–237.

44. Sarah Kofman, *The Enigma of Woman,* trans. Catherine Porter (Ithaca: Cornell University Press, 1985), pp. 109–110.

45. "Thus *heimlich* is a word the meaning of which develops towards an ambivalence, until it finally coincides with its opposite, *unheimlich."* In light of the one-sex model, with its insistence on the vagina as an internal penis, this all becomes still stranger: "This *unheimlich* place, however," writes Freud, "is the entrance to the former *heim* [home] of all human beings, to the place where everyone dwelt once upon a time and in the beginning." Freud, "The 'Uncanny,'" (1919), *Studies in Parapsychology,* ed. Philip Rieff (New York: Collier, 1963), pp. 30, 51.

46. See Evelyn Fox Keller, *Reflections on Gender and Science* (New Haven: Yale University Press, 1985), pp. 177–179.

47. François Jacob, *The Logic of Life: A History of Heredity,* trans. Betty E. Spillmann (New York: Pantheon, 1973; 1970 in French), p. 16. Jacob won the Nobel Prize for his work in molecular genetics. I use the term "narratives" to mean all those contexts in which the body figures, all those stories told about it. I once used the more limited term "metaphors," which in its strict sense is too limiting.

48. Auguste Comte, *Cours de philosophie positive,* in G. Lenzer, ed., *Auguste Comte and Positivism* (New York: Harper and Row, 1975), p. 178; italics mine. Positivism, first used systematically as a term by Saint-Simon and taken up by Comte in the 1830s, is the immensely influential view that an objective, scientific knowledge of nature was not only possible but could be the basis for social regeneration.

49. Emile Durkheim, *The Division of Labor in Society,* trans. W. D. Halls (New York: Free Press, 1984), p. 14. I am grateful to my student Paul Friedland for this and the preceding reference.

50. Barbara Johnson, *The Critical Difference,* quoted in Elizabeth Abel, ed., *Writing*

and Sexual Difference (Chicago: University of Chicago Press, 1982), p. 1. My understanding of this epigram is indebted to Jane Gallop's exegesis, *"Writing and Sexual Difference:* The Difference Within," in Abel, pp. 283–291.

51. I here accept and turn on its head Elizabeth Abel's comment in her introduction to *The Critical Difference.*

52. Charles Darwin, "On the Two Forms, or Dimorphic Condition, in the Species of *Primula,* and on Their Remarkable Sexual Relations," in Paul H. Barrett, ed., *Collected Papers of Charles Darwin* (Chicago: University of Chicago Press, 1980, 2 vols. in one), 2.61.

53. See the review of the literature on the division of sexual labor in Michael T. Ghiselin, *The Economy of Nature and the Evolution of Sex* (Berkeley: University of California Press, 1974), pp. 99–137.

54. George Ewart Evans and David Thomson, *The Leaping Hare* (London: Faber, 1972), pp. 24–25; Gilbert H. Herdt, *Guardians of the Flute* (New York: McGraw-Hill, 1981), p. 154. The Sambia are a tribe in the New Guinea highlands; their men appear to believe that ingesting semen is a necessary step toward becoming masculine, and they engage in fellatio with other males as part of a prolonged period of transition to adulthood.

55. Claude Lévi-Strauss, *The Savage Mind* (Chicago: University of Chicago Press, 1966), p. 46 and chap. 2 generally. See also Edmund Leach's illuminating "Anthropological Aspects of Language: Animal Categories and Verbal Abuse," in Eric H. Lenneberg, ed., *New Directions in the Study of Language* (Cambridge: MIT Press, 1964).

56. Leonore Davidoff's and Catherine Hall's *Family Fortunes* (Chicago: University of Chicago Press, 1987) is a model of the studies I have in mind.

57. Frederic Harrison, "The Emancipation of Women," *Fortnightly Review,* 298 (October 1, 1891), 442, 448. Harrison, the leading British positivist, gave this lecture on the anniversary of Comte's death. Below I discuss Millicent Fawcett's reply in this debate among progressives on the woman question.

58. On gender in slasher films, see Carol C. Clover, "Her Body, Himself: Gender in Slasher Film," *Representations,* 20 (Fall 1987), 187–228. For "the triumph of contract and the 'individual' over sexual difference" in de Sade, see Carole Pateman, *The Sexual Contract* (Stanford: Stanford University Press, 1988), p. 186. Contract theory actually works on a no-sex model, which I discuss below. Pateman's is the best treatment I know of the implications of liberal individualism for theories of sexual difference.

59. In addition to Bleier's *Science and Gender* and Fausto-Sterling's *Myths of Gender,* see Lynda Birke, *Women, Feminism, and Biology* (New York: Methuen, 1986).

60. Elizabeth Fee, "Nineteenth Century Craniology: The Study of the Female Skull," *Bulletin of the History of Medicine,* 53 (1979), 433. On the question of bias in science, see Sandra Harding and Jean F. O'Barr, eds., *Sex and Scientific Inquiry* (Chicago: University of Chicago Press, 1987).

2. Destiny Is Anatomy

1. Galen, *On the Usefulness of the Parts of the Body,* trans. Margaret Tallmadge May, 2 vols. (Ithaca: Cornell University Press, 1968), 2.628–629; hereafter abbrevi-

ated *UP.* Denis Diderot, *Rameau's Nephew and Other Works,* trans. Jacques Barzun and Ralph H. Bowen (Indianapolis: Bobbs-Merrill, 1964), p. 135.

2. Galen, *On the Natural Faculties,* trans. Arthur John Brock, Loeb Classical Library (Cambridge: Harvard University Press, 1952), 3.2, pp. 227–229. Cophonis' *Anatomia porcis,* an apocryphal Galenic text produced at the famous medical school in Salerno during the twelfth century, begins the discussion of the matrix as an organ contrived so that whatever superfluities a woman generates during the month, her menstrual flow, could be sent there "like the bilge water of the entire body *(tanquam ad sentinam totius corporis).*" It is primarily a storage space. As if an afterthought, the writer says it is also the field of generation. See George W. Corner, *Anatomical Texts of the Earlier Middle Ages* (Washington: Carnegie Institute, 1927), pp. 50, 53.

3. See Isidore of Seville, *Etimologias,* ed. José Oroz Reta and Manuel A. Marcos Casquero (Madrid: Biblioteca de Autores Christianos, 1983), 12.1.134, for *uterum* in relation to *caulis;* the Latin text in this edition of the *Etymologiarum* is identical to that in the standard edition of W. M. Lindsay (Oxford, 1911). The force of the proposition is somewhat dulled when Isidore goes on to say that the uterus resembles a little stalk *(cauliculus);* this word, a cognate of the Latin and Greek *caulis,* was the important medical writer Celsus' preferred term for penis and was used metaphorically for the male organ in Petronius, *Satyricon* 132.8. See J. N. Adams, *The Latin Sexual Vocabulary* (London: Duckworth, 1982), pp. 26–27.

   Perhaps the ancient association of the uterus with the stomach/belly explains what would seem the bizarre claim, given then current anatomical knowledge, that the wandering womb pressing upward from the abdomen caused the choking and general feeling of constriction characteristic of hysteria. If one interprets this literally, there would be no explanation for male hysteria or for how the ancients thought that the womb made its way up through the various organs and divisions above it. But if one construes the womb as retentive space/belly, hollow/stomach, the source of hysteria is properly localized. My sense is that ancient medicine is less interested in specific organic causes than in corporeal metaphors that correlate with symptoms.

4. Isidore is making much of the roots of *uterum* meaning belly, but he does have a separate discussion of *agualiculus* (stomach) at 11.1.136. This word also has the sense of any vessel, hence belly. See Adams, *Latin Sexual Vocabulary,* pp. 100–101. We retain this in the way we speak to young children—"Mommy has a baby in her belly"—when we wish to be anatomically vague. On vulva–vagina–gateway to the belly, see Pseudo-Albertus Magnus, *De Secretis mulierum* (1665 ed.), pp. 12, 19, or *Anatomia Magistri Nicolai Physici,* in Corner, *Anatomical Texts,* p. 85.

5. Isidore of Seville, *Etymologiarum* 11.1.139.

6. It does not help matters that *sinus*–bosom–vagina or womb, as in *sinus muliebris,* could also, as in Lactantius' use *(sinus pudendus),* mean penis. Adams, *Latin Sexual Vocabulary,* pp. 90–91.

7. On the nature of heat and the difference between its quantity and quality, see Everett Mendelsohn, *Heat and Life: The Development of the Theory of Animal Heat* (Cambridge: Harvard University Press, 1964), pp. 17–26, esp. n. 58.

8. *UP* 2.629. Galen did not invent the trope of the mole's eyes as the paradigmatic case of the imperfect version of a more perfect structure found elsewhere. See Aristotle, *Historia animalium,* 1.9.491b26ff and 4.8.533a1–13; hereafter abbreviated *HA.*

9. Aristotle [?], *Economics,* 2.3.1343b25–1344a8. I have throughout this book used the translation in Jonathan Barnes, ed., *Complete Works of Aristotle,* 2 vols. (Princeton: Princeton University Press, 1984), but have checked terms and arguments critical to my exposition in the standard Greek texts.

10. On generation and Aristotle's theory of causality, see Anthony Preus, "Galen's Criticism of Aristotle's Conception Theory," *Journal of the History of Biology,* 10 (Spring 1977), 78, and more generally his "Science and Philosophy in Aristotle's *Generation of Animals,*" idem, 3 (Spring 1970). *Generation of Animals* (hereafter abbreviated *GA*) not only begins (1.1.715a3) but ends (5.5.789b3) with discussion of cause. A. L. Peck points out the importance of a theory of causality in Aristotle's thought and gives an extremely clear exposition of how he develops such a theory in his work on generation; see the introduction to *GA,* Loeb Classical Library (Cambridge: Harvard University Press, 1958), pp. xxxviii–xliv.

11. *GA* 1.2.716a13–14, 716a20–22; 4.3.768a25–28. Male and female are "contrary," in *Metaphysics,* 10.9.1058a29–30. I take this formulation for the relationship between biology and a model of filiation from Giulia Sissa, "Subtle Bodies," in *Fragments for a History of the Human Body,* part 3, ed. Michel Faher et al., *Zone,* 5 (1989), 154, n. 6.

12. *GA* 4.1.765b35ff. For *perineos* used to refer to the female genitalia, see *HA* 1.14.493b9–10. Female genitals are called *aidoion* at *HA* 1.14.493b2; male genitals are referred to by the same term at *HA* 2.1.500a33–b25. See also Peck, *GA,* p. 388, n. c; for *pudenda* see Adams, *Latin Sexual Vocabulary,* p. 66.

13. *GA* 1.2.716a19–b1; *HA* 1.13.493a25. At *HA* 1.2.489a10–14 Aristotle defines "male" as emitting into another and "female" as emitting into itself—a suitably ambiguous effort to ground difference in anatomy and physiology.

14. *HA* 9.50.632a22. I put "ovaries" in quotation marks because Aristotle does not recognize the existence of female testicles, and no writer before the late seventeenth century construed the organ we now call the ovary as the source of an egg. The organ whose excision Aristotle referred to was "cut from the place where the boars have their testicles and adhere to the two divisions of the womb."

15. This sentence is necessarily awkward because the relationship between genitals and gender is so complicated, as the studies of Robert Stoller on cases of ambiguous or "misassigned" sex suggest. See his *Sex and Gender* (New York: Science House, 1968), and Richard Green and John Money, eds., *Transexualism and Sex Reassignment* (Baltimore: Johns Hopkins University Press, 1969).

16. *GA* 1.7.718a23. This works because "that which is carried too far is cooled."

17. Eva Keuls, *The Reign of the Phallus* (New York: Harper and Row, 1985), pp. 68–69.

18. *GA* 1.4.717a26–30. Aristotle's linking of the reproductive with the digestive system is based on the commonplace that generative products and products of the digestive system are both residues. Thus at *GA* 1.20.728a201–24 Aristotle

argues that just as diarrhea is caused by insufficient concoction of the blood in the bowels, "so are caused in the blood vessels all discharges of blood, including the menstrual blood," though the former condition is morbid and the latter is not. Still, menstrual discharge is the result of a failure, of the woman's not being as hot as the man and thus unable to concoct residue for the last time and to produce sperma.

19. Aristotle uses the highly specialized word *kapria* (female pig part) for the organ whose removal produces the dramatic results he describes. *Kapria* is the "sow virus," a liquid from the female pig related to the spermlike substance (*gones*, generative material) that oozes out of the sexual organs of mares in heat. The latter substance, the hippomanes, apparently a version of the black matter on a newborn foal's head, "resembles the sow virus (*kapria*), and is in much demand amongst women who deal in drugs," Aristotle says (*HA* 6.18.572a21–23). In the Renaissance the hippomanes was still considered an aphrodisiac. Aristotle seems to suggest that the hippomanes, qua liquid, is produced exclusively by mares impregnated by the wind but that the word also refers to the caul on foals, however they were conceived. The standard Greek term for ovaries was *orcheis* (testicles), or *didymoi* (twins); the Latin version *orchis* referred to a flower. The ovaries are said to have been discovered by Herophilus of Alexandria in the third century B.C. See Staden, *Herophilus,* pp. 167–168. Neither the word "ovary" nor "ovum" for its content was used until the late seventeenth century.

20. *GA* 1.3.716b33 and more generally *HA* 1.17.497a30–31. This simile works because the two suspensory ligaments, including presumably what are now called the Fallopian tubes, are imagined as "horns of the uterus"; the ovaries then become visual analogues to the testes so that the body of the uterus becomes the female scrotum of Galen's description.

21. See Soranus, *Gynecology,* trans. Owsei Temkin (Baltimore: Johns Hopkins University Press, 1956), 9.1.16, p. 14, and p. 10, n. 6, where Temkin points out that the word for tube is also the word for penis. Celsus writing in the first century B.C. used *caulis* (stalk), which he got from the Greek *kaulos,* as his standard term for the penis. Caelius Aurelius used *kaulos* as the equivalent of *aidoion,* which was a common word for penis as well as for female pudenda. He and other Latin medical writers regarded *aidoion* as meaning *ueretrum,* another common Latin word for penis. See Adams, *Latin Sexual Vocabulary,* pp. 26–27, 52–53.

22. Julius Pollux, *Onomasticon* (Vocabulary), ed. Eric Bethe (Leipzig: Teubner, 1900), 2.171. Pollux was little known in antiquity, but the 1502 printing of his text and subsequent Latin-Greek editions were immensely important during the Renaissance as a source for new, non-Arabic anatomical nomenclature.

23. *HA* 10.4.636a6–7. If this writer had Soranus' image in mind, it would commit him to having the womb ejaculate into its own foreskin. The genuine Aristotle frequently writes about the womb breathing in material—it draws it up like a cupping glass—but does not believe that the womb itself ejaculates semen (for example, *GA* 2.4.739b1–20 and *HA* 7.3.583a15–16).

24. G. E. R. Lloyd, *Science, Folklore, and Ideology* (Cambridge: University Press, 1984), pp. 107–108.

25. Aristotle holds that though women and men are "contrarities," they are not separate species because they differ only in matter and not in formula, much as a
black man differs from a white man only incidentally, in color. Women differ
from men not as a circle does from a triangle but as a circle or a triangle of one
material does from a circle or triangle of another. See *Metaphysics* 10.1058a29ff
and *HA* 5.11.538a13.

26. Pseudo-Aristotle, *Problems,* 1.50.865a33f. Phlegm also has a complicated relationship to heat and inflammation as well as to the theory held by Plato, Hippocrates, and others that semen derives from the brain and spinal matter rather
than from the blood.

27. See the learned notes in Iain M. Lonie, *The Hippocratic Treatises: "On Generation," "The Seed," "On the Nature of the Child," "Diseases IV,"* in the series *Ars
Medica: Texte und Untersuchungen zur Quellenkunde der Alten Medizine* (Berlin:
Walter de Gruyer, 1981), pp. 124–132, 102–103, 277–279, which emphasize
the openness of fluid boundaries.

28. Isidore, *Etymologiarum,* 11.1.77. Galen discusses the convertibility of blood and
milk in clinical detail in *UP* 2.639. See also Hippocrates, *Aphorisms,* 5.37, 52.

29. *The Seven Books of Paulus Aegineta,* trans. Francis Adams (London, 1844),
3.609–614; Aetios of Ameda, *Tetrabiblion,* trans. James V. Ricci (Philadelphia:
Blakiston, 1950), chaps. 4 and 26; Soranus, *Gynecology,* pp. 18–19. These observations are quite commonplace and I cite Paulus Aegineta, Aetios, and Soranus
as general medical authorities only because they provide readily accessible, coherent accounts. They are also clinically astute, though not for the reasons then
supposed. For example, on modern thinking as to why exercise, obesity, and
severe weight loss result in amenorrhea, see Leon Speroff et al., *Clinical and
Gynecological Endocrinology and Infertility* (Baltimore: Willimans and Wilkins,
1983), chaps. 1 and 5, esp. pp. 171–177.

30. *HA* 10.5.637a18–19. On fig. 2 see Zelda Boyd, "'The Grammarian's Funeral'
and the Erotics of Grammar," *Browning Institute Studies,* vol. 16, ed. Robert
Viscusi (Browning Institute, Southwestern College, 1988), p. 5. On the throat/
neck of the womb, vagina, or cervix, see Ann Hanson and David Armstrong,
"The Virgin's Voice and Neck: Aeschylus, *Agamemnon 245* and Other Texts,"
British Institute of Classical Studies, 33 (1986), 97–100; and Lloyd, *Science, Folklore,* pp. 326–327. Galen in *De uteri dissectione 7* says that "Herophilus likens
the nature of the uterus [cervix?] to the upper part of the windpipe"; Staden,
Herophilus, p. 217.

31. Hippocrates, *Aphorisms* 32 and 33 and *Epidemics* 1.16, in *The Medical Works of
Hippocrates,* ed. John Chadwick and W. N. Mann (Oxford: Oxford University
Press, 1950). These clinical observations would be repeated for two thousand
years. A Renaissance doctor reports, for example, that a woman who was suffering from headaches because her menses did not flow was temporarily relieved
when "these were at length ejected by vomiting." The complaint reappeared and
was permanently eliminated when the doctor let blood from her ankle, which
"compelled the menses to flow regularly from the natural place." Antonio Beni-
veni (1443–1502), *De abditis nonnullis ac mirandis morborum et sanationum causis,*
trans. Charles Singer (Springfield: Charles C Thomas, 1954).

32. Soranus, *Gynecology*, p. 19. He notes also that the amount of menstrual flow is less in "teachers of singing and in those journeying away from home." The interconnection between fluids seems endless. Thus Albertus Magnus held that sexual stimulation of both men and women produced an ejaculate halfway between sperm and sweat. James R. Shaw, "Scientific Empiricism in the Middle Ages: Albertus Magnus on Sexual Anatomy and Physiology," *Clio Medica*, 10.1 (1975), 61.

33. *GA* 1.19.727a11–15; *HA* 7.10.587b32–588a2; this passage follows Aristotle's account of why lactating women do not menstruate.

34. *GA* 1.19.727a31ff; *HA* 7.2.582b30–583a4; for the connection of milk and sperm, see *HA* 3.20.521b7; on milk, blood, and sperm, *GA* 4.4.771a4ff. I cite Aristotle here because of his importance in western thought on the subject, but these views are commonplace throughout ancient and later writings, even those not directly in the Aristotelian tradition.

35. See below, Chapter 5, as to why these discoveries made more plausible, but did not entail, a two-sex model and why it would be anachronistic to use the modern terms "sperm" and "egg" for what seventeenth-century scientists saw.

36. Two-seed theories, like those of Hippocrates and Galen, hold that "seeds" from both parents are necessary to vivify the matter provided by the mother. One-seed theories, of which Aristotle's was the most influential, hold that the male provides the *sperma* (the efficient and, more problematically, the formal cause) in generation while the female provides the *catamenia* (the material cause). The female ejaculate in this model had no purpose because by definition the female provides no seed. See Michael Boylan, "The Galenic and Hippocratic Challenges to Aristotle's Conception Theory," *Journal of the History of Biology*, 17 (Spring 1984), 85–86, and Preus, note 10 above.

37. Aline Rousselle, *Porneia*, trans. Felicia Pheasant (Oxford: Blackwell, 1988), pp. 24–26, argues that, in the absence of opportunities for male doctors to examine women dead or alive, the quite precise observations regarding female pleasure and physiology were given to the doctors by midwives or female patients. Though there is no direct evidence for this view, I would like it to be true since it suggests that much of what I will say in this book reflects not just a high, male, medical tradition but the imaginative worlds of women as well. I would disagree with Rousselle, however, in attributing to Aristotle a fundamentally different view of the phenomenological aspects of reproduction from that of the Hippocratic writer. I use the phrase "Hippocratic writer" to suggest that the corpus of works attributed to Hippocrates are now thought to be written by many writers in his tradition. It is awkward to use this locution consistently, and so I revert to calling these writers by the name of one of them: Hippocrates.

38. "On Generation," Lonie ed., 6.1 and 6.2, p. 5, and the illuminating commentary on pp. 124–132.

39. Ibid., 7.2; again at 8.2.

40. Ibid., 6.2. The existence of male and female sperm in each parent is adduced to explain why some women produce male offspring with some men and female offspring with a later husband. Since the Hippocratic tradition is pangenesist, holding that each part of the body produces part of the sperm, each feature of

the child is a result of the same sort of battle that determines sex. (See *GA* 1.17.725b13ff for the classical attack on this position.) "On Generation" simply asserts that no child can resemble only one parent, which is another way of saying that men are necessary and women cannot simply clone themselves (see 8.1 and 8.2). On pangenesis and ancient theories of inheritance generally, see Erna Lesky, *Die Zeugungs und Vererbungslehre der Antike und ihr Nachwirken* (Mainz: Akademie der Wissenschaften und der Literatur, 1950).

41. Hippocrates offers no account of why there are not, as this model might suggest there would be, a large number of creatures whose genital configuration would be "in between," making them difficult to classify socially. Nor does he address the question, which vexed others, of why the female needs the male at all if she is indeed capable of producing a strong malelike sperm.

42. The case is made explicitly in Galen, *Peri spermatos* (On the Seed), Kuhn ed., 4.2.4, p. 622. He argues elsewhere in this text that "females have seminal ducts and testes full of semen." If males had milk in their mammary ducts, there would be no reason to inquire what it was for. "Likewise, since females have semen there is no need to wonder whether they excrete it" (2.1, p. 600).

43. Avicenna, *Canon* (Venice, 1564), 3.20.1.3. At 3.31.1.1 Avicenna, like Galen, makes the case that the female organ of generation, the womb, is "as it were, the male organ reversed." The Latin translation of Avicenna's Arabic uses *sperma* for both the male and female ejaculate, and Avicenna is at pains to criticize those who equate the female seed with the menstrual fluid. Generally speaking, Avicenna maintains an Aristotelian position on generation while reproducing almost verbatim the Galenic system of anatomical isomorphisms. See Danielle Jacquart and Claude Thomasset, *Sexuality and Medicine in the Middle Ages* (Princeton: Princeton University Press, 1988), pp. 36ff.

44. See Boylan, "Galenic Challenge." On other occasions Aristotle uses *gonimos* (generative, productive) to refer to sperm. He uses the same word to refer to the female contribution.

45. *GA* 1.21.729b17ff; 2.1.734b20ff, which discusses the complicated relation of the soul(s) to sperma generally; 2.3.737a10–16. Rennet is the mucous lining of a calf's stomach which contains rennin, an enzyme used to curdle milk. Fig juice serves a similar function; *HA* 6.18.572a15.

46. Biological and intellectual conception are closely related, as Aristotle's seventeenth-century proponent William Harvey noted.

47. The medieval text (*De secretis mulierum*) of the Pseudo-Albertus Magnus uses *menstruum* to refer to the female seed and *sperma* to the male seed in the discussion of conception, in which the two seeds (*duo semine*) meet in the *vulva* (vagina). See Charles Wood, "The Doctors' Dilemma: Sin, Salvation, and the Menstrual Cycle in Medieval Thought," *Speculum*, 56 (1981), 716, and John F. Benton, "Clio and Venus: An Historical View of Medieval Love," *The Meaning of Courtly Love,* ed. F. X. Newman (Albany: State University of New York Press, 1969), p. 32, on *menstruum* as seed and *sanguinis menstruus* as menstrual blood. Aquinas' concern is to have the Virgin be both a material and a formal cause for the human Christ; see esp. *Summa theologica*, 3a.31.5, and Wood, p. 27. Clearly more than biology is at stake in the question of whether *menstruum* is called a seed. In claiming a lack of clear distinctions between a one-seed and a two-seed

model, I am arguing against the position put forward by Anne-Liese Thomasen, "'Historia animalium' contra 'Gynaecyia' in der Literatur des Mittlealters," *Clio Medica,* 15 (1980), 5–23, where she describes two distinct and mutually exclusive traditions.

48. *GA* 4.8.776b10. See Boylan, "Galenic Challenge," p. 94, where he concludes, rightly I think, that the uterus does engage in an inferior form of the "fourth concoction of pepsis" which the spermatic ducts in men do better. More generally on how food is heated to produce blood and generative material, see Michael Boylan, "The Digestive and 'Circulatory' Systems in Aristotle's Biology," *Journal of the History of Biology,* 15 (1982), 89–118. The fact that *HA* 10.1.634b30ff and 10.6.637b32, for example, use *sperma* for both the male and female generative products is one reason why scholars doubt the authenticity of book 10. Whether by Aristotle or not, this linguistic equation seems to move in the direction taken in genuine Aristotelian texts.

49. *GA* 1.19.726b5ff; on old men and boys see *GA* 1.18.725b20. The semen of drunkards, says the pseudo-Aristotelian *Problems* 50.865a33, is infertile because it is too moist and produces too liquid a residue.

50. It is no wonder, Peter Brown suggested to me, that both the Gnostic and the Manichean traditions emphasize ejaculating sperm as the final step in delivering light/spirit from base matter.

51. Paul Delany, "Constantinius Africanus' *De Coitu:* A Translation," *Chaucer Review,* 4.1 (1969), 59. Constantinius Africanus was an eleventh-century physician, steeped in ancient medical learning, who taught in the medical school at Salerno. For more on this point and on the widely varying and often contradictory advice proffered by doctors, see Jacquart and Thomasset, *Sexuality,* pp. 53ff, 87–96. As will become clear, I differ from them in that I argue against the sharp division they wish to make between male and female reproductive physiology.

52. *The Divine Comedy: Purgatory,* trans. Dorothy L. Sayers (Harmondsworth: Penguin, 1955), 25.37–45, p. 264. Obviously the reference is to the male, but his refined blood is sprayed on the refined blood of the female, which has been concocted by an identical process.

53. Pseudo-Albertus Magnus, *De secretis mulierum,* 1.19. This twelfth-century text was widely copied and later printed first in Latin and then in various vernaculars. (There is an English edition as late as 1745.) See Lynn Thorndike, "Further Consideration of the *Experimenta, Speculum Stromiac,* and *De Secretis Mulierum* ascribed to Albertus," *Speculum,* 30 (1955), 413–443.

54. The story of Tiresias is in Ovid, *Metamorphoses,* 3.323–331. One might translate the question more specifically as "which sex had the better orgasm (*maior voluptas*)." See also Leonard Barkan, *The Gods Made Flesh: Metamorphosis and the Pursuit of Paganism* (New Haven: Yale University Press, 1986), pp. 41–42; and his discussion of how the act of love in Ovid and other poets "blurs distinctions by transforming the lovers into hermaphrodites" (p. 57). The story of Narcissus in *Metamorphoses* follows immediately after the brief account of Tiresias.

55. *UP* 2.651. By "genital areas" Galen means here the inner organs and their male equivalents. Note once again the association of parts: scrotum/uterus as well as digestive organs/genital organs.

56. *The Parts of Animals,* 4.9.689a5ff, in *Complete Works,* Ross ed.

57. Galen explains, correctly by modern standards, that the vessel from the right kidney, today called the internal spermatic vessel, passes directly to the uterus. He thought that this gave the serous, exciting, residue a straight shot to its sensitive target (*UP* 2.641). "Right" here is from the viewer's perspective.

58. Pseudo-Aristotle, *Problems,* 1.26.879a36–880a5. P. H. Schrijvers, the editor of Caelius Aurelianus' *De Morbis Chronicus IV.9: Eine medizinische Erklarung der mannlichen Homosexualitat aus der Antike* (Amsterdam: B. R. Gruner, 1985), comments on this passage and argues that the passive homosexual, the *mollis,* is therefore a "bisexual" with excessive desire (an excess of semen). The connections between these organs are mirrored in language: *vagina* as a sheath was a metaphor for anus. Adams, *Latin Sexual Vocabulary,* pp. 20, 115. See Jacquart and Thomasset, *Sexuality,* pp. 124–125, for an account of a long and technical discussion comparing the anal sphincter with the muscles of the uterus (vagina, cervix, etc.) in al-Samau'al ibn Yahyâ (d. 1180), *Book of Conversation with Friends on the Intimate Relations Between Lovers in the Domain of the Science of Sexuality.*

59. *UP* 2.622–623, 658–659, 660–661. The *nympha* (2.661), by which Galen seems to mean the clitoris, is said to be like the uvula, which gives protection to the throat. Here again reproduction and breath, breathing and ejaculation, the throat and the genital passages, are linked.

60. See Shaw, "Albertus Magnus," p. 60.

61. Avicenna, *Canon,* 3.20.1.3, 25. Avicenna in his accounts of reproduction combines an essentially Galenic physiology with Aristotelian metaphysics.

62. "On Generation," Lonie ed., 1.1, 4.1.

63. Galen, *UP* 2.640–643. The citation from Democritus to which Galen refers is probably the following: "Coition is a slight attack of apoplexy: for man gushes forth from man, and is separated by being torn apart with a kind of blow. See Herman Diels and Walther Kranz, *Die Fragmente der Vorsokratiker* (Berlin: Weidmann, 1951–52), p. 68b22. Though Aristotle was explicitly opposed to Democritus' interpretation of this explosion as evidence for pangenesis, he too regarded the intense pleasure of orgasm as being due to a sudden blast of *pneuma* in both men and women (*GA* 1.20.728a10, 2.4.738b26–32). The image of coitus as a version of epilepsy remained in currency for centuries; see, for example, the first major Christian educational guide, Clement of Alexandria's *Paedagogus,* 2.10. In the early 1960s the Vatican censored Alberto Moravia's novel *Empty Canvas* for its "sexual realism" because a love scene was likened to epilepsy; the woman and not the bored lover had the seizure.

64. Pseudo-Aristotle, *Problems,* 4.1.876a30–35.

65. Tertullian, *A Treatise on the Soul,* in *The Ante-Nicene Fathers,* ed. Alexander Roberts and James Donaldson, 3 vols. (Grand Rapids: Erdmans, 1976 reprint), 3.208; the sentence before the ellipsis is from a translation by Peter Brown which he has kindly allowed me to use. I substitute his language to emphasize that *both* sexes are caught up in the pleasure of sex no matter who contributes true semen. See J. H. Waszink's commentary, pp. 342–348, in his edition of Tertullian's *De anima* (Amsterdam: J. M. Meulenhoff, 1947), on the nature of the contributions of each sex to a new life which makes manifest the difficulties of determining what ancient authors actually meant.

66. Lucretius, *The Nature of the Universe,* trans. Ronald Latham (Penguin: Harmondsworth, 1951), pp. 165, 168.

67. *GA* 2.4.739a27–30. He wants to hold that even when a woman emits, it is not semen but "merely proper to the part concerned" (*GA* 1.20.727b35–728a1).

68. *GA* 2.4.739a20–35; also 1.19.727b34–728a24. Aristotle is willing to admit that men can emit semen without feeling it, as in wet dreams.

69. *GA* 1.18.723b33. This argument works, in Aristotle's view, against the pangenesist position that female orgasm is evidence for her producing semen and that semen comes from all parts of the body of both sexes.

70. *GA* 1.20.728a11–21. This is where Aristotle argues that a woman is an impotent man, or like a boy.

71. *HA* 10.638a5ff. At *GA* 2.739a20–26 Aristotle argues that, even though women also have wet dreams, their discharge does not contribute to the embryo, because boys who have no semen and men who seem infertile also have nocturnal dreams. Again the move is to shield the maleness of generativity from empirical investigation.

72. Aetios, *Tetrabiblion,* 16.1, Ricci trans., pp. 19, 36. It is implicit in the Hippocratic "On Generation," 5.1, which describes how the womb contracts when it has received the seeds. Experienced women could supposedly tell from this contraction the precise day of conception. See Lonie's commentary, p. 124, for other references to the womb's sucking in its own and the male ejaculate.

73. "On Generation," 4.2. When the hierarchy of heat does not work, the hierarchy of activity takes over. Thus man's sperm arriving at the womb before the woman's orgasm extinguishes "both the heat and pleasure for woman," just as pouring cold water into boiling water cools the latter. Again, one must not construe "hot" and "cold" in the medical literature as meaning what they would today. Thus most of the Hippocratic corpus regards men as hotter and hence more perfect than women, whereas *Regime* holds that men are colder and more perfect. No empirical dispute divides these positions.

74. *HA* 10.3.635b19–24. This sweating is also likened to the tears that come in bright light or as a response to cold or great heat. For my purposes, again, it does not matter that this book is probably not by Aristotle. The specificity of its reference to preorgasmic lubricity, as opposed to the emission of female sperm at orgasm, may indicate that the passage is in the voice of women as transmitted by an anonymous ancient physician. See note 37 above.

75. *HA* 10.5.636b12ff; see also 10.1.634b28ff and 10.1.634b3, regarding optimal conditions of dryness or wetness.

76. Rhazes, *Liber ad almansorum* (1481), 5.73.

77. *Canon,* 3.20.1.44. One might imagine this better in a basically polygymous society where wives are prized either for the pleasure they give or for their capacity to bear sons. Abandoned by their husbands, they seek pleasure among themselves. Perhaps the point is to enforce the norm that men should keep trying to give women pleasure since reproduction, of sons, is as much their responsibility as the women's.

78. This may seem totally implausible. But Soranus had an escape. Just as a grieving widow might not know that she has an appetite and both needs and will make

good use of food, so a woman might not know that she actually does desire intercourse. Certain feelings can be masked by others. *Gynecology,* Temkin ed., p. 36. I discuss the implications of this view in the debate over the possibility of conception in rape cases (Chapter 5).

79. Soranus, *Gynecology,* pp. 34–35, 38–39.

80. Polemo, *Physiognomonika,* 1.112, 1.10.36, cited by Maud Gleason, "The Semiotics of Gender: Physiognomy and Self-Fashioning in the Second Century A.D.," in Halperin et al., eds., *Before Sexuality.*

81. On the naturalness of homosexuality, see K. J. Dover, *Greek Homosexuality* (New York: Vintage Books, 1980), pp. 60–68. Specifically on the naturalness of a man's genital response to young boys, see Dover, p. 170, and Caelius Aurelianus, *On Acute Diseases,* 3.180–181, in Schrijvers, *Eine Medizinische,* pp. 7–8.

82. Plato, *Symposium,* ed. Alexander Nehamas and Paul Woodruf (Cambridge, Eng.: Hackett, 1989), 189e–193a, pp. 25–29; Aristotle, *Rhetoric,* 1371b15–16. I do not want to suggest that in Greek culture generally homosexuality was thought to be natural; indeed, while Aristophanes seeks to give a natural history of men's love for men, Pausanius in the *Symposium* maintains a sort of cultural relativism: "the customs regarding Love" might be easy to understand in most cities, but "in Athens (and in Sparta as well) they are remarkably complex" (182a–182b, p. 15).

83. On recognizing the *cinaedus* from even the slightest gesture, see Gleason, "The Semiotics of Gender." On honor and appropriate exchanges generally in male same-sex relationships, see David J. Cohen, "Law, Society and Homosexuality in Classical Athens," *Past and Present,* 117 (November 1987), 3–21; David Halperin, "One Hundred Years of Homosexuality," *Diacritics* (Summer 1986), 34–45, and a fuller version of this piece, "Paederasty, Politics, and Power in Classical Athens," forthcoming in George Chauncey et al., *The New Social History of Homosexuality* (New American Library).

84. See Dover, pp. 182–184, on the question of the aggressive, masculine "lesbian"; and Schrijvers, *Eine Medizinische,* p. 8, for the equivalence of *mollis* and *tribade.*

85. Vicky Spellman, "Aristotle, Females, and Women." I am grateful to Spellman for allowing me to read this typescript.

86. *Republic,* 454e, in *The Collected Dialogues,* ed. Edith Hamilton and Huntington Cairns (Princeton: Princeton University Press, 1963), p. 693. Plato of course does not maintain this view of sexual equality in other contexts, as in the *Laws* or the myth of the origin of women in the *Timaeus.* I have profited greatly in understanding the context of Plato's arguments on this subject from Monique Canto, "The Politics of Women's Bodies: Reflections on Plato," in Susan Rubin Suleiman, ed., *The Female Body in Western Culture* (Cambridge: Harvard University Press, 1986), pp. 339–353. Whereas my reading emphasizes Plato's rejection of the biology of reproduction as a relevant political difference, Canto makes the positive case that Plato is arguing for a "communal" account of procreation that neutralizes the effects of difference; raising children communally, as is proposed elsewhere in the *Republic,* is a continuation of this political strategy. The highly contextual quality of Plato's view of women generally is made clear in

Gregory Vlastos, "Was Plato a Feminist?" *Times Literary Supplement*, March 17–23, 1989, pp. 276, 288–289.

87. He insists also at *GA* 2.1.734b20–735a10 that heat alone makes neither an ax nor flesh. The sword is made by movements containing the principles of art, and the same is true for what the male parent contributes to the flesh.

88. *On the Heavens*, 2.7.289a29–30. See *GA*, Appendix A, n. 7, Peck ed., and Mendelsohn, *Heat and Life*, pp. 11–13, for an account of Aristotle's and other ancient writers' views of heat.

89. On the political and biological uses of the same terms, see Mary Cline Horowitz, "Aristotle and Women," *Journal of the History of Biology*, 9 (Fall 1976), 183–213.

90. See R. Howard Bloch, *Etymologies and Genealogies: A Literary Anthropology of the Middle Ages* (Chicago: University of Chicago Press, 1983), and the extremely useful account of how Isidore's etymologies worked in Jacquart and Thomasset, *Sexuality*, pp. 8–14.

91. Isidore, *Etymologiarum*, 9.6.4 ("Semen") and 4.5.4 ("Blood").

92. Ibid., 9.5.24. *Vidua* is translated "spouseless mother" because in a previous section Isidore has already dealt with the case of a posthumous child legitimately born to a widow. Lewis and Short give *spurium* as meaning female pudenda. For Plutarch see Adams, *Latin Sexual Vocabulary*, p. 96.

93. Ibid., 11.1.145.

94. It is by no means clear that Regnier de Graaf discovered the mammalian egg, since he identified it with what we now know as the Graafian follicle. Similarly, the sperm seen by Leuwenhoek and Ham was thought by them to be wholly different from what we now take it to be.

95. Aeschylus, *The Eumenides*, trans. Richmond Lattimore, in David Greene and Lattimore, eds., *Greek Tragedies*, vol. 3 (Chicago: University of Chicago Press, 1960), lines 606ff, 653, 657ff, pp. 26–28. For "mounts" Aeschylus uses *throsko*, which in its usual intransitive forms means to leap or spring. This passage is the only one given in Liddell and Scott for the transitive form meaning to mount or impregnate. It is also the *locus classicus* of what Michael Boylan has called the "furrowed field" theory of generation, the view that the male provides all the relevant causes for generation. See his "Galenic Challenge," pp. 85–86.

96. Sigmund Freud, *Moses and Monotheism* (1939), in *The Standard Edition of the Complete Psychoanalytical Works*, ed. James Strachey (London: Hogarth Press), 23.113–114. I have somewhat amended the translation based on the German version in Freud, *Gesammelte Werke*, ed. Marie Bonaparte et al. (London: Imago, 1950), 14.220–221. *Geist*, and hence *Geistigkeit*, is notoriously difficult to translate since "spirit" and "spirituality" have too religious a connotation and a neologism like "intellectuality" means little. But Freud's argument, which continues in the next section where he defends his ranking of *Geist* over *Sinn* (senses), emphasizes both the cultural and the intrapsychic superiority of spirit, reason, reflectivity, and restraint over the materially present, immediacy, and instinct.

97. See Nancy G. Siraisi, *Taddeo Alderotii and His Pupils: Two Generations of Italian Medical Learning* (Princeton: Princeton University Press, 1981), pp. 197–199.

98. On wind eggs, see *GA* 2.3.737a28ff, 3.1.749a34–749b7; *HA* 4.2.559b20–

560a17. Mola, the bits of unformed flesh and hair sometimes found in women, were not thought to be exact equivalents to wind eggs in birds because they supposedly never occurred without prior intercourse with a male. This actually is not the case since the molas the ancients observed in women were probably dermatoid cysts that form parthenogenically from primordial germ cells. They also occur in men, though rarely. But the point is that, in proportion to the monumental task of forming the flesh of hotter animals, the female had to be understood as proportionately less potent in relation to the work that had to be done; *UP* 2.630. The advantage is that women are warm enough to nurture the conceptus but not so warm as to burn it up. If women were men, the new life would fall onto a desert and perish.

99. Plutarch, *Advice to Bride and Groom* in *Moralia,* vol. 2, trans. F. C. Babbitt, Loeb Classical Library (Cambridge: Harvard University Press, 1927), 48.145e, p. 339. See also 33.142e, p. 323; 4.138f. p. 303; 42.144b. pp. 331–332.

100. On the Christianization of the body, see Brown, *Body and Society.*

101. Brown, "Julian of Eclanum," p. 70.

102. Aristotle argued that erection, like changes in the rhythms of the heart, was involuntary and thus not susceptible to moral blame or praise. *De motu animalium,* 703b5–7, trans. Martha Nussbaum (Princeton: Princeton University Press, 1978). It was precisely the incapacity of the will to control erection which made it, and more tellingly still impotence, so deeply revelatory of man's fallen state.

103. Augustine, *The City of God,* trans. Henry Bettenson (Harmondsworth: Penguin, 1984), 14.24, pp. 588–589.

104. Thomas Tentler, *Sin and Confession on the Eve of the Reformation* (Princeton: Princeton University Press, 1977), p. 181; Innocent III, *On the Misery of the Human Condition,* trans. Margaret Mary Dietz (Indianapolis: Bobbs-Merrill, n.d.), p. 8.

105. Brown, *Body and Society,* p. 69.

106. See, for example, G. E. R. Lloyd, "Right and Left in Greek Philosophy," *Journal of Hellenistic Studies,* 82 (1962), 55–66; O. Kember, "Right and Left in the Sexual Theories of Parmenides," idem, 91 (1971), 70–79; and for a more general discussion of the categories in relationship to sex/gender, Carol P. MacCormack, "Nature, Culture, and Gender: A Critique," in MacCormack and Marilyn Strathern, eds., *Nature, Culture, and Gender* (Cambridge: University Press, 1980), pp. 1–24.

3. New Science, One Flesh

1. Guillaume Bouchet, *Les Sérées de Guillaume Bouchet,* ed. C. E. Roybet, six vols. (Paris, 1873–1882), 1.96; Christopher Wirsung, *Ein Neues Artzney Buch Darinn fast alle eusserliche und innerliche Glieder des Mennschlichen leibs . . . beschriben werden* (1572), p. 416; Thomas Vicary, *The Anatomy of the Bodie of Man* (1548 as reissued in 1577), ed. F. J. and P. Furnivall (Oxford: Early English Text Society, 1988), p. 77.

2. Similarly, "tail" could refer not only to the posterior extremity but to both the penis and the female pudenda, although this is a slang usage I have not encountered in medical texts.

3. *Auslegung und Bescreibung der Anathomy oder warhafften abcontersetung eines inwendigen corpers des Manns und Weibs* (1539), section "von der mutter" (on the mother), no pagination. For the connection between uterus and scrotum via words for bag, and for associations with other organs as well—womb as "breeding gut," for example, to take up again the uterus/intestine connection—see Torild W. Arnoldson, *Parts of the Body in Older Germanic and Scandinavian* (Chicago: University of Chicago Press, 1915), pp. 160–175, and *Parts of the Body in the Later Germanic Dialects* (Chicago: University of Chicago Press, 1920), pp. 104–121.

4. Pseudo-Albertus Magnus, *De secretis mulierum* (1655 ed.), p. 19. The context is a discussion of male and female ejaculation; when the two seeds have been received by the womb, it "is shut like a purse (*matrix mulieris clauditur tanquam bursa*)." The next paragraph repeats this phrase and gives as the reason for the closure, on the authority of Avicenna, that the womb "rejoices in the heat it has received and does not want to lose it (*quia guadet ex calido recepto nolens perdere*)."

5. *Aristotle's Masterpiece* (1684), p. 28.

6. Laevinius Lemnius, *The Secret Miracles of Nature* (London, 1658), p. 19, which was originally published as *De occultis naturae miraculis* in 1557.

7. Columbus, *De re anatomica* (Venice, 1559), 11.16, pp. 447–448. Matteo Realdo Colombo (1516–1559?)—I retain the Latin form of his name in my text—was the distinguished successor of Vesalius as lecturer in surgery at Padua.

8. Ibid., pp. 444–445. The idea of the seven-cell uterus is not found in Galen or in the major Arabic authors, but appears first in the writings of the twelfth-century anatomical school of Salerno. On this point see Robert Reisert, *Der seibenkammerige uterus: Studien zur mittlealterlichen Wirkungsgeschichte und Enfaltung eines embryologischen Gebärmutermodells* (Hanover: Wurzburger medizinshistorische Forschungen, 1986).

9. Fallopius, *Observationes anatomica* (Venice, 1561), p. 193. These are said to be the lecture notes of Fallopius (Gabriello Fallopio, 1523–1562), the anatomist who did discover the oviducts.

10. *Bartholinus' Anatomy, Made from the Precepts of His Father, and from Observations of All Modern Anatomists, Together with His Own* (London, 1668), p. 75. This book is a translation of revisions in 1641 by Thomas Bartholin (the discoverer of the lymphatic system) of his father Kaspar's famous text, *Institutiones anatomicae* (1611). It was Thomas' son Kaspar II (1655) who gave his name to the greater vestibular glands that lubricate the lower end of the vagina during coitus.

11. Jane Sharp, *The Midwives Book, or the Whole Art of Midwifery Discovered Directing Childbearing Women How to Behave Themselves in Their Conception, Breeding, Bearing and Nursing Children* (London, 1671), pp. 40, 42. Mrs. Sharp says that her book is based on thirty years' experience, that it is aimed at a broad female audience (hence no Latin), and that she has incurred great costs in translating the latest French, Dutch, and Italian sources into English.

12. Columbus, *Anatomica*, pp. 447–448.

13. I take Jacqueline Rose's argument that "there can be no work on the image, no challenge to its powers of illusion and address, which does not simultaneously challenge the fact of sexual difference," to mean that facts of sexual difference do not exist independently of forms of allusion and address. *Sexuality in the Field of Vision* (London: Verso, 1987), p. 226. She is commenting on a footnote in Freud's account of Leonardo's immensely ambiguous depiction of sexual intercourse, which is not, as Freud suggests, an idiosyncratic result of Leonardo's bisexuality, but a commonplace example of Renaissance depictions of the genital organs.

14. What I mean by "from a modern perspective" is that contemporary texts would not make this sort of case. There is obviously an enormous problem, which I discuss briefly in my first chapter, about using modern research as the standard. Even when someone today argues that women's secretions during orgasm are histochemically like male prostatic fluid or that the neurology of orgasm is similar in both sexes or that negative pressures during female orgasm aid conception, they are not making the same kinds of claims that Renaissance observers were making. The problem of theoretical translation is, in my view, more acute in biology than in the physical sciences.

15. Columbus, *Anatomica,* pp. 448, 453–454.

16. *GA* 2.4.739a29–30; 1.19.727b6–11.

17. M. Anthony Hewson, *Giles of Rome and the Medieval Theory of Conception* (London: Athlone Press, 1975), p. 87. The case cited by Averroës, used by Giles to make even stronger claims, was well known in the Renaissance.

18. William Harvey, *Disputations Touching the Generation of Animals* (1653), trans. Gweneth Whitteridge (Oxford: Blackwell Scientific Publications, 1981), p. 165.

19. On the popularity of early printed medical works in Tudor England, see Paul Slack, "Mirrors of Health and Treasures of Poor Men," in Charles Webster, ed., *Health, Medicine, and Mortality in the Sixteenth Century* (Cambridge: University Press, 1979), pp. 237–273.

20. I have relied on the survey of data on this subject in Lisa Lloyd's manuscript "Evolutionary Explanations of Human Female Orgasm," which she has kindly let me read.

21. Herman W. Roodenburg, "The Autobiography of Isabella De Moerloose: Sex, Childbearing and Popular Belief in Seventeenth Century Holland," *Journal of Social History,* 18 (Summer 1985), 517–540. (I discuss aspects of this diary below in note 83.) A woman writing about conception in her diary in the nineteenth century still speaks largely in the language of Hippocrates.

22. The best direct evidence for the absence of radically divergent views between doctors and their patients are the casebooks of Johann Storch, a small-town physician practicing in early eighteenth-century Eisenach, which have been brilliantly analyzed by Barbara Duden, *Geschichte unter der Haut* (Stuttgart: Klett-Cotta, 1987). On the creation of popular culture by the pulling away of a high tradition, see Natalie Z. Davis, "Proverbial Wisdom and Popular Errors," *Society and Culture in Early Modern France* (Stanford: Stanford University Press, 1975), pp. 227–267. I suggest below that, in matters relevant to this book, the differences between the new medicine based on purified classical texts and direct ob-

servation, on the one hand, and traditional views on the other were minimal. See also Paul-Gabriel Bouché, "Imagination, Pregnant Women, and Monsters in Eighteenth-Century England and France," in G. S. Rousseau and Roy Porter, eds., *Sexual Underworlds of the Enlightenment* (Chapel Hill: University of North Carolina Press, 1988), pp. 86–100, for an account of how it was not until the eighteenth century that doctors came to attack, and then not with one voice, the commonplace view that the behavior of pregnant women could cause monstrosities.

23. See Emily Martin, *The Woman in the Body* (Boston: Beacon Press, 1987).

24. Robert J. Smith and Ella Lury Wiswell, *The Women of Suye Mura* (Chicago: University of Chicago Press, 1982): "She demonstrated with her hands how the womb opens up when receptive" (pp. 63–64). The book is based entirely on Wiswell's field notes.

25. Françoise Héritier-Augé, "Semen and Blood: Some Ancient Theories Concerning their Genesis and Relationship," *Zone,* 5 (1989), 160–161. It is in fact unclear whether the anthropologist interrogated both male and female Samo, but she presents the evidence as if it spoke for generally accepted views. See also the surveys of women's views on menstruation and fertility cited in the introduction to T. Buckley and A. Gottlieb, eds., *Blood Magic: The Anthropology of Menstruation* (Berkeley: University of California Press, 1988), pp. 42–43.

26. Willard van Orman Quine, "Two Dogmas of Empiricism," *From a Logical Point of View* (New York: Harper and Row, 1963), pp. 42–43; see also the formulation in Quine and J. S. Ullian, *The Web of Belief* (New York: Random House, 1978, 2nd ed.). Thomas Kuhn in *The Structure of Scientific Revolutions* argues the same case historically.

27. I make this case based on many different authors. For an immensely rich study of the logic of the body, the relationships between its various structural, metaphoric, and macrocosmic aspects, see Marie Christine-Pouchelle's study of Henri de Mondeville, *Chirugia, corps et chirugie à l'apogée du moyen-age* (Paris: Flammerion, 1983).

28. This does not mean that Vesalius and his successors escaped the influence of classical learning in general or of Galen in particular. All of Galen's works were edited and translated in numerous vernaculars; Vesalius himself was involved with producing the great *Opera Galeni* published in Venice (1541–42) and regarded Galen as "the prince of physicians and preceptor of all." See Richard J. Durling, "A Chronological Census of Renaissance Editions and Translations of Galen," *Journal of the Warburg and Courtauld Institutes,* 24 (1961), which enumerates 630 items between 1473 and 1600, excluding long citations in the work of others. J. B. deC. Saunders and Charles D. O'Malley, *The Anatomical Drawings of Andreas Vesalius* (New York: Bonanza, 1982), p. 13. For reasons discussed below, Aristotle, who was not an anatomist or physician, was much less influential in writings about the body. But there is much of Aristotle in Avicenna, who was a major influence in Renaissance medical teaching. See Nancy Siraisi, *Avicenna in Renaissance Italy: The Canon and Medical Teaching in Italian Universities after 1500* (Princeton: Princeton University Press, 1987). His philosophical influence was enormous. See also Charles B. Schmitt, "Towards a Reassessment of

Renaissance Aristotelianism," *History of Science,* 11 (1973), 159–193, and more generally *Aristotle and the Renaissance* (Cambridge: Harvard University Press, 1983).

29. Preface to *The Fabric of the Human Body,* trans. Logan Clendening, *Source Book of Medical History* (New York: Dover, 1942), p. 136.

30. Fallopius, *Observationes,* p. 195.

31. On theater and public anatomies, see Giovanna Ferrari, "Public Anatomy Lessons and the Carnival in Bologna," *Past and Present,* 117 (1987), 50–107.

32. Harvey Cushing, *A Bio-Bibliography of Andreas Vesalius* (Hamden: Archon Books, 1962, 2nd ed.), pp. 81–82. It is usually said that the young man in the chair in fig. 3 is the professor and that the dissecters below are his assistants. But the man in the chair was more likely a junior assistant whose job it was to read the text while the professor—the older man bent over the body—dissected. See Jerome J. Bylebyl, "The School of Padua: Humanistic Medicine in the Sixteenth Century" in Webster, ed., *Health, Medicine,* pp. 335–371. Vesalius' epistemological claim on the title page and the evidence of the images themselves remain intact, in my view.

33. My reading of the body in this picture is heavily indebted to W. S. Heckscher's account of Rembrandt's "Anatomy of Dr. Nicolaas Tulp" in his *Rembrandt's "Anatomy"* (New York: New York University Press, 1958). "Anatomies" as a literary genre were predicated on the process of penetrating representations and getting at the "real" truth. See Devon L. Hodges, *Renaissance Fictions of Anatomy* (Amherst: University of Massachusetts Press, 1985), pp. 6–17. For the use of classical sculpture to contain human anatomy, see Glenn Harcourt, "Andreas Vesalius and the Anatomy of Antique Sculpture," *Representations,* 17 (Winter 1987), 28–61.

34. The decline, during the scientific revolution, of an idea of nature as a nurturing mother to whom humanity is organically bound, and the rise of a conception of nature as a feminine object to be studied and exploited by men, is the theme of Carolyn Merchant's *The Death of Nature: Women, Ecology and the Scientific Revolution* (New York: Harper and Row, 1980).

35. I have not counted reuses or reworkings of a plate in new editions of the original work or in altogether different works. This is by no means a proper survey, but I would be surprised if such a study altered the results significantly. Because more men than women were executed, more male cadavers were undoubtedly available for dissection. Still there was ample opportunity for doctors to examine women. Vesalius dissected at least seven. Autopsies, as Katherine Park argues in *Doctors and Medicine in Early Renaissance Florence* (Princeton: Princeton University Press, 1985), pp. 52–53, were routinely performed, and even noble women had no qualms about being examined either while alive or in contemplation of death. She cites the case of a patrician woman who was suffering from a uterine flux and asked that she be autopsied so that doctors might better treat her daughters should they develop the same condition. Anecdotal evidence, as in *Beloved Son: The Journal of Felix Platter, a Medical Student in Montpellier in the Sixteenth Century,* trans. Sean Jennett (London: Frederick Muller, 1961), p. 90, suggests that women's bodies were made available by grave robbing.

36. Samuel Y. Edgerton, *Pictures and Punishments: Art and Criminal Prosecution during the Florentine Renaissance* (Ithaca: Cornell University Press, 1985), pp. 215–217, and chap. 5 passim, notes that in this picture the anatomist is cast as an exalted, almost priestlike figure. The corpse may resemble the dead Christ of Pietà paintings, but it is the anatomist who seems to be making a godlike claim.

37. See R. K. French's informative "Berengario da Carpi and the Use of Commentary in Anatomical Teaching," in A. Wear, R. K. French, and I. M. Lonie, eds., *The Medical Renaissance of the Sixteenth Century* (Cambridge: University Press, 1985), pp. 42–74, esp. 54–62.

38. On illustration in medieval texts, see Karl Sudhoff, *Eine Beitrage zur der Geschichte der Anatomie in Mittelalter, speziell der Anatomischen Graphik nach Handscriften des 9. bis 15 Jahrhunderts,* in *Studien zur Geschichte der Medizin*, 4 (1908), 1–94 and 24 plates, in which he argues for the schematic nature of the illustrations, the difficulty in demonstrating their connection with a particular text, and their reliance on one another—especially in the case of the skeletons (pp. 28–51)—rather than on nature. There are no known anatomical illustrations from antiquity, and the earliest gynecological drawing (of a uterus) is from the ninth century. See Fritz Weindler, *Geschichte der Gynaekologisch-Anatomomischen Abbildung* (Dresden: Zahn und Jaensch, 1908), pp. 14–15 and pp. 81–89 on Berengario as the great pre-Vesalian innovator. The most comprehensive history of anatomical illustration is Johann Ludwig Choulant, *A History and Bibliography of Anatomic Illustration,* trans. Mortimer Frank (New York: Hafner, 1945, 1962 reprint). I have also consulted R. Herrlinger, *History of Medical Illustration from Antiquity to 1600* (New York: Editions Medicina Rara, 1970). The manifestly new relationship between print and text is difficult to characterize precisely because it is not, as the history of science literature suggests, simply a matter of more naturalistic illustrations replacing schematic ones. Nor is it the case, as Geoffrey Lapage, *Art and Scientific Illustration* (Bristol: John Wright, 1961), argues, that truth in illustration somehow lies in the attainable goal of simply avoiding distortion when a print is produced from a scientist's observations. All anatomical illustration is necessarily schematic in relation to an infinitely less clear and more crowded body. Moreover, so-called naturalistic anatomical illustrations, though they might be drawn from nature, are still heavily dependent on artistic conventions and even ideological imperatives (see Chapter 6). On the power of convention, see E. H. Gombrich's account of the longevity of Durer's largely fanciful though conventionally naturalistic drawing of a rhinoceros, "Truth and the Stereotype," in *Art and Illusion: A Study in the Psychology of Pictorial Representation* (New York: Pantheon, 1960), pp. 81–82.

39. On the Michelangelo self-portrait in the skin of St. Bartholomew, see Leo Steinberg, "Michelangelo and the Doctors," *Bulletin of the History of Medicine*, 56 (1982), 543–553, esp. 549–551. On its relation to Valverde's text, see Edgerton, *Pictures and Punishments,* pp. 217–219 and n. 53.

40. See French, "Berengario," pp. 43–49, and L. R. Lind, *Studies in Pre-Vesalian Anatomy: Biography, Translations, and Documents* (Philadelphia: American Philosophical Society, 1975).

41. Jacopo Berengario da Carpi, *A Short Introduction to Anatomy* [*Isagoge brevis*],

trans. L. R. Lind (Chicago: University of Chicago Press, 1959), p. 80. The *Isagoge* is a kind of summary of Berengario's far larger *Commentary on Mondino* (1521), which was the first anatomy book to integrate illustrations with text.

42. Both the male and female genitals, Vesalius directs, are in the first instance to be attached to a "figure we have drawn to act chiefly as a base for all the others . . . the figure representing a nude female." The nude in fig. 19c made of blood vessels is, as it were, the inside of the chaste classical nude woman (fig. 19d) included in a special chapter devoted to the terminology of surface anatomy.

43. Despite Gombrich's argument in *Art and Illusion* that all art originates in the human mind and that stylistic convention determines the mode of representation, he remains, as Svetlana Alpers points out, committed to the notion that a perfect representation is possible and that certain schemata are more likely than others to produce truth in pictures. See Alpers, "Interpretation without Representation, or the Viewing of *Las Meninas,*" *Representations,* 1 (February 1983), 31–42. Without arguing these points generally, I simply want to hold that inflexible conventions are not the cause of the peculiar form of seeing suggested by these figures.

44. The classic instance is Vesalius' insistence that there is a network of blood vessels at the base of the human brain, the *rete mirabile,* when in fact such a structure does not exist in humans. Seeing something on the basis of an authority is a commonplace in the history of anatomy and in the modern anatomy laboratory.

45. John Dryander was professor of medicine and mathematics at the new Protestant university at Marburg. I take the illustrations and text from his *Der Gantzen Artzenei Spiegel* (Frankfurt, 1542), pp. 17–19, a book intended, we are told in a long title, for doctors, barber surgeons, and others who needed to know about the body. Much of his text is taken from Mondino, many of the illustrations from Vesalius. His nomenclature comes straight from the Latin: *testes* (literally witness) becomes in German *Zeuglin* from *Zeuge* or *Zeugin* (witness). The other word used in Renaissance German texts for both testicles and ovaries is *Hode*. Note also the image of the ovaries and testicles as producers. *Zeug* means stuff, material, in German; *erzeugen* is to produce. Dryander translates the Latin *pudenda,* which derived from terms for shame or disgrace, to the German *Scham* and uses it to refer only to the female's external genitalia. But in Latin *pudenda* was used to refer to "private parts," the genital organs of both sexes (see Adams, *Latin Sexual Vocabulary,* p. 55). In other German texts *Scham* refers to both male and female external organs. See, for example, Wirsung, *Neues Artzney,* p. 260, which regards an unnaturally early appearance of hair around the male *Scham* as an indication of excess heat and hence infertility. For *Hode* and *Zeugin* see Jacob and Wilhelm Grimm, *Deutsches Wörterbuch* (Leipzig: S. Hirzes, 1965).

46. Saunders and O'Malley, *Anatomical Drawings,* p. 170, note that others have called the drawing from *Fabrica* plate 20 "monstrous" or the result of a "Freudian quirk," but they explain what they take to be its peculiarities by the haste with which Vesalius had to perform the particular dissection from which it was derived. Charles Joseph Singer, *A Short History of Anatomy from the Greeks to Harvey* (New York: Dover, 1957), pp. 119–120, attributes its peculiarities, and

Vesalius' many "errors" in female anatomy, to the fact that he had opportunities to dissect only seven women. As I have argued, Vesalius' image is not due to such circumstances, nor is it in any way out of the ordinary.

47. Charles Estienne, *De dissectione partium corporis humani* (Paris, 1545), 3.7, p. 289. Estienne was the scion of a distinguished family of printers and was court anatomist to Francis I. This work also appeared in French translation. According to Singer, *A Short History,* p. 102, Estienne had plenty of material for his dissections and claims to have seen everything he describes. The major gross-anatomical difficulty with the proposed thought experiment I quote is that the female testes are not attached to the Fallopian tubes, which in Renaissance illustrations are construed as both the ovarian/testicular arteries and as the deferent ducts of the testes.

48. Helkiah Crooke, *Microcosmographia: A Description of the Body of Man* (London, 1615), p. 250. Crooke is basing these arguments on the work of Gaspard Bauhin and on Andreas Laurentius, of Jewish descent, professor of medicine at Montpellier, and physician to Henry IV.

49. Estienne, *De dissectione,* 3.7, p. 289.

50. *Bartholinus' Anatomy,* pp. 62–63. This book was published in England, perhaps out of sympathy for Bartholin's egalitarian views, by Nicholas Culpepper and Abadiah Cole. Culpepper was extremely active in the political reform of medicine during the English revolution; in his own works, however, he expounded the old relationships between male and female organs.On the important role of Culpepper in producing vernacular literature in defiance of the medical establishment, see Charles Webster, *The Great Instauration: Science, Medicine and Reform, 1626–1660* (London: Holmes and Meier, 1975), pp. 268–271. The prostate was described in detail as early as 1536 by the Venetian Niccolo Massa. Its secretions are now used to argue for the essential similarity of male and female sexuality because of the histochemical properties they share with the secretions of Bartholin's glands.

51. *Bartholinus' Anatomy,* pp. 71–72.

52. Jacques Duval, *Traité des hermaphrodits* (Rouen, 1612; reprint Paris, 1880), pp. 342–349. By *vulve* he means what we would call the vulva, vagina, and cervix with the corpus and fundus of the uterus attached. This is a holdover from the classical use of *vulva* to mean what we would call the uterus with its outer parts, as in Celsus, *De medicina,* trans. W. G. Spencer (London: Heineman, 1935) 4.1.12, pp. 14–15. I am puzzled by Duval's reference to Aristotle rather than to Galen as the author of the inversion exercise.

53. William Harvey, "On Parturition," in *The Works of William Harvey* (London, 1847), pp. 537–538.

54. The cod was literally the scrotum, so a codpiece was the bag that held the bag that held the testes. A codpiece could also be an appendage to the female attire worn on the breast.

55. François Rabelais, *The Histories of Gargantua and Pantagruel,* trans. J. M. Cohen (Harmondsworth: Penguin, 1982), 1.8, p. 55. See the OED for "cod."

56. The carnation was "generally recognized as a token of betrothal in fifteenth and

sixteenth century Northern European painting." Metropolitan Museum of Art Exhibition Catalogue, *Liechtenstein: The Princely Collections* (New York, 1985), p. 239.

57. I am grateful to Paul Alpers for the Gascoigne poem.

58. I have not studied the nomenclature for the male reproductive anatomy thoroughly, and I know of no general study of the subject. There are to be sure many different words for penis, testicle, or scrotum, but in my reading the referents of these terms are unambiguous. Perhaps this is the linguistic correlative of the corporeal telos generally: the male body is stable, the female body more open and labile.

59. Columbus, *Anatomica,* p. 443. No such metaphorical excursus is exercised on the male organs. *Bartholinus' Anatomy,* p. 65 (chap. 28, "Of the Womb in General"), spends a paragraph explaining how for Pliny *vulva* meant particularly a pig's womb, a "delicate dish" for the Romans, but that other writers, such as Celsus, used it to mean the womb of any animal. *Vulva,* Bartholin speculates, is a corruption of *bulga,* which means bag but also refers to the "satchel or knapsack hanging from a Mans Arm."

60. Columbus, *Anatomica,* p. 445. *Mentula* was an obscene word for penis in antiquity (Adams, *Latin Sexual Vocabulary,* p. 9) but became the standard term in the Renaissance. *Vagina* was not used in Latin in its modern sense but referred to a tube or sheath, usually for a sword. It seems to have been used humorously as "anus" (Adams, pp. 20, 115).

61. Columbus, *Anatomica,* pp. 447–448. Columbus, like all other Renaissance anatomists, refers to the ovaries as testes that are slightly larger and firmer than the male's and are contained within rather than pendant.

62. Fallopius, *Observationes,* pp. 193, 195–196. He bases the distinction on what he takes to be the use of Soranus and Galen who, he says, refer to the vagina as a female "kolpos" and distinguish it from the true cervix. They are not so consistent. Singer, *A Short History,* p. 143, claims that Fallopius was the first to use the term *vagina* in a modern sense, but I have not found this usage. Fallopius offers no theory of the function of his "tubes," but he observes that they do not touch the ovaries, which in turn do not produce semen.

63. Gaspard Bauhin, *Anatomes* (Basel, 1591–92), 1.12, pp.101–102. *Porcus* (pig) was apparently a Roman nursery word for the external pudenda of girls (Adams, p. 82). Perhaps the allusion is to a perceived resemblance between the part in question and the end of a pig's snout.

64. Jacquart and Thomasset, *Sexuality,* p. 34, quoting al-Kunna al-Maliki. Consulting the French edition of this book I could not tell what Arabic word was translated as *clitoris.* But the authors do give *lèvres* as an alternative translation, and in the context it is clear that the labia minora are the referent.

65. *The Anatomy of Mundinus,* in Singer, ed., *Fasciculo,* p. 76 and n. 64.

66. Berengario, *Isagoge brevis:* "at the end of the cervix little skins are added at the sides; these are called prepuces" (p. 78); and in referring to the penis, "a certain soft skin surrounds this glans; it is called the prepuce" (p. 72). Josef Hyrtl, *Onomatologia Anatomica: Geschichte und Kritik der Anatomischen Sprache der Gegen-*

wart (Vienna, 1880), gives "nymphae" as meaning both labia and prepuce; see entry "nymphae und myrtiformis."

67. John Pechy, *The Complete Midwives Practice Enlarged* (London, 1698, 5th ed.), p. 49, and *A General Treatise of the Diseases of Maids, Bigbellied Women* (1696), p. 60.

68. Vicary, *Anatomy,* p. 77. Albucasim uses *tentigo* in his *Chirurgia,* 2.71; see Hyrtl, *Onomatologia,* entry "clitoris"; Adams, *Latin Sexual Vocabulary,* pp. 103–104, and OED, entry "tentigo." By the seventeenth century, *tentigo* meant quite precisely the clitoris. See, for example, the Jena dissertation of André Homberg, *De tentigine, seu excrescentia clitoridis* (1671), listed as a reference in the long entry "clitoris" in *Dictionnaire des sciences medicales* (Paris, 1813), vol. 5.

69. It occurs in *De anima,* 3.9.432b21; or "God and nature create nothing that is pointless," in *De caelo,* 1.4.271a33.

70. Nathaniel Highmore, *The History of Generation* (London, 1651), pp. 84–85.

71. Lemnius, *The Secret Miracles,* pp. 8–9. In general Aristotle was in somewhat low repute. The sixteenth century, as Jerome Bylebyl puts it, was "the golden age of Galenism" ("School of Padua," p. 340). Ian Maclean, *The Renaissance Notion of Woman* (Cambridge: University Press, 1980), agrees with this assessment in his examination of specific theories of generation. But Aristotle, even if he was in some circles the major representative of outmoded scholastic learning, remained influential and well worth attacking.

72. Vicary, *The Englishe Man's Treasure* (London, 1586), p. 55. This is a version of his 1548 *Anatomy.*

73. Sherman J. Silber, *How to Get Pregnant* (New York: Scribners, 1980), in addition to giving a useful layman's account of the statistics of fertilization, says that half of the married women who have not become pregnant after a year of planned trying become pregnant during the following six months with no therapeutic intervention. A pat on the head would thus appear to work for fully half of the supposedly infertile population. A considerable literature supports the view that this happens in a high proportion of cases.

74. René Bretonnayan, *La Generation de l'homme et le temple de l'aime* (Paris, 1583), section entitled "De la conception et sterilité." Pleasure here and in all my texts refers to heterosexual procreative intercourse. Though the manuals I have seen may also have been used as guides to sexual pleasure for its own sake, they are all couched in terms of procreation. Many of these works also point out that defects that make conception impossible—atresia of the vagina, absence of a womb, malformed penis—do not necessarily interfere with pleasure.

75. Gabriello Fallopio, *De decoratione* in *Opuscula* (Padua, 1566), p. 49, "De praeputii brevitate corrigenda." This and most other works, except for the *Anatomical Observations* (1561), were probably written by Fallopius' students or others who traded on his name. God ordained circumcision among the Jews, this text says, so that they might concentrate on his service rather than on pleasures of the flesh. The notion that circumcision reduces pleasure and hence the chance of conception is fairly widespread.

76. Lorenz Fries, *Spiegel,* p. 129; Avicenna, *Canon,* 3.20.1.44.

77. I take this example from Wirsung, *Neues Artzney,* p. 258.

78. Guillaume de la Motte, *A General Treatise of Midwifery,* trans. Thomas Tomkyns, Surgeon (London, 1746), p. 12. He is identified as a surgeon and male midwife at Valognes, a small city in northwestern France.

79. All of this is commonplace, but there is a particularly thorough discussion of the problem of heat and barrenness in Trotula of Salerno, *The Diseases of Women,* ed. Elizabeth Mason-Huhl (Los Angeles: Ward Ritchie Press, 1940), pp. 16–19. This text is in all likelihood not by the female healer named Trotula to whom it is usually attributed. But it was among the most widely circulated medieval works on gynecology (Chaucer cites it), was translated during the Renaissance into various vernaculars, and was included in the many editions of Caspar Woolf's massive encyclopedia of gynecology (which first appeared in 1566). See John Benton, "Trotula, Women's Problems, and the Professionalization of Medicine in the Middle Ages," *Bulletin of the History of Medicine,* 59 (Spring 1985), 30–54.

80. One of the most complete discussions of the physiology and clinical treatment of barrenness is Lazarus Riverius, *The Practice of Physick* (London, 1672), pp. 502–509. More generally see Nicholas Fontanus, *The Woman's Doctour* (London, 1652), pp. 128–137; Leonard Sowerby, *The Ladies Dispensatory* (London, 1652), pp. 139–140, for materials that "cause standing of the yard"; Jacob Rueff, *The Expert Midwife* (London, 1637), p. 55.

81. I have found nothing on women's using lascivious talk to influence men, but generally male impotence and the inability to engender a child are dealt with pharmacologically—occasionally magically—in the tracts and treatises I have consulted, much in the same way they consider the problem in women.

82. John Sadler, *The Sicke Woman's Private Looking Glass* (London, 1636), p. 118. Since the advice Sadler offers in English is quite explicit, it is curious that the sentence about foreplay is in Latin: "Mulier praepari ac disponi debet molli complexu, lascivis verbis oscular lasciviora miscenda."

83. The sentence on titillation to bring down the seeds is in the French edition, *Oeuvres* (Paris, 1579), bk. 22, chap. 4; the remainder is in "Of the Generation of Man," in *The Workes of the Famous Chirurgion,* trans. [from the Latin version of the French original] Thomas Johnson (London, 1634), bk. 24, pp. 889–890. These excerpts serve to remind us of the complexity of Renaissance metaphors of generation. "If you find her hard to the spur [a horseman's metaphor, perhaps a play on the two meanings of venery, from *venari,* to hunt, and *vener,* sexual pleasure], and the cultivator will not enter into [plow] the field of nature freely," says the French edition, mixing images of the hunt with what seem to be Aristotelian images of the womb as a field. But then Paré shifts to the Galenic two-seed model when, during orgasm, both sexes produce seeds that mix.

This mixing of metaphors is not confined to medical tracts. A Dutch clergyman's wife, for example, complains in her diary of her husband's penchant for coitus interruptus. It is, Isabella De Moerloose laments, no better than masturbation. Indeed, it is worse because in such truncated intercourse she too casts seed on barren ground: "If it had been on one side only, then it is still acceptable, but two seeds which are discharged at the same time must certainly be a child." Now she shifts to an Aristotelian metaphor: "just as rumen makes the milk curdle, the male makes the female seed curdle." Here and throughout her diary she mixes images of female activity and passivity, ideas chosen from contradictory sources as the moment dictated. See Roodenburg, "The Autobiography of Isabella De Moerloose," pp. 530–531.

84. Peter of Spain's *Thesaurus pauperum,* a major medieval source, gives 34 prescriptions for aphrodisiacs, 26 for contraceptives, and 56 to ensure fertility not counting those designed to bring on the menses which can be seen as abortifacients (Jacquart and Thomasset, *Sexuality,* pp. 91–92 and chap. 3). See also John Scarborough in A. C. Crombie and Nancy Siraisi, eds., *The Rational Arts of Living* (Northampton: Smith College Studies in History, no. 50, 1987). Two of the largest sixteenth-century herbals refer to more than 40 plants that were thought to be sexually stimulating. See Thomas G. Benedek, "Beliefs about Human Sexual Function in the Middle Ages and Renaissance," in Douglas Radcliffe-Umstead, ed., *Human Sexuality in the Middle Ages and Renaissance* (Pittsburgh: University of Pittsburgh Press, 1978), p. 108.

85. G. R. Quaife, *Wanton Wenches and Wayward Wives: Peasants and Illicit Sex in Early Seventeenth England* (London: Croom Helm, 1979), p. 172. A young girl claimed in early eighteenth-century Yorkshire that a clergyman tried to seduce her with the promise that he was too drunk for her to conceive. Another prostitute claimed to swallow some spices to prevent conception. Drenching oneself with water to cool the body was said to have a similar effect. Almost all of the texts I have cited have long sections on heating (and cooling) drugs in relation to infertility and menstrual dysfunction. Dryander, *Der Gantzen Artzenei,* chap. 7 on barrenness and chap. 19 on the mother, is particularly rich in medications, as is Michael Baust the Elder, *Wunderbarliches Leib und Wund Artzneybuch* (Leipzig, 1596), pp. 109–113. Book 1 of Baust's *Der Ander Theil des Wunderbarliches* (Leipzig, 1597) is devoted entirely to human blood and makes clear the extent to which a single economy of fluids is common to both sexes. See Nicholas Culpepper, *School of Physick* (London, 1696), p. 245, for calamint explicitly as an abortifacient and for "bringing on the termes."

86. Vesalius, *Epitome,* p. 84; on p. 85 he says that the same holds for women. Cushing, *Bio-Bibliography,* pp. 44–45, gives credence to Guenther of Andernach's open letter (1536) praising his pupil Vesalius for discovering the asymmetrical insertion of the two seminal veins. Singer and Rabin in *A Prelude,* pp. lxii–lxiii, argue that this fact was known to Mondino, who in turn cites Avicenna, who himself cites Galen (the cite given is *De. ven. art. diss.,* Kuhn, 2.808). *UP* 2.635 (see Chapter 2, note 1, above) seems to make the point as well, though Galen reverses, from a modern perspective, right and left.

87. On the issue of the epigastric vessels, see Charles Singer's introduction to Joannes Ketham, *The Fasciculo di Medicina* (Florence, 1925), 1.104, n. 59. Some writers also argued explicitly for a close connection between the genitals and the chest in men—too much sex leads to spitting blood, for example. See Jacquart and Thomasset, *Sexualité,* p. 123.

88. Laurent Joubert, *Erreurs populaires* (Bordeaux, 1579, 2nd ed.) pp. 451, 157; he was also chancellor of the faculty of medicine at the University of Montpellier. On this important writer and class of texts see Davis, "Proverbial Wisdom," esp.

pp. 258–262; Paré, *Workes,* trans. Johnson, p. 547. Joubert's account echoes that of Isidore of Seville.

89. Nicolo Serpetro, *Il Mercato delle maraviglie della natura, overo istoria naturale* (Venice, 1653), p. 23. I am grateful to Paula Findlen for this material.

90. Wirsung, *Neues Artzney,* p. 440; Crooke, *Microkosmographia* (1615), bk. 3, chap. 20. One might have thought that the publication of Harvey's *Essay on the Motion of the Heart and Blood in Animals* in 1628, which among other things argued that the heart was a pump and not a furnace, would have made views like Crooke's immediately obsolete. But they remained hearty throughout the seventeenth century. This is true of many other discoveries. Aranzio, for example, in 1564 found that the mother's blood supply did not anastamose directly with the fetus' via the "cotyledons," but this did not alter the view that the mother's blood nurtured the child and that there was therefore no extra blood to be expelled as menses. On this discovery see Howard B. Adelmann, *Marcello Malpighi and the Evolution of Embryology* (Ithaca: Cornell University Press, 1966), 2.754; Adelmann's introduction is a magisterial history of theories of generation from antiquity to Malpighi.

91. John Bulwer, *Anthropometamorphosis* (London, 1653), p. 390, says that the cuts provided an alternative drain for the body's plethora; Joubert, *Erreurs,* pp. 159–160; Culpepper, *Directory for Midwives,* p. 68. I am interested in the logic of these claims in relation to the themes of this book. I do not here, or elsewhere, want to commit myself to their truth or falsehood. It is quite possible that because of a high level of exercise, diet, low body fat, extended lactation, and such, Indian women menstruated less or less regularly than European women. Generally speaking, almost nothing is known about the cross-cultural nature or even existence of the menstrual cycle. See Buckley and Gottlieb, *Blood Magic,* pp. 44–47.

92. Cardanus is cited in Crooke, *Microcosmographia,* pp. 193–194.

93. Serpetro, *Il Mercato,* p. 24.

94. A. R. [Alexander Ross], *Arcana Microcosmos, or the hidden secrets of man's body discovered* (London, 1652), p. 88; Joubert, *Erreurs,* pp. 474–475 (his source in Aristotle may be *HA* 3.20.522a13ff).

95. On these themes see Caroline Bynum, *Holy Feast,* and her *Jesus as Mother* (Berkeley: University of California Press, 1982).

96. Wirsung, *Neues Artzney,* p. 427, "white stuff" (*weiss gesicht,* literally white-appearing). Note the assumption that one needs to designate which semen is at issue. There is a fascinating account of how an eighteenth-century German doctor and his female patients understood the interconvertibility of milk and other fluids in Duden's *Geschichte,* pp. 127–129.

97. Albrecht von Haller, *Physiology: Being a Course of Lectures* (London, 1754), 2.293; Hermann Boerhaave, *Academical Lectures on the Theory of Physic* (London, 1757), p. 114. Haller was one of the giants of eighteenth-century biology, and Boerhaave was arguably the most important clinical teacher of the late seventeenth and early eighteenth centuries. For further clinical notes on the relationship between bleeding generally and menstruation, see John Locke, *Physician and Philosopher . . . with an Edition of the Medical Notes* (London: Wellcome History of Medicine Library, 1963), pp. 106, 200.

98. I am grateful to Natalie Zemon Davis for this information on Louise Bourgeois.

99. Maclean, *Renaissance Notion of Woman*, p. 3.

100. I have used the translation by Benjamin Jowett in Hamilton and Cairns, eds., *Collected Dialogues*. After a discussion of how the "animated" substances of procreation are created in men and women, Plato says that in men "the organ of generation, becoming rebellious and masterful, like an animal disobedient to reason . . . seeks to gain absolute sway, and the same is the case with the so-called womb or matrix in women" (p. 1210).

101. See Walther von Wartburg, *Französisches Etymologisches Wörterbuch* (Tubingen: J. C. B. Mohr, 1948).

102. Quoted in Ilza Veith, *Hysteria: The History of a Disease* (Chicago: University of Chicago Press, 1965), p. 39; see pp. 28–29 for Soranus' argument that the womb is not an animal.

103. Galen, *On Anatomical Procedures,* trans. Charles Singer (New York: Oxford University Press, 1956), 6.5.561, p. 159.

104. Smollett's attack on Nihell is in *Critical Review,* 9 (1760), 187–197. The passage in question is in Elizabeth Nihell's *A Treatise on the Art of Midwifery* (London, 1760), p. 98, and her response is in *An Answer to the Author of the Critical Review . . . by Mrs. Elizabeth Nihell, a Professed Midwife* (London, 1760). I am grateful to Lisa Cody for these references.

105. Recall that some writers in the Hippocratic corpus thought that women were warmer than men, but then of course the values were reversed. Hot and cold may have meant more than good and bad, but they certainly meant that as well.

106. Columbus, *Anatomica,* pp. 446–447. I owe this novel grammatical analysis of Columbus entirely to my research assistant, Mary McGarry.

4. Representing Sex

1. Stephen Greenblatt, "Fiction and Friction," in *Shakespearean Negotiations* (Berkeley: University of California Press, 1988), p. 68. "Bias," from the game of bowls, refers to the curved path imparted by an off-center lead weight to a ball when it is rolled.

2. Angus Fletcher, *Allegory: The Theory of a Symbolic Mode* (Ithaca: Cornell University Press, 1964), pp. 110, 115–116. Foucault in *The Order of Things* makes much the same point.

3. Andreas Vesalius, *De humani corporis fabrica* (Basel, 1543), 5.12, pp. 519–520.

4. *Signatures of Internal Things: or A True and Lively Anatomy of the Greater and Lesser World* (London, 1669), pp. 5–6. This book is explicitly Paracelsian, but the system of belief outlined here extends well beyond any one tradition, as Keith Thomas shows in *Religion and the Decline of Magic* (New York: Scribner's, 1971).

5. John Tanner, *The Hidden Treasures of the Art of Physick Fully Discovered in Four Books* (London, 1659), pp. 36–37. The OED gives the following from Sir Walter

Raleigh for the use of "perspective glasse": "A worthy astrologer now living [Galileo] who by the help of perspective glasses has found in the stars many things unknowne to the Ancients."

6. Robert Bayfield, *Enchiridion medicum* (London, 1655), introduction, n.p.
7. Anon., *Anthropologia Abstracted: or the Idea of Humane Nature Reflected* (London, 1655), p. 74. According to its preface, the book is by a "Doctor of Physick" at a great university who died young more than twelve years prior to its publication. On Arachne, see Ovid, *Metamorphosis,* trans. Mary M. Innes (Harmondsworth: Penguin, 1955), pp. 134–138.
8. Christopher Wirsung, *Ein Neues Artzney Buch* (1572), p. 417. Nicholas Culpepper, *Directory for Midwives* (1696), pp. 67–68, argues that the menses is called the flowers in English because they go before conception "as flowers go before fruit."
9. *The Faerie Queene,* 3.6.3–8, in *The Poetical Works of Edmund Spenser,* ed. J. C. Smith and E. De Selincourt (Oxford: Oxford University Press, 1912; paperback 1977), pp. 171–172.
10. It would take more than two centuries of experimental evidence before the link between heat and reproduction was finally broken and the possibility of spontaneous generation put to rest. From a modern vantage, the very first results should have demonstrated its impossibility.
11. Hildegard quoted in Peter Dronke, *Women Writers of the Middle Ages* (Cambridge: University Press, 1984), p. 176.
12. Lorenz Fries (Laurentius Phryssen), *Spiegel der Artzney* (Strasbourg, 1518, 1546), pp. 127–128. "Brosam" is a curious simile for the fetus' being protected by the amniotic sac/crust. It is used in the sense of "crumb" by Luther in translating Luke 16.21, where dogs lick the beggar Lazarus' sores as he is "desiring to be fed with the crumbs [brosam] which fell from the rich man's table."
13. Mikhail Bakhtin, *Rabelais and His World* (Cambridge: MIT Press, 1968), p. 318.
14. Ibid., pp. 317–318, 320–323. It is curious that, since he includes pregnancy in the functions of the grotesque body, Bakhtin fails to mention the womb as one of its central organs.
15. *Winthrop's Journal: History of New England, 1630–1649,* ed. James Kendall Hosmer (New York: Scribner's, 1908; 1966 reprint), 1.266–269. For a general account of the creation of monsters which reviews earlier theories, see Paul-Gabriel Bouce, "Imagination, Pregnant Women, and Monsters in Eighteenth Century England and France," in G. S. Rousseau and Roy Porter, eds., *Sexual Underworlds of the Enlightenment* (Chapel Hill: University of North Carolina Press, 1988), pp. 86–100.
16. Norbert Elias, *The History of Manners: The Civilizing Process,* trans. Edmund Jephcott (New York: Pantheon Books, 1978).
17. Leah Marcus, "Shakespeare's Comic Heroines, Elizabeth I, and the Political Uses of Androgyny," in Mary Beth Rose, ed., *Women in the Middle Ages and the Renaissance* (Syracuse: Syracuse University Press, 1986), pp. 141–142. See also Carla Freccera, "The Other and the Same: The Image of the Hermaphrodite in Rabelais," in Margaret W. Ferguson, Maureen Quilligan, and Nancy J. Vickers, *Rewriting the Renaissance* (Chicago: University of Chicago Press, 1986), pp. 145–158.

18. See Roberto Zapperi, *L'Homme encient* (Paris: Presses Universitaires de France, 1983).

19. "The Lady That Was Castrated," in *Bawdy Tales from the Courts of Medieval France,* trans. Paul Brians (New York: Harper and Row, 1973), pp. 24–35.

20. Michel Foucault, introduction to *Herculine Barbin* (New York: Pantheon, 1980), pp. vii–viii. Ivan Illich makes much the same point when he distinguishes "economic sex" from "vernacular gender." The former is, I think, what sex generally means in the modern world, a "complementary duality"; the latter means "the polarization of a common characteristic," which is roughly how I regard sex in the one-sex model. Both sex and gender, Illich says, "are social relations with only tenuous connection to anatomy." *Gender* (New York: Pantheon, 1982), p. 14.

21. There are of course other traditions in which this debate is carried out. In addition to Maclean, *Renaissance Notion of Woman,* see Manfred Fleischer, "'Are Women Human?' The Debate between Valens Acidalius and Simon Gediccus," *Sixteenth Century Journal,* 12.2 (1981), 107–120. Much of this sounds like classical concern about bodily adornment and effeminacy, for which see Maud Gleason, Chapter 2, note 80.

22. Castiglione, *The Book of the Courtier* (1561), trans. Thomas Hoby (London: Dent, Everyman's Library, 1966), p. 39.

23. Ibid., p. 200. I cannot find this view in Aristotle from the citation the editor Thomas Hoby gives.

24. Ibid., pp. 193–194.

25. Ambroise Paré, *On Monsters and Marvels,* trans. Janis L. Pallister (Chicago: University of Chicago Press, 1982), pp. 31–32.

26. Michel de Montaigne, *Travel Journal,* trans. Donald Frame (San Francisco: North Point Press, 1983), pp. 5–6. See also *The Complete Essays of Montaigne,* trans. Donald Frame (Stanford: Stanford University Press, 1965), 1.2, p. 69.

27. Gaspard Bauhin, *Theatrum anatomicum* (Basel, 1605), p. 181, cited in William Harvey, *Lectures on the Whole Anatomy* [*Prelectiones anatomiae universalis,* 1616], trans. C. D. O'Malley, F. N. L. Poynter, and K. F. Russell (Berkeley: University of California Press, 1961), p. 132 and n. 467.

28. Pliny, *Natural History,* trans. H. Rackham, Loeb Classical Library (London: Heinemann, 1942), 7.4.36–38, vol. 2, p. 531.

29. Sir Thomas Browne, *Pseudodoxia Epidemica: or, Enquiries into Vulgar and Common Errors* (1846). The seventeenth-century pornographic work by J. B. Sinibaldi, *Rare Verities: The Cabinet of Venus Unlocked and Her Secrets Laid Open* (London, 1658), has a chapter that answers affirmatively the question "whether females may change their sex." See Roger Thompson, *Unfit for Human Ears* (Ottawa: Rowman and Littlefield, 1979), pp. 168–169.

30. Ovid, *Metamorphoses,* 9.794. See also Barkan on Iphis in *Gods Made Flesh,* pp. 70–71.

31. The story of Marie is not in the "A" text of the *Essays* but was added subsequently by Montaigne. This may account for why the comments about the imagination seem to apply more immediately to the Iphis story than to the new interpolation. See *Oeuvres complètes* (Paris: Gallimard, 1962), pp. 96, 1453.

32. In book 1, chap. 8, "On Idleness," Montaigne seems to regard the imagination

as an external force that can act upon the body. Fertile ground brings forth all manner of weeds unless it is subjected and properly sown. Similarly, women bring forth "lumps of shapeless flesh" unless "manured with another kinde of seede." (See Chapter 2 above on mola). So it is, he continues, with minds that, unless busied with some subject, will "slatter themselves through the vaste field of imagination."

33. See Nancy J. Vickers, "The Mistress in the Masterpiece," in Nancy K. Miller, ed., *The Poetics of Gender* (New York: Columbia University Press, 1986), pp. 36 and 19–41. I have also consulted Vickers' manuscript "Blazon," which discusses in detail this new courtly genre that "assumed its definitive elaboration in a collective volume of 1543 entitled *Blasons anatomiques du corps feminin.*" On the efforts of women to be heard among these male voices, see for example Ann Rosalind Jones, "City Women and Their Audiences: Louise Labé and Veronica Franco," in *Rewriting the Renaissance,* pp. 299–316.

34. Charles Estienne, *La Dissection des parties du corps humain* (Paris, 1546), 3.41, in the context of explaining how to organize, essentially stage, a dissection. I presume that the potentially attractive privities are those of women, but the term *partie honteuse,* though feminine, is used throughout to refer to the "shameful part" of both sexes.

35. Jacques-Louis Binet and Pierre Descargues, *Dessins et traités d'anatomie* (Paris, 1980), pp. 39–40.

36. Susan Koslow, "The Curtain-Sack: A Newly Discovered Incarnation Motif in Rogier van der Weyden's *Columba Annunciation,*" *College Art Association Proceedings,* February 1985. *Vas* in its classical Latin usage was more commonly used in a sexual sense to refer to the penis and testicles (Adams, *Latin Sexual Vocabulary,* pp. 41–43, 88); sure enough, Estienne includes a similar object in an engraving of a man. Once again, however, nomenclature for the reproductive system blurs the boundaries of difference.

37. Estienne, *Dissection,* 3.7.

38. *Aretino's Dialogues,* trans. Raymond Rosenthal (New York: Stein and Day, 1971), pp. 169–170, quoted in Laura Walvoord, "'A Whore's Vices Are Really Virtues': Prostitution and Feminine Identity in Sixteenth Century Venice," unpublished research paper, Berkeley, 1987. Walvoord makes the case that shifting symbolic systems were played out by way of prostitutes.

39. Columbus, *De re anatomica* (1559), "Concerning Things Which Rarely Happen in Anatomy," 15, pp. 494–495.

40. Ibid. Columbus was obviously fascinated by this "woman" but did not intervene clinically in what even over the distance of centuries is a sad and disturbing case. "The poor woman wished that I would cut off her penis with a knife, which penis she said was an impediment when she desired intercourse with a man. She also asked me to increase the opening of her vulva so as to be capable of bearing a man. But I, who frequently longed to perceive the distinctions between these implements, put her off with words. For I did not dare to undertake the satisfaction of this desire, since I did not think it could be done without a risk of life."

41. Ambroise Tardieu, *Questions médico-légales de l'identité dans les rapports avec les vices de conformation des organes sexuels* (Paris, 1874, 2nd ed.), pp. 18–32. I am grateful

to Vanessa Schwartz for this reference. The case that evoked Tardieu's indignation seems to be that of Herculine Barbin, though Foucault reads it differently (see note 20).

42. Jacques Duval, *Traité des hermaphrodites* (1612). Actually Marie was accused of sodomy, which involved putting the right organ in the wrong place or the wrong organ in the right place or the wrong organ in the wrong place. This means that she was alleged to have put her clitoris in any of her partners' orifices, since none would have been appropriate. Neither one of two women just rubbing their genitals together would have been guilty of sodomy but of a lesser offense.

43. The question, like that in Natalie Zemon Davis' *The Return of Martin Guerre* (Cambridge: Harvard University Press, 1983), is not so much who is the real Martin—the impostor seems to have been a better Martin than the original—but who by what set of criteria gets to play the part.

44. For a more extensive discussion of this case, see my "Amor Veneris," *Zone*, 5 (1989).

45. See ibid., and Paré, *On Monsters*, p. 188, n. 35.

46. Montaigne, *Travel Journal*, pp. 5–6; *Oeuvres complètes*, p. 1118. I have modified the Frame translation slightly and noted the personal pronoun only when it is actually in the text: "Il devint amoureaux" or "elle avoit esté condamnée."

47. Paolo Zacchia, *Questionum medico-legalium* (Basel, 1653). Zacchia in this text is concerned with a wide variety of medical-legal issues: how to detect poisoning, distinguishing real from apparent death, establishing paternity, cataloguing monsters, and of course establishing sex in difficult cases.

48. Zacchia is writing very much in the tradition of Gaspard Bauhin: see Katherine Park and Lorraine J. Daston, "Unnatural Conceptions: The Study of Monsters in France and England," *Past and Present*, 92 (1981), 20–54.

49. Zacchia, para. 22, p. 494. Zacchia is at pains to argue against genitals being a proof of sex, which he regards as a function of heat. Thus he points out, in accordance with common medical knowledge, that while testes take their name from "testifying" to virility, "those selfsame parts in women are also called *testes* even if they are hidden in them." Even external testes are not a sure sign. Some male animals and birds have testes inside, and "it is clear from very trustworthy accounts that even women have a genital projecting externally" (para. 23).

50. Ibid., para. 8, p. 492. The argument here is not functional. Zacchia cites two cases of women with a clitoris so large that they could play the man's role in intercourse, and in one case the woman even claimed to have emitted through her clitoris (para. 15, p. 502).

51. Ibid., para. 42, p. 498; para. 13, p. 493.

52. Ibid., para. 28, pp. 494–495. This comes after a long discussion, paras. 26–27, of putative men becoming women and of creatures, human or otherwise, changing back and forth. Zacchia's basic view is that in cases in which men seem to become women—such as Daniel, a married soldier who became pregnant when he was lying with his wife and was impregnated by a comrade—the original male designation was mistaken (para. 13, p. 493). Daniel may have appeared to be a man but his "valid" sex was female (para. 28).

53. Llewellyn, "Dedication to Harvey," cited in Elizabeth B. Gasking, *Investigations*

into Generation, 1651–1828 (Baltimore: Johns Hopkins University Press, 1967), p. 16.

54. I have used the translation of *Exercitationes de generatione animalium* (1651) by Gweneth Whitteridge (Oxford: Blackwell, 1981).

55. On the substantive contributions of Harvey in this area, see Adelmann, *Marcello Malpighi,* 2.762–765, and the learned account in Gasking, *Investigations into Generation,* pp. 16–35.

56. Carolyn Merchant, *The Death of Nature: Women, Ecology, and the Scientific Revolution* (San Francisco: Harper and Row, 1980), pp. 156, 159.

57. The phrase just before, however, reaffirms the continuity implicit in the one-sex model and in the chain of perfection: "thus the organ of generation begins as male and is completed as female." See also Harvey, *Lectures on the Whole Anatomy,* p. 127.

58. The female is also the final or first cause, since the male is driven to venery by her presence. At times Harvey seems to want the female alone to be the efficient cause, ordered into action by the sperm. At other times, pp. 162–163 for example, he argues "that both the male and the female are the efficient causes of generation."

59. Gasking, *Investigations into Generation,* p. 16; Walter Pagel, *William Harvey's Biological Ideas* (Basel, N.Y.: Karger, 1967), p. 44. See also Pagel's *New Light on William Harvey* (Basel: Karger, 1976), which sets Harvey's invocation of epigenesis against radical atomists like Highmore as well as against Galenists. For a brief summary of Harvey's views in the context of contemporary writing on the subject, see Charles Bodemer, "Embryological Thought in Seventeenth Century England," in *Medical Investigation in Seventeenth Century England* (Los Angeles: William Andrews Clark Memorial Library, 1968), pp. 3–25.

60. *Disputations,* pp. 4–10. I do not want to exaggerate Harvey's Baconianism or his belief in a transparently readable nature. On p. 9 he gives, as his own, Aristotle's account of the relationship of universals to particulars: "Knowledge is acquired by reasoning from universals to particulars" (*Physics,* 184a16–25). He also regards science as an enterprise casting light into the darkness: "conception is indeed a dark business . . . full of shadows" (p. 443). I am obviously not the first to suggest that Harvey and his contemporaries were still deeply involved with the philosophical problems, biases they used to be called, of ancient science.

61. Harvey, like Boyle, believed that—contra the Duhem-Quine thesis—it was possible to construct a crucial experiment to prove or disprove a theory. See on this question Steve Shapin and Simon Schaffer, *Leviathan and the Air-Pump: Hobbes, Boyle, and the Experimental Life* (Princeton: Princeton University Press, 1985).

62. *Disputations,* pp. 352–353. Harvey does not tell us how long after coition the dissection was undertaken. Since his previous chapter discusses rutting in September, and since we are told that the isolation of the does began in early October, there seems to have been a period of some weeks between mating and dissection. After such an interval, there would be no evidence of semen in the womb. Harvey makes much of the image of fertility as a heightened and more noble version of the way "epidemic, contagious and pestilential diseases scatter their seeds . . . and so quietly multiply themselves" (pp. 189–190).

63. I have this account of Ruysch from David Davis, *The Principles and Practices of Obstetric Medicine* (London, 1836), 2.830.

64. *Disputations,* pp. 165–166. Most of his book is an argument for the creative power of male semen. Contrary to Aristotle, Harvey regards the female as well as the male as an efficient cause of generation, since on command of the male she actually produces the new life. Having been made fecund "by no perceptible corporeal agent," she—Harvey is puzzled as to whether the uterus itself or the whole female is the locus of power—"exercises the formative power of engendering and procreates her own like, not otherwise than does a plant which we see is empowered with the force of both sexes" (p. 443).

65. Ibid., pp. 182–183, 189, 452, 351–352.

66. Ibid., pp. 150–151, 125 (48r). I have retained Harvey's punctuation.

5. Discovery of the Sexes

Chapter epigraph: Victor Jozé, "Le Féminisme et le bons sens," *La Plume,* 154 (September 15–30, 1895), 391–392; quoted in Deborah Silverman, *Art Nouveau in Fin-de-Siècle France* (Berkeley: University of California Press, 1989), p. 72.

1. Claude Martin Gardien, *Traité complèt d'accouchements, et des maladies des filles, des femmes et des enfants,* 2nd ed. (Paris, 1816), 1.2–3, quoted in Erna Olafson, "Women, Social Order, and the City: Rules for French Ladies, 1830–1870" (diss., Berkeley, 1980), p. 97.

2. Jacques L. Moreau, *Histoire naturelle de la femme,* 2 vols. (Paris, 1803), 1, chap. 2. Moreau says that all organs, genital and others, mark sexual difference. But he also claims to be following Pierre Roussel who, as Michèle le Doeuff argues, genitalizes the entire body except in the genitals. Le Doeuff's theoretical evidence for this claim is that the phallocentric point of view has to see difference between the sexes everywhere but cannot see it at the genital level. Her substantive case is based on Roussel's—and Moreau's—position that menstruation is not a natural function of the female reproductive system but a product of modern luxury. Thus what we might take to be a specific reproductive function is expressly somaticized. See "Les Chiasmes de Pierre Roussel" in Michèle le Doeuff, *Recherches sur l'imaginaire philosophiques* (Paris: Pagot, 1980), p. 190 and passim. I discuss later the role of egg and sperm in understanding difference.

3. Cited in V. C. Medvei, *A History of Endocrinology* (Cambridge: MIT Press, 1982), p. 357. An eighteenth-century clergyman in Holme, Yorkshire, who tried to seduce one of his parishioners after a Christening service by telling the said Martha Haight "that she might venture to suffer him to have the pleasure of her body for that he was drunk and would do her no harm," was still working on the ancient theory that extra heat dried up the generative elements (Borthwick Institute MS RVII.I.360.1716). Heating elixirs to cure sterility, to induce abortions, or to do whatever more heat was supposed to do were still widely marketed in the London newspapers of the mid-eighteenth century.

4. Dr. Paul G. Donohue, syndicated column, November 10, 1987. I am grateful to Bonnie Smith for sending me this clipping. The doctor's answer misses the point. The question until the 1930s, and even to some extent today, is whether orgasm in women plays a significant role in ovulation as it does in some mammals. The

so-called gender-choice system notes that "female orgasm is not necessary but will further increase your chances of having a boy." Female orgasm is strongly counterindicated to conceive a girl. See *Mother Jones,* December 1986, p. 16.

5. See Ursula Heckner-Hagen, "Women White Collar Workers in Imperial Germany, 1889–1914: *Des Verband für weibliche Angestellte,*" (M.A. thesis, University of California, Davis, 1978), p. 62.

6. The two explanations are obviously related. The success of doctors at the expense of priests as experts on public morality is the consequence of political developments made possible by the epistemological revolution.

7. Michel Foucault, *The Order of Things: An Archaeology of the Human Senses* (New York: Pantheon, 1971), pp. 32, 54–55. I want to see this as a more general development than Foucault does; the new classical episteme continued to underlie nineteenth-century science.

8. Unlike Peter Gay, *Education of the Senses* (New York: Oxford University Press, 1984), for example, I have no stake in debating which position was more prevalent or which better described reality.

9. On the political background for Wolstenholme's claim, see Sheila Jeffreys, *The Spinster and Her Enemies* (London: Pandora, 1985), pp. 28–35, esp. 34–35.

10. This idea of heightened genital sensitivity goes back to antiquity. The contrary view is part of a new racist discussion about why black men are supposedly sexually insatiable, about the relationship of white men to black women, and so on. See the somewhat pornographic *Untrodden Fields of Anthropology* by an anonymous Dr. Jacobus (New York: Falstaff Press, n.d., ca. 1900), pp. 125, 238–239. In general, there are important parallels between post-eighteenth-century discussions of sexual and racial differences, since both seek to produce a biological foundation for social arrangements.

11. Comte George Louis Leclerc de Buffon, *Natural History* (London, 1807, orig. in French, 44 vols., 1749–1804), 4.34.

12. On the decline of Galenism as a model for organizing knowledge about the body, see Oswei Temkin, *Galenism: Rise and Decline of a Medical Philosophy* (Ithaca: Cornell University Press, 1973), chap. 4. Galenic therapeutics did not suffer a comparable decline. In late eighteenth-century London, as the Westminster Coroner's Inquests make clear, bleeding was still the first aid of choice in cases ranging from drowning and suicide by hanging to profusely bleeding head wounds. Essentially Galenic therapy to restore natural balance still dominated American medicine in the first two thirds of the nineteenth century, and Hippocrates enjoyed a substantial revival in early nineteenth-century France. On America, see John Harley Warner, *The Therapeutic Perspective* (Cambridge: Harvard University Press, 1986), pp. 83–92.

13. For an account of this changing imagery in popular medical literature, see Robert A. Erickson, "'The Books of Generation': Some Observations on the Style of the English Midwife Books, 1671–1764," in Paul-Gabriel Bouche, ed., *Sexuality in Eighteenth-Century Britain* (Manchester: Manchester University Press, 1982).

14. For the relation between "generation" and "reproduction," see François Jacob, *The Logic of Life: A History of Heredity,* trans. Betty Spillman (New York: Pantheon, 1974), chap. 1. The quote is from Bernard de Fontanelle, *Lettres galantes:*

Oeuvres, 1.322–323, in Jacob, p. 63. Actually the dominant preformationist views did not strictly entail reproduction: in a sense, nothing was either reproduced or generated in this scheme but merely grew from an already existing thing. The term "reproduction" first applied to the capacity of polyps and other such creatures to reproduce a lost appendage.

15. See Philip Curtin, *The Image of Africa* (Madison: University of Wisconsin Press, 1964), pp. 28–57.

16. I take this claim from S. T. von Soemmerring, *Über die Köpperliche Verschiedenheit des Negers vom Europäer* (Frankfurt, 1785), p. 67, who cites, in addition to his own anatomical studies—the various parts of the Negroes he discusses are available in his collections for verification—one Father Charlevoir, who reports on the severely restricted mental capacities of the New Guinea Negro: some are dumb, and some can count only to three.

17. François Poullain de la Barre, *The Woman as Good as the Man: or, the Equality of Both Sexes,* "written originally in French [*De l'égalité des deux sexes: Discours physique et moral,* 1673] and translated into English by A. L." (London, 1677), pp. 2–4.

18. For Aristotle there is nothing that could cause one to be in error in holding that the efficient cause (that which defines the male) is superior to the material cause (that which defines the female).

19. In my view, the Enlightenment's new valuation of nature as applied to women was not, as Bloch and Bloch suggest, always or even usually conservative, but open to a wide range of uses. See Maurice Bloch and Jean H. Bloch, "Women and the Dialectics of Nature in Eighteenth-Century French Thought," in MacCormack and Strathem, eds., *Nature, Culture and Gender,* pp. 25–41.

20. John Locke, *Two Treatises on Government,* ed. Peter Laslett (Cambridge: University Press, 1960), 1, para. 47, pp. 209–210.

21. This is Carole Pateman's reconstruction (*Sexual Contract,* p. 49) of the extremely obscure argument in the *Leviathan.*

22. For Locke see Lorenne M. G. Clark, "Women and Locke: Who Owns the Apples in the Garden of Eden?" in Clark and Lynda Lange, eds., *The Sexism of Social and Political Theory* (Toronto: University of Toronto Press, 1979), pp. 16–40. I obviously do not agree with Clark in seeing Locke's project as simply a version of earlier efforts to establish the inferiority of women; indeed, I think the useful evidence she adduces suggests the novelty of Locke's arguments for a very old case.

23. Jean-Jacques Rousseau, *A Discourse on Inequality,* trans. Maurice Cranston (Harmondsworth: Penguin, 1984), p. 104.

24. Alexis de Tocqueville, *Democracy in America,* ed. Phillips Bradley (New York: Knopf, 1945), 2.223.

25. See Londa Schiebinger, *The Mind Has No Sex?* (Cambridge: Harvard University Press, 1989), pp. 191–200.

26. See John Mullen, "Hypochondria and Hysteria: Sensibility and the Physicians," *The Eighteenth Century,* 25.2 (1984), 141–174, esp. 142. See also Michel Foucault, *Madness and Civilization* (New York: Vintage, 1988), pp. 153ff, for the relation between sympathy and nerves and between sympathy and order.

27. Attributed to Charles Cotton, *Erotopolis: The Present State of Betty-land* (London, 1684); Thomas Stretzer, *Merryland*(orig. 1740, New York: Robin Hood House, 1932), 45–65. I am grateful to Lisa Cody for these references.

28. Robert B. Todd, *Cyclopedia of Anatomy and Physiology* (London, 1836–1839) 2.685–686, 684–738. The major French medical encyclopedia of the period gives a similar account.

29. Lazzaro Spallanzani, *Experiences pour servir à l'histoire de la generation des animaux et des plantes* (Geneva, 1785), para. 123.

30. R. Couper, *Speculations on the Mode and Appearances of Impregnation in the Human Female* (Edinburgh, 1789), p. 41.

31. On Hunter, see Evard Home, "An Account of the Dissection of an Hermaphrodite Dog, to Which Are Prefixed Some Observations on Hermaphrodites in General," *Philosophical Transactions,* 69 (1799), part 2. More generally on artificial insemination, though without reference to its value for understanding how conception works in the female, see F. N. L. Poynter, "Hunter, Spallanzani, and the History of Artificial Insemination," in Lloyd G. Stevenson and Robert P. Multhauf, eds., *Medicine, Science and Culture* (Baltimore: Johns Hopkins University Press, 1968), pp. 99–113.

32. Samuel Farr, *The Elements of Medical Jurisprudence* (London, 1785), pp. 42–43.

33. Soranus, *Gynecology,* trans. Temkin, p. 36. Soranus makes this claim seem commonsensical by pointing out that "Similarly in women who mourn, appetite for food often exists but is obscured by grief."

34. Richard Burn, *Justice of the Peace* (London, 1756), p. 598. He cites a line of lawyers who subscribed to this view but then quotes Hawkins to the effect that their legal opinions are dubious: "the previous violence is in no way extenuated by the present consent," because if one followed this rule the trial of the offender would have to wait until pregnancy was determined, and finally the "philosophy of this notion may be very well doubted of."

35. Cited in J. S. Forsyth, *A Synopsis of Modern Medical Jurisprudence* (London, 1829), pp. 499–500.

36. Nevertheless, as late as 1865 a leading forensic physician reported that rapists' attorneys were using the fact of pregnancy in their clients' defense and worried that, if juries actually believed these arguments, great injustice would result. Susan Edwards, *Female Sexuality and the Law* (Oxford: Robertson, 1981), p. 124.

37. Matthew Hale (1609–1676), *Historia placitorum coronae,* p. 631; first American ed., *History of the Pleas of the Crown* (Philadelphia, 1847).

38. On the prosecution of rape, see Anna Clark, *Women's Silence, Men's Violence* (London: Pandora, 1987).

39. John Mason Good, *The Study of Medicine* (Boston, 1823), 4.100.

40. J. A. Paris and J. S. M. Fontblanque, *Medical Jurisprudence* (London, 1823), 1.436–437.

41. T. R. Beck, *Elements of Medical Jurisprudence* (London 1836, 6th ed.), p. 109. Beck admits that "we do not know, nor shall probably ever know, what is necessary to cause conception."

42. Couper, *Speculations,* p. 40; E. Sibley, *Medical Mirror* (London, n.d. but ca. 1790), p. 15.

43. Thomas Denman, *An Introduction to the Practice of Midwifery* (London, 1794), 1.73–74.
44. I base these observations on the 27th, or 100th anniversary, edition of Henry Gray, *Anatomy of the Human Body,* ed. Charles Mayo Goss (Philadelphia: Lea and Febiger, 1959), figs. 74, 77, 90, 827, among many others.
45. S. T. von Soemmerring, *Abildung des menschlichen Auges* (Frankfurt, 1801), preface, unpaginated.
46. Bernard Albinus, *Table of the Skeleton and Muscles of the Human Body,* quoted in Schiebinger, *The Mind Has No Sex?,* p. 203. My account of the making of the perfect female skeleton is indebted to Schiebinger, pp. 200–211.
47. Leon Battista Alberti, *On Painting,* trans. J. R. Spencer (New Haven: Yale University Press, 1966), bk. 3, p. 93. This was already a very old quote when Alberti recorded it.
48. Quoted in Schiebinger, p. 200.
49. It should be pointed out that while physical anthropologists can generally determine the sex of a skeleton, it is very difficult to do so from pictures without the gross exaggerations used by late eighteenth- and nineteenth-century anatomists. Skeletal specimens in the anatomy lab do not make sexual difference readily apparent, as any student will attest. For illustrations of these various skeletons, see Schiebinger, pp. 204–205.
50. See, for example, the table showing the embryological origins of the male and female reproductive system in Rudolph Wagner, *Handwörterbuch der Physiologie* (Braunschweig, 1853), 3.763, which is essentially identical to that in any modern developmental anatomy text.
51. See Desmond Heath, "An Investigation into the Origins of a Copious Vaginal Discharge during Intercourse: 'Enough to Wet the Bed—That Is Not Urine'," *Journal of Sex Research,* 20 (May 1984), 194–215.
52. William Cowper, *The Anatomy of Humane Bodies* (London, 1737), introduction, no pagination. Note that Cowper still thought it necessary to specify *masculine* sperm; the word did not mean what it does today but refers rather to the whole of the male ejaculate, to what we would call semen.
53. Hartsoeker, *Essai de dioptrique* (Paris, 1694), chap. 10, sec. 89, quoted in Jacob, *Logic of Life,* p. 59.
54. The word *reproduction* came to be finally distinguished from the older term *generation* only during the course of the nineteenth century, when the production of new parts of individuals (regeneration) was understood as fundamentally different from the making of new individuals.
55. I rely heavily here on Frederick B. Churchill, "Sex and the Single Organism: Biological Theories of Sexuality in Mid-Nineteenth Century," in William Coleman and Camille Limoges, eds., *Studies in the History of Biology* (Baltimore: Johns Hopkins University Press, 1979), pp. 139–177, and on the excellent summary of eighteenth-century embryological theories in Shirley A. Roe, *Matter, Life, and Generation: 18th-Century Embryology and the Haller-Woolf Debate* (Cambridge: University Press, 1981), pp. 1–22.
56. Rudolph Jakob Camerarius, *De sexu plantarum epistola* (Tubingen, 1694), p. 20, cited in Delaporte, *Nature's Second Kingdom,* p. 94.

57. Carolus Linnaeus, *Species plantarum* (1753), vol. 1, with an introduction by W. T. Stearn (facsimile of first edition, printed for the Ray Society, London, 1957), pp. 32–33.

58. Even then, of course, their meanings were not fixed as further studies of the nucleus made their comparative size less significant. Continued research into fertilization continues to change views of what aspects of egg and sperm matter.

59. Roe, *Matter, Life*, pp. 44, 70–73, 77–79, and n. 24, p. 178.

60. See Churchill, "Sex and the Single Organism," pp. 142ff; Gasking, *Investigations into Generation,* pp. 63–65 and chap. 5 generally.

61. Roe in *Matter, Life* makes it clear that, however much one side or the other wanted to base some claim about gender on the nature of sperm and egg, the scientists who argued futilely about ovism and animalculism debated on different and more technical grounds. Haller's move from spermaticist to epigenesist, holding that the embryo was preformed in the egg, had to do with the importance granted to certain observations and to the politics of science, not of gender.

62. On sperm, see F. J. Cole, *Early Theories of Generation* (Oxford: Clarendon Press, 1930), chaps. 1 and 2. On Spallanzani, see Gasking, *Generation,* pp. 132–136.

63. Buffon, *Natural History,* 3.228–229. The arguments made against Buffon by Haller were similar to those made by Aristotle against the pangenesists. See Roe, pp. 28–29.

64. Pierre de Maupertuis, *The Earthly Venus* (orig. in French, 1745 and 1756), p. 6. See also chap. 2 in which he discusses the vexed question of whether the male semen actually touches the egg. He, like many before and after, denied that such contact was necessary for fertilization. He writes that "nine months after a woman has surrendered to the pleasure which perpetuates mankind, she brings into the world a small creature" (p. 4). He marvels, in a culturally fraught passage: "how will this blissful area [the womb] become a dark prison for a formless and senseless embryo. How can it be that the cause of such pleasure, the origin of so perfect a being, is nothing more than flesh and blood" (p. 6). Preformationist theories arose precisely because there seemed no other plausible account of how matter organized itself into new forms.

65. Achille Chereau, *Memoires pour servir a l'étude des maladies des ovaires* (Paris, 1844), p. 91. Still the best article on nineteenth-century views of the ovary is Carol Smith-Rosenberg and Charles Rosenberg, "The Female Animal: Medical and Biological Views of Women in Nineteenth-Century America" *Journal of American History,* 60 (September 1973). See also Carol Smith-Rosenberg, "Puberty to Menopause: The Cycle of Femininity in Nineteenth-Century America," reprinted in her book *Disorderly Conduct* (New York: Oxford University Press, 1985), pp. 182–196.

66. Hysteria in women is entirely of ovarian and not of uterine origin, says the authoritative late nineteenth-century *Dictionnaire encyclopédique des sciences médicales,* under "sexe."

67. One must not be too quick to condemn all gynecological surgery. As the distinguished American physician and researcher Mary Putnam Jacobi wrote Elizabeth Blackwell, the first woman to take a medical degree in Great Britain, "when you

shudder at 'mutilations' it seems to me you can never have handled a degenerated ovary or a suppurating Fallopian tube,—or you would admit that the mutilation had been effected by disease . . . or neglect . . . before the surgeon intervened." Letter dated December 25, 1888, Library of Congress, Blackwell MS Box 59. I am grateful to Regina Marantz Sanchez for providing this material.

68. See Brown, *Body and Society,* pp. 67–68, and Rousselle, *Porneia,* pp. 121–128.

69. *History of Animals,* 9.50.632a22.

70. Percival Pott, *The Chirurgical Works* (London, 1808, new ed.), case 24, "An Ovarian Hernia," pp. 210–211. A modern gynecologist interprets this as a rare instance of bilateral prolapse of the ovaries into the inguinal sacks. Their removal would lead to the masculination Pott describes. I am grateful to Roger Hoag, M.D., for his diagnosis.

71. Theodor von Bischoff, *Beweis der von der begattung unabhängigen periodischen reifung und belosung der eier der säugethiere unter des Menschen als der ersten Bedingung ihrer fortpflanzung* (Giessen, 1844), pp. 41–42, citing G. Roberts, *Fragments d'un voyage dans les provinces intérieures de l'Inde en 1841.* Edward John Tilt, the well-known English gynecologist who was one of the main proponents of the view that the ovaries controlled women's sexual urges and were in turn affected by them—"excessive use of sexual intercourse is not infrequently a cause of sub-acute ovaritis"—also cites Roberts as evidence for the fact that the ovaries produce "the characteristic luxuriance" of the female form. He claims, in one of the most bizarre of all nineteenth-century orientalist fantasies, that the operation was carried out "to serve the lascivious propensities of Eastern despots." *On Diseases of Menstruation and Ovarian Inflammation* (London, 1850), p. 53. Roberts' description seems consistent with what would ensue from prepubertal removal of the ovaries but also with a pituitary disorder.

72. L. Hermann, *Handbuch der Physiologie* (Leipzig, 1881), vol. 6, pt. 2: section by V. Hensen, "Physiologie der Zeugung," pp. 69ff. Menstruation after removal of the ovaries appeared because, not knowing what the ovary did, physicians were not careful to remove all ovarian tissue and left some remaining on the pedicle or mesovarium to which the ovary is attached.

73. A. Charpentier, *Cyclopedia of Obstetrics and Gynecology,* trans. Egbert H. Grandin (New York, 1887; orig. in French, 1882), pp. 95–96.

74. Cited in George Corner, "The Early History of Estrogenic Hormones," *Journal of Endocrinology,* 31 (1964–65), iv. His comments are in the context of writings about ovarian pathology.

75. Hegar uses the term "castration" advisedly. Some people, he writes, want to use ovariotomy to refer to the excision of diseased ovaries, whereas castration is reserved for the removal of healthy ones. No, says Hegar. The ovaries he removes may appear from clinical examination to be healthy, but one cannot deny that his patients suffer. To claim that the ovaries he removes are healthy is like saying that the sclerotic arteries of a man about to die of a stroke are healthy just because his physician could not diagnose them as diseased. Hegar is determined, in short, to see the ovaries as guilty until proven innocent. "Zur Begriffsbildung der Kastration," special reprint from the *Centralblatt für Gynäkologie* (1887), pp. 44, 6–7 (available in Hegar's collected papers at the Crerar Library, Chicago). Re-

sponding to criticism, he also denies that he routinely prescribes castration for hysteria, which he acknowledges has a broad range of causes; only in the rarest case can it be shown to originate in the gonads. But there are other diseases of a neurotic nature that do stem from the genital organs. These have a physical quality ("das leiden"), which sometimes goes away during pregnancy and menstruation. Hegar is clearly uncomfortable under attack and tries to display himself as a responsible physician. See "Für Castration bei Hysterie," *Berlin Klinischen Wochenschrift*, 26 (1880).

76. Alfred Hegar, *Die Castration der Frauen von physiologischen und chirugischen Stanpunkte aus* (Leipzig, 1878), pp. 41ff.

77. MS journal of Mabel Loomis Todd, Yale University Library, May 15, 1879. I am grateful to Peter Gay for providing this material.

78. I do not mean to suggest that these questions are easy to answer. Competent scientists come to very different conclusions from the same, by now rather large, body of data from humans and primates. See Donald Symons, *The Evolution of Human Sexuality* (New York: Oxford University Press, 1979), and the review by Sarah Blaffer Hrdy in *Quarterly Review of Biology*, 54 (September 1979), 309–313.

79. I shall return to the history of the vexed question of what causes ovulation in the next chapter. As it happens, rabbits and a few other relatively exotic creatures—ferrets, minks, short-tailed shrews—are coitally induced ovulators. Humans and most other mammals ovulate "cyclically" or "spontaneously." The distinction was not clear until the twentieth century. Regnier de Graaf, *De mulierum organis generationi inservientibus*, trans. George W. Corner in *Essays in Biology in Honor of Herbert Evans* (Berkeley: University of California Press, 1943), pp. 55–92. Regarding coitally induced versus spontaneous ovulation, see A. V. Nalbandov, *Reproductive Physiology of Mammals and Birds* (New York: Freeman, 3rd ed. 1976), pp. 132–133, and R. M. F. S. Sadleir, *The Reproduction of Vertebrates* (New York: Academic Press, 1973), pp. 127–129. Current thinking tends to blur the rigid distinction between coitally induced and spontaneous ovulators and regards animals on a continuum. For an application of this approach to humans, see J. H. Clark and M. X. Zarrow, "Influence of Copulation on Time of Ovulation in Women," *American Journal of Obstetrics and Gynecology*, 109 (April 1971), 183–185.

80. William Smellie, *A Treatise on the Theory and Practice of Midwifery* (London, 1779), 1.90.

81. Albrecht von Haller, *Physiology: Being a Course of Lectures*, vol. 2 (London, 1754), pp. 300–303. Haller was an ovist when he wrote this, but there are almost identical accounts by spermaticists. Thus Henry Bracken writes, "In the act of generation, the pleasure is so exquisite, as to alter the Course of the Blood and Animal Spirits, which at this time move al those parts which before lay still." *Midwife's Companion* (London, 1737). The prominent William Smellie gave essentially the same account (*Treatise*, 1.92).

82. This autopsy is reported in Pierre Dionis, *The Anatomy of Humane Bodies* (London, 1716, 2nd ed.), p. 237.

83. W. C. Cruickshank: "Experiments in which, on the third day after impregnation,

the ova of rabbits were found in the Fallopian tubes." *Philosophical Transactions,* 87 (1797). As von Baer recognized, Cruickshank came very close to identifying the mammalian ova.

84. This is in some ways curious, since von Baer makes much of the development of the Graafian follicle, which he calls the egg, and of the actual egg, the little egg or "eichin" within it, as part of the natural history of the ovary. "On the Genesis of the Ovum of Mammals and of Man," trans. C. D. O'Malley, *Isis,* 47 (1956), 117–153, esp. 119.

85. John Pulley, *Essay on the Proximate Causes of Animal Impregnation* (Bedford, 1801), pp. 9–10. This claim is made in the context of an argument with Haighton (see below).

86. J. G. Smith, *The Principles of Forensic Medicine* (London, 1827, 3rd ed.), p. 483.

87. J. F. Blumenbach, *The Elements of Physiology,* trans. from the 4th Latin ed. by John Elliotson, M.D. (London, 1828), n. "i" by translator, p. 468.

88. Henry John Todd, ed., *Cyclopedia of Anatomy and Physiology* (London, 1836–1839), "Generation," p. 450.

89. For less ambivalent statements of this position, see David Davis, *Principles and Practice of Obstetric Medicine* (London, 1836), 2.830–831, who argues strongly that a corpus luteum is evidence of impregnation. Haighton in the *Philosophical Transactions* (1797), p. 164, says that it furnishes "incontestable proof" of prior pregnancy. An instance of the practical significance of this issue is related in G. F. Girdwood, "Mr. Girdwood's Theory of Menstruation," *Lancet* (1842–43), p. 829, where a dead woman who had long been married but childless was suspected of marital infidelity. Her large corpus luteum was taken as proof by some of impregnation, but Girdwood denied the inference and posthumously saved the woman's honor.

90. The analogy here and above is with the beasts: the corpus luteum has "also been found in the female quadruped after a state of periodical lasciviousness, where no copulation had taken place." Smith, *Principles,* p. 482.

91. Physicians recognized the existence of a large corpus luteum in the ovaries of women who died during early pregnancy, but were at a loss to explain either their function or their relation to the many "miniature corpora lutea" they also found there. See, for example, Robert Knox, "Contributions to the History of the Corpus Luteum, Human and Comparative," *Lancet* (May 9, 1840), pp. 226–229. Robert Lee, a major teacher of gynecology and obstetrics, was still trying to sort this out in 1853. See his *Clinical Reports on Ovarian and Uterine Diseases* (London, 1853), pp. 16–20.

92. Blumenbach, *Elements,* pp. 483, 485.

93. Johannes Müller, *Handbuch der Physiologie des Menschen,* vol. 2 (Coblenz, 1840), pp. 644–645. The 1848 supplement to the 1843 English translation of this work berates Müller for having dealt with the production of the ovum inadequately and cites a great deal of literature in this very active field. See William Baly, *Recent Advances in the Physiology of Motion, the Senses, Generation* (London, 1848), pp. 43–61.

94. John Bostock, *An Elementary System of Physiology* (Boston, 1828), p. 25.

95. Davis, *Principles,* p. 831. Todd, in *Cyclopedia of Anatomy* under "Generation," had

already argued that the ovaries "become unusally vascular during sexual union." It is not clear what he means by "venereal orgasm," but probably not a process that includes an affective component. "Orgasm" in nineteenth-century medical writing usually refers simply to some form of turgidity or to a state of great pressure.

96. Erna Lesky, *The Vienna Medical School in the 19th Century* (Baltimore: Johns Hopkins University Press, 1976), pp. 106–116.

97. Girdwood to Grant, *Lancet* (1842–43), pp. 825, 826; *Lancet* (1840–41), p. 295; F.-A. Pouchet, *Théorie positive 'de ovulation spontanée* (Paris, 1847), pp. 125ff.

98. Davis, *Principles,* pp. 57–58.

99. Todd, *Cyclopedia,* pp. 439, 447, 443.

100. Havelock Ellis, *Studies in the Psychology of Sex,* vol. 3 (Philadelphia: F. A. Davis, 1920, 2nd ed.), pp. 193–194.

101. Adam Raciborski, *De la puberté et de l'âge critique chez la femme* (Paris, 1844), p. 486; Acton, *Functions,* (4th ed., 1865), p. 112.

102. Davis, *Principles,* p. 830. These speculations are cited by later physicians, and I have added the italics to emphasize the assumption, common in the nineteenth century, that menstruation is the human equivalent of heat and that women are more sexually responsive then.

103. Cited in Peter Gay, *The Bourgeois Experience,* p. 161.

104. R. D. Owen, *Moral Physiology* (New York, 1828), p. 44.

105. Josef Ignaz von Dollinger, "Versuch einer Geschichte der Menschlichen Zeugung," trans A. W. Meyer, *Human Generation* (Stanford: Stanford University Press, 1956), p. 37.

106. Giuseppe Pitre, *Sicilian Folk Medicine,* trans. Phyllis Williams (Lawrence, Kans.: Coronado Press, 1971), introduction.

107. Henry Campbell, *Differences in the Nervous Organization of Man and Woman: Physiological and Pathological* (London, 1891), pp. 200–201.

108. Carl Degler, "What Ought to Be and What Was," *American Historical Review,* 79 (December 1974), 1467–90.

109. Rosalind Rosenberg, *Beyond Separate Spheres* (New Haven: Yale University Press, 1982), p. 181, n. 6.

110. Matthews Duncan, *On Sterility in Women,* Gulstonian Lecture delivered at the College of Physicians, February 1883 (London, 1884), pp. 96–100.

111. E. Heinrich Kisch, *Die Sterilitat des Weibes* (Vienna and Leipzig, 1886), pp. 5, 16–17. Kisch was professor of medicine at Prague and during summers the chief physician at Marienbad.

112. Barton C. Hirst, *A Textbook of Obstetrics* (Philadelphia and London: W. B. Saunders, 1901), p. 67.

6. Sex Socialized

1. *Encyclopédie* (1751), 5.471, "Femme, droit nat." See also p. 469 for an explicit attack on the Galenic view that the penis was only a prolapsed uterus and more generally that woman was only a lesser man.

2. Dorinda Outram, *The Body and the French Revolution* (New Haven: Yale University Press), p. 156.

3. Marquis de Condorcet, "On the Admission of Women to the Rights of Citizenship," in *Selected Writings,* ed. Keith Baker (Indianapolis: Bobbs-Merrill), pp. 98, 99.

4. Olympe de Gouges, "Declaration of the Rights of Woman and Citizen" (1791), in Susan Groag Bell and Karen Offen, eds., *Woman: The Family and Freedom* (Stanford: Stanford University Press, 1983), p. 105.

5. Necker de Saussure, *L'Education progressive,* 2.274, quoted in Hellerstein, pp. 184–185; see also Leonore Davidoff and Catherine Hall, *Family Fortunes* (Chicago: University of Chicago Press, 1987).

6. Auguste Debay, *Hygiene et physiologie du mariage* (Paris, 1850 ed.), part 1, "Philosophie du mariage," pp. 88–90, 39–48, 55. See on doctors versus clergy, Angus McLaren, "Doctor in the House: Medicine and Private Morality in France, 1800–1850," *Feminist Studies,* 2.3 (1974–75), 39–54.

7. William Acton, *Functions . . .* (1857).

8. Susanna Barrows, *Distorting Mirrors* (New Haven: Yale University Press, 1981), chap. 1.

9. Susan Sleeth Mosedale, "Science Corrupted: Victorian Biologists Consider 'The Woman Question,'" *Journal of the History of Biology,* 11 (Spring 1978), 1–55; Elizabeth Fee, "Nineteenth-Century Craniology: The Study of the Female Skull, *Bulletin of the History of Medicine,* 53 (Fall 1979), 915–933; Lorna Duffin, "Prisoners of Progress: Women and Evolution," in Sara Delamont and Lorna Duffin, eds., *The Nineteenth-Century Woman: Her Cultural and Physical World* (New York: Barnes and Noble, 1978), pp. 915–933. For two contemporary English articulations of these themes, see Grant Allen, "Plain Words on the Woman Question," *Fortnightly Review,* 46 (October 1889), 274; and W. L. Distant, "On the Mental Differences Between the Sexes," *Journal of the Royal Anthropological Institute,* 4 (1875), 78–87.

10. Pateman, *The Sexual Contract,* p. 41.

11. See on this point Jean Bethke Elshtain, *Public Man, Private Woman: Women in Social and Political Thought* (Princeton: Princeton University Press, 1981), chap. 3.

12. Millicent Fawcett, "The Emancipation of Women," *Fortnightly Review,* 50 (November 1891), a response to Frederic Harrison's article with the same title in the previous month's issue, which argued that emancipated women would become like men; italics mine.

13. Joan B. Landes, *Women and the Public Sphere* (Ithaca: Cornell University Press, 1988), p. 11. More generally see chap. 3 on Rousseau's "Reply to Public Women" and chaps. 1 and 2 on new women's voices and symbolic politics.

14. For a recent account of the centrality of sexual difference in Rousseau's account of the origins of sociability and dependence, see Joel Schwartz, *The Sexual Politics of Jean-Jacques Rousseau* (Chicago: University of Chicago Press, 1984), pp. 3 and 1–40 passim. Elshtain in *Public Man, Private Woman* also argues for the centrality of sexual difference in the political philosophy of Rousseau.

15. Jean-Jacques Rousseau, *A Discourse on Inequality,* trans. Maurice Cranston (Har-

mondsworth: Penguin, 1984), pp. 102–104. Rousseau also argues against Locke that in the state of nature there is no reason for any one woman to seek any one man, or vice versa. The bonds of family as well as passion are the creation of civilization (pp. 165–166, n.L).

16. Pufendorf quoted in Schwartz, *Sexual Politics,* p. 19. These passages, which do not of course deal with the ultimate difference of male and female passion, are explicitly concerned with establishing another divide: that between humans and beasts.

17. Rousseau, *Discourse,* pp. 102–103, 110; *Emile,* trans. Allan Bloom (New York: Basic Books, 1981), book 5, pp. 359 and 362n.

18. *Emile,* pp. 357–358, 362–3; italics mine.

19. *Encyclopédie,* "Jouissance," 5.889. I have used the translation with some modifications in Stephen J. Gendzier, trans., *The Encyclopedia* (New York: Harper and Row, 1967), p. 97.

20. For a general account of the condition of women as markers of progress in Millar's four stages of civilization, see Paul Bowles, "John Millar, the Four-Stages Theory, and Women's Position in Society," *History of Political Economy,* 16 (Winter 1984), 619–638. Bowles rightly sees economics as the engine of change in Millar but plays down the active mediation of women in translating structural changes into new cultural norms. See also Ignatieff's article in Istvan Hont and Michael Ignatieff, eds., *Wealth and Virtue: The Shaping of Political Economy in the Scottish Enlightenment* (Cambridge: University Press, 1983), and Sylvana Tomaselli, "The Enlightenment Debate on Women," *History Workshop,* 20 (1985), 101–124.

21. John Millar, *Origin of the Distinction of Ranks* (Basel, 1793), pp. 14, 32, 86, 95–96.

22. In my discussion of Wheeler I rely heavily on Barbara Taylor, *Eve and the New Jerusalem: Socialism and Feminism in the Nineteenth Century* (New York: Pantheon, 1983), esp. chap. 2.

23. Catherine Gallagher, "The Body versus the Social Body in the Works of Thomas Malthus and Henry Mayhew," in Gallagher and Thomas Laqueur, eds., *The Making of the Modern Body* (Berkeley: University of California Press, 1987), pp. 83–106.

24. Anna Wheeler and William Thompson, *An Appeal of One-Half the Human Race, Women, Against the Pretensions of the Other Half, Men, To Retain Them in Political and Thence in Civil and Domestic Slavery* (London, 1825), pp. 60–61.

25. Ibid., p. 145 and part 2, question 2.

26. Mary Wollstonecraft, *Thoughts on the Education of Daughters* ... (London, 1787), p. 82. Mary Poovey, *The Proper Lady and the Woman Writer* (Chicago: University of Chicago Press, 1984), pp. 80–81, 48–81 passim. See also Zillah Eisenstein, *The Radical Future of Liberal Feminism* (New York: Longmans, 1981), pp. 89–112.

27. Theodor G. von Hippel, *On Improving the Status of Women* (1792), trans. Timothy F. Sellner (Detroit: Wayne State University Press, 1979), pp. 66, 143, 147, and chap. 5 passim. Hippel's term for "opposite sex" is "anderes geschlecht" and might be translated simply as "the other sex," but in German as in English it is

almost always used by a male writer or speaker to refer to the female or to his own sex in relation to the other. The sense of opposition rather than superiority/inferiority is part of the linguistic making of incommensurability. I do not know when this sense of the term entered German, but the OED gives the first English use in the *Spectator* (1711); "Nothing makes a woman more esteemed by the opposite sex than chastity." Again, the capacity for sexual control defines opposition.

28. Mary Wollstonecraft, *Female Reader* (London, 1789), p. vii; Taylor, *Eve,* pp. 47–48.

29. Davidoff and Hall, *Family Fortunes,* p. 179 and chap. 3. Domestic ideology may be defined as the belief that the domestic sphere is the primary arena for teaching morality and proper conduct, that this sphere is dominated by women, and that women therefore exercise enormous public influence by their efforts in the home.

30. Sarah Ellis, *The Wives of England* (London, n.d.), p. 345; *The Daughters of England: Their Position in Society, Character and Responsibilities* (London, 1842), p. 85. Mitzi Myers, "Reform or Ruin: A Revolution in Female Manners," *Studies in the Eighteenth Century,* 11 (1982), 199–217, makes a persuasive case for considering writers as far apart politically as the domestic ideologist and Wollstonecraft as engaged in a similar moral enterprise.

31. Elizabeth Blackwell, *The Human Element in Sex* (London, 1884), pp. 52, 57, 16.

32. Ibid., pp. 54, 21, 26, 44, 31.

33. Elizabeth Blackwell, *A Medical Address on the Benevolence of Malthus, contrasted with the Corruptions of Neo-Malthusianism* (London, 1888), pp. 17, 25, 34, 32.

34. For a more general discussion of this theme, see Sheila Jeffreys, *The Spinster and Her Enemies* (London: Pandora, 1985).

35. Aldous Huxley, "Literature and Science" (1963), quoted in Peter Morton, *The Vital Science: Biology and the Literary Imagination* (London: Allen and Unwin, 1984), p. 212.

36. Evolutionary theory can be, and of course was, interpreted to support the notion of an infinitely graded scale, reminiscent of the one-sex model, on which women were lower than men, childlike in the same way that blacks were childlike with respect to whites. I do not want to argue that Darwin himself consistently held any particular view on this subject or that any particular view can be derived from Darwinian theory. As with the debate about the nature of competition in society, any view—and thus no view—about sexual difference can be derived from evolutionary biology. My reading of Darwin has been influenced by Morton, *The Vital Science,* and by Gilian Beer, *Darwin's Plots* (London: Routledge, 1983).

37. Charles Darwin, *The Origin of Species* (1859) (Garden City: Doubleday, 1958), pp. 96–97. There is actually among animals considerable evidence against the idea of the coy female. See Sarah Blaffer Hrdy, "Empathy, Polyandry, and the Myth of the Coy Female," in Ruth Bleier, ed., *Feminist Approaches to Science* (New York: Pergamon Press, 1986), pp. 118–146.

38. Havelock Ellis argues explicitly that "it is the spontaneous and natural instinct of the lover to desire modesty in his mistress." *The Evolution of Modesty,* part 1, in *Studies in the Psychology of Sex* (1900, 1920), 1.45, quoted in Ruth Yeazell, "Nature's Courtship Plot in Darwin and Ellis" (unpublished MS), who argues for

the narrative generalization of Darwin's views. Like Diderot and Rousseau, Ellis regards modesty as engendering sexual desire and thinks that it diminishes after marriage and notes: "The difference in ticklishness between the unmarried woman and the married woman corresponds to their difference in degree of modesty." *Sexual Selection in Man,* in *Studies,* 6.18.

39. *The Descent of Man and Selection in Relation to Sex* (1871) (Princeton: Princeton University Press, 1981), 2.402 (part 2, chap. 21); 2.329–330 (chap. 19).

40. O. S. Fowler, *Practical Phrenology* (New York, n.d.), p. 59; also p. 67. I purchased my copy of this book from a workingman's club library in Aylesbury, Bucks. For a general overview of these issues, see Robert M. Young, *Mind, Brain, and Adaptation in the Nineteenth Century* (London: Oxford University Press, 1970), pp. 47–49. The cerebellar locus of sexuality is expounded most fully in George Combe, *On the Functions of the Cerebellum by Dr. Gall* (Edinburgh, 1838), a translation of the relevant parts of F. J. Gall and J. C. Spurzheim, *Anatomie et physiologie du systeme nerveaux* (Paris, 1810–1819). J. Chitty, *A Practical Treatis on Medical Jurisprudence* (London, 1834), p. 270, treats it as a commonplace.

41. The history of physiology in the nineteenth and twentieth centuries generally is underdeveloped, and the history of both human and animal reproductive physiology is even less well explored. There simply is not enough work on the day-to-day practice of nineteenth-century scientists working in reproductive biology to understand in detail how social issues structured their practice.

42. Theodor von Bischoff, *Beweis der von der Begattung unabhängigen periodischen Reifung und Loslösung der Eier der Säugethiere und des Menschen* (Giesen, 1844), pp. 28–31.

43. F. A. Pouchet, *Théorie positive de l'ovulation spontanée et de la fécondation des mammifères et de l'espèce humaine* (Paris, 1847), pp. 452, 104–167. Pouchet was a distinguished naturalist, a corresponding member of the French Academy of Science, and a man of considerable independent judgment and courage. In the famous debate between Pouchet, a believer in spontaneous generation, and his opponent Pasteur, he and not Pasteur worked against the grain of religious and political orthodoxy. See John Farley and Gerald Giesen, "Science, Politics and Spontaneous Generation in Nineteenth Century France: The Pasteur-Pouchet Debate," *Bulletin of the History of Medicine,* 48 (Summer 1974), 161–198.

44. Jules Michelet, *L'Amour* (Paris, 1859), p. xv.

45. Bischoff, *Beweis,* p. 43.

46. V. Hensen in L. Hermann, *Handbuch der Physiologie* (Leipzig, 1881), 6.2.69.

47. Q. U. Newell et al., "The Time of Ovulation in the Menstrual Cycle as Checked by Recovery of the Ova from the Fallopian Tubes," *American Journal of Obstetrics and Gynecology,* 19 (February 1930), 180–185.

48. By the twentieth century, pathologists were better able to tell the age of the corpus luteum and could therefore put ovulation somewhere in the middle of the menstrual cycle. But even this evidence showed wide variation, with several major researchers locating mean time of ovulation just after or within days of the end of the menses. See the summary of this research in Carl G. Hartman, *Time of Ovulation in Women* (Baltimore: Williams and Wilkins, 1936).

49. Paget cited in R. J. Tilt, *Diseases of Menstruation* (London, 1850), p. xxvii.
50. Ibid., pp. 141–155; R. L. Dickenson, *Human Sex Anatomy* (Baltimore: Williams and Wilkins, 1933), fig. 42. We know now that days 12–14 in the cycle are by far the most likely for conception.
51. George H. Napheys, *The Physical Life of Woman* (Walthamstow, 1879), pp. 69–70.
52. Carl Capellmann, *Facultative Sterilität ohne Verletzung der Sittengesetze* (Aachen, 1882).
53. Marie Stopes, *Married Love,* p. 148. It is thus little wonder, as Dr. Bessie Moses said in her report on the first five years of the Baltimore Contraception Bureau, that almost all of her patients who claimed to be using the rhythm method abstained from intercourse just before, during, and just after the menstrual flow, that is, during what they took to be the fertile period. See Hartman, *Time of Ovulation,* p. 149.
54. A. S. Parkes, "The Rise of Reproductive Endocrinology, 1926–1940," *Journal of Endocrinology,* 34 (1966), xx–xxii; Medvei, *History,* pp. 396–411; George W. Corner, "Our Knowledge of the Menstrual Cycle, 1910–1950," *Lancet,* 240 (April 28, 1951), 919–923.
55. Pouchet, *Théorie positive,* p. 227.
56. Augustus Gardiner, *The Causes and Curative Treatment of Sterility, with a Preliminary Statement of the Physiology of Generation* (New York, 1856), p. 17. *Lancet* (January 28, 1843), 644, states flatly: "The menstrual period in women bears a strict physiological resemblance" to the heat of "brutes."
57. Cited as the epigraph of chap. 3, "The Changes That Take Place in the Non-Pregnant Uterus During the Oestrous Cycle," in F. H. A. Marshall's classic *The Physiology of Reproduction* (New York, 1910), p. 75.
58. Bischoff, *Beweis,* pp. 40, 40–48.
59. *GA* 738b5ff, 727a21ff; see also Chapter 2 above.
60. Pliny, *Natural History,* 7.13.15.63; Loeb ed., 2.547.
61. Haller, *Physiology,* p. 290 (419 of the 1803 English ed.).
62. Blumenbach, *Elements,* pp. 461–462.
63. Robert Remak, "Über Menstruation und Brunst," *Neue Zeitschrift für Geburtskunde,* 3 (1843), 175–233, esp. 176.
64. Müller, *Handbuch,* 2.640.
65. Jean Borie, "Une Gynaecologie passionée," in Jean-Paul Aron, ed., *Misérable et glorieuse: La Femme du XIX siècle* (Paris: Fayard, 1980), pp. 164ff.
66. Pouchet, *Théorie positive,* pp. 12–26 (on the use of logic in the absence of hard evidence, see his discussion of the first law, esp. p. 15); pp. 444–446 for a summary of his program.
67. G. F. Girdwood, "On the Theory of Menstruation," *Lancet* (October 1844), 315–316.
68. Adam Raciborski, *Traité* (Paris, 1868), pp. 43–47. His *De la puberté et de l'âge critique chez la femme* (1844) was often cited, along with Bischoff, as having established spontaneous ovulation in women.
69. Ibid., pp. 46–47. "L'orgasme de l'ovulation," incidentally and again inexplicably, was not a moment of pleasure but an orgasm in the sense of "an increase in the

vital action" of the organ being considered. See Littré, *Dictionnaire,* "Orgasme." This heightened activity in turn resulted in nervous irritation, which was somehow communicated to the uterus and caused it to be fattened with blood. Then, with the bursting of the ovarian follicle, the dam was broken, the egg was released, and the womb gave up its extra blood. Alternatively, the pressure affected the uterus, which began to bleed somewhat before the release of the egg.

70. Nägele, *Erfahrungen und Abhandlungen* . . . (Mannheim, 1812), p. 275. See p. 270 regarding estrus of domesticated animals occurring at times other than when they are fertile.

71. Blumenbach, *Physiology,* p. 455.

72. Heape had argued explicitly in an earlier paper that heat and menstruation were analogous, with differences due to the general conditions affecting higher mammals. "The Menstruation of *Semnopithecus entellus,*" *Philosophical Transactions,* 185.1 (1894). It is unfair to discuss this man only in the context of his political views because his work on menstruation and ovulation in primates is of considerable scientific importance.

73. Walter Heape, "Ovulation and Degeneration of Ova in the Rabbit," *Proceedings of the Royal Society,* 76 (1905), 267.

74. Walter Heape, *Sex Antagonism* (London, 1913), p. 23.

75. L. Adler and H. Hitschmann, "Der Bau der Uterusschleimhaut des geschlechsreifen Weibes mit besonderer Berucksichtigung der Menstruation," *Monatschrift für Geburtshulfe und Gynäkologie,* 27.1 (1908), esp. 1–8, 48–59.

76. This is actually Marshall's summary in his immensely popular *Textbook,* p. 92, of Heape's account in "Menstruation of *Semnopithecus entellus.*"

77. Havelock Ellis, *Man and Woman: A Study of Human Secondary Sexual Characteristics* (London, 1904), pp. 284, 293.

78. Rudolf Virchow, *Der püpurele Zustand: Das Weib und die Zelle* (1848), quoted in Mary Putnam Jacobi, *The Question of Rest for Women During Menstruation* (New York, 1886), p. 110.

79. According to Michelet, *L'Amour,* p. 393, the ovary was not the only source of women's fundamental sickness. The nineteenth was the century of the uterus: "Ce siècle sera nommé celui des maladies de la matrice," he argues, having identified the fourteenth century as that of the plague and the sixteenth as that of syphilis (p.iv). For discussion see Thérèse Moreau, *Le Sang de l'histoire: Michelet, l'histoire et l'idée de la femme au XIXe siècle* (Paris: Flammarion, 1982).

80. Charpentier, *Cyclopedia of Obstetrics and Gynecology,* trans. Grandin (1887), part 2, p. 84.

81. Quoted in Hans H. Simmer, "Pflüger's Nerve Reflex Theory of Menstruation: The Product of Analogy, Teleology and Neurophysiology," *Clio Medica,* 12.1 (1977), 59.

82. Elie Metchnikoff, *The Nature of Man: Studies in Optimistic Philosophy,* trans. P. Chalmers Mitchel (New York: Putnam's, 1908). Metchnikoff, after 1883 a professor at the Pasteur Institute, was according to his translator "a votive of the new religion who has left everything for science" (p. 91). He thought that menstrual periods were a result of progress and culture, particularly a high marriage age: "In these circumstances it is not wonderful that menstruation should appear

so abnormal and even pathological." See below for others who believed that menstruation was a sign of civilization.

83. Jacobi, *Question,* pp. 1–25, 81, and 223–232. Section 3, pp. 64–115, is devoted to laying out and criticizing the so-called ovarian theory of menstruation.

84. Ibid., pp. 98–100. Jacobi was generally an opponent of what she took to be sentimental or romantic views of women's role in the world.

85. Ibid., pp. 83, 165.

86. Ibid., pp. 99, 167–168.

87. W. F. Ganong, *Review of Medical Physiology,* 8th ed. (Los Altos: Lang, 1977), p. 332.

88. Edward Westermarck, *The History of Human Marriage* (New York, 1891). Westermarck assumes "that marriage existed among primitive men," taking as a premise that which he wishes to conclude.

89. The poem "Ellis Ethelmer," in *Woman Free* (Congleton: Women's Emancipation Union, 1893), pp. 10–17. I am grateful to Susan Kent for sending me a copy of this poem. See her *Sex and Suffrage in Britain, 1860–1914* (Princeton: Princeton University Press, 1987), for the general context of Wolstenholme's attack on the notion of complementary and cooperative separate spheres.

90. Ellis, *The Phenomena of Sexual Periodicity,* in *Studies,* 1.85–160, summarizes the vast nineteenth-century literature. He was so committed to the menses-estrus connection that when he was able to study "directly" the cycles of desire in two women—in one case based on a diary of erotic dreams, in another on a diary of masturbatory episodes—he interpreted a second and to him surprising peaking of desire evident at the middle point of the cycle. This was the result of *mittelschmerz,* now regarded as a symptom of ovulation but considered by Ellis as *nebenmenstruation,* a secondary menstruation, a "minor or abortive menstruation" that might be the first sign of a future breaking up of the menstrual cycle into two. Modern studies find no consistent cycle of desire in relation to ovulation. The following literature on behavioral aspects of the menstrual cycle is especially useful: Robert Snowden et al., *Patterns and Perceptions of Menstruation* (New York: St. Martin's Press, 1983); Lorraine Dennerstein: "Hormones and Female Sexuality" and "The Menstrual Cycle-correlating Biological and Psychological Changes," in Dennerstein and Myriam de Senarclens, eds., *The Young Woman: Psychosomatic Aspects of Obstetrics and Gynaecology* (Princeton: Excerpta Medica, 1983); Naomi W. Morris and J. Richard Udry, "Epidemiological Patterns of Behavior in the Menstrual Cycle," and Gregory D. Williams and Ann Marie Williams, "Sexual Behavior and the Menstrual Cycle," in Richard C. Friedman, ed., *Behavior and the Menstrual Cycle* (New York: Marcel Dekker, 1982).

91. This story is from Nancy Burley, "The Evolution of Concealed Ovulation," *American Naturalist,* 114 (December 1979), 835–858. For an argument on the other side, correlating the social and endocrine elements in primate sexual behavior, see M. J. Baum, "Hormonal Modulation of Sexuality in Female Primates," *BioScience,* 33.9 (1983), 578–582. Sarah Blaffer Hrdy argues that hidden ovulation in primates, and by extension in humans, is a way of making a number of males feel that they might have been the father of a child and hence obliged to care for it; the certainty of paternity is clearly not necessary to bind father to

child. For a popular summary of this argument, see "Heat Loss," *Science*, 83 (October 1983), 73–78, and a more technical account in Barbara B. Smuts et al., eds., *Primate Societies* (Chicago: University of Chicago Press, 1986), "Patterning of Sexual Activity, pp. 370–384.

92. This view is widespread, but I quote here Peter Wagner's description of the new literature on masturbation in *Eros Revived: Erotica of the Enlightenment in England and America* (London: Secker and Warburg, 1988), p. 16.

93. M. A. Petit, *Medium of the Heart,* cited in M. Larmont, *Medical Advisor and Marriage Guide* (New York, 1861), p. 325. Petit was supposedly a physician in Lyon.

94. Joseph W. Howe, M.D., *Excessive Venery, Masturbation, and Continence* (New York, 1896), p. 67.

95. Foucault, *History of Sexuality,* vol. 1. *Onania* appears in newspaper advertisements during the first decade of the eighteenth century and went through countless editions over the next two centuries. Tissot's work was written in 1760 and translated into English in 1766. *The Silent* came out in Birmingham around 1840.

96. I disagree in my account with Schwartz, *Sexual Politics,* who distinguishes these episodes, pp. 105–106; Rousseau, *Confessions,* Modern Library ed., p. 111; Rousseau, *Emile,* pp. 4, 334–335. Rousseau's account is an early version of the modern adage, "Masturbation is sex with someone you love."

97. Henry Thomas Kitchener, *Letters on Marriage . . . and on the Reciprocal Relations between the Sexes* (London, 1812), 1.22. He cites Rousseau's *Emile* at this point. The title of course speaks to the alternative, social sexuality.

98. Goss and Co., *Hygeiana* (n.d., ca. 1840), pp. 59–60. The horror stories regarding females are worse than those involving males in this book: spasm, hysteria, rickets, painfully enlarged clitoris, vaginal discharge, and much more. Using the cordial produced by this company renders such poor creatures once again fit mothers with regular menses.

99. Owen, *Moral Physiology,* pp. 34–35.

100. Samuel Sullivan, *A Guide to Health, or Advice to Both Sexes in Nervous and Consumptive Complaints, Scurvy, Leprosy, Scrofula: also on Certain Disease and Sexual Debility* (London, 66th ed., n.d. but sold in New York in 1847), p. 207. I give the full title to show the company that masturbatory debility keeps.

101. Richard Carlile, *Every Woman's Book or What Is Love containing Most Important Instructions for the Prudent Regulation of the Principle of Love and the Number of a Family* (London, 1828), esp. pp. 18, 22, 26–27, 37–38. I consulted a 1892 reprint of the 1828 edition published by the Malthusian League; the tract was originally published in Carlile's ultraradical *Red Republican*.

102. The connection between the unleashing of desire and its valuation in classic economic thought, as brilliantly discussed by Albert Hirschman in *The Passions and the Interests* (Princeton: Princeton University Press, 1977), has never been studied in relation to the new differentiation of desire in which men produce and desire sex while women reproduce and desire goods. They, after all, are the new consumers. Isabel Hull is exploring these issues in her studies of sexuality and the making of civil society in eighteenth-century Germany.

103. Mothers are cautioned to warn their daughters that the solitary vice will make them unfit to fulfill their normal functions and leave them with something they will not be able to share with their virtuous husband without shame. Eliza Duffy, *What Women Should Know* (London, 1873). Old views of producing if not monsters then certainly deformities through social practices were alive and well in the eighteenth century. See the opening of Sterne's *Tristram Shandy* and, for a general account, Paul-Gabriel Boucé, "Imagination, Pregnant Women, and Monsters in Eighteenth Century England and France," in G. S. Rousseau and Roy Porter, eds., *Sexual Underworlds of the Enlightenment* (Chapel Hill: University of North Carolina Press, 1988), pp. 86–100.

104. Tilt, *Diseases of Menstruation*, p. 54; Ryan, *Philosophy of Marriage*, p. 168. Ryan, like most other nineteenth-century authorities, continued to believe in so-called moral causes of sterility and held that "reserve and frigidity during the approach of the sexes" can lead to barrenness within marriage (p. 157). See also, for example, Frederick Hollick, *The Marriage Guide or Natural History of Generation* (London, 1850), p. 72; Campbell, *Differences*, pp. 211–212; Ryan, *Jurisprudence*, p. 225; Napheys, *Physical Life*, pp. 77–78.

105. See Fleetwood Churchill's classic *Outlines of the Principal Diseases of Females* (Dublin, 1838), in which he greatly admires Parent-Duchâtlet's work but nevertheless maintains what he presumes to be the "general opinion": "scarcely any organ [as the clitoris] is so liable to enlargement from frequent excitation, and thus in turn prompts to repetition of the excitement." Perhaps in his view prostitutes do not engage in clitoral stimulation. On prostitution and exchange, the classic text is George Simmel, "Prostitution," in Donald Levine, ed., *On Individuality and Social Forms* (Chicago: University of Chicago Press, 1971).

106. Lucretius, *The Nature of the Universe*, trans. Ronald Latham (Harmondsworth: Penquin, 1951), p. 170. No one, as far as I can tell, cited any evidence for this claim between its articulation in the twelfth century and its going out of favor in the late nineteenth.

107. Regarding excessive moisture as a cause of barrenness, see, for example, R. B. [R. Buttleworth?], *The Doctresse: A Plain and Easie Method of Curing Those Diseases Which Are Peculier to Women* (London, 1656), p. 50. A variant on the heat argument is that ordinary women experience two orgasms, one from the alteration in her cold state caused by the inflow of hot sperm from the male and another from her own emission. Harlots, whose wombs are already hot from excessive intercourse, lack the first. On this claim see Helen R. Lemay, "William of Saliceto on Human Sexuality," *Viator*, 12 (1981), 172. She attributes it to William of Conches or some twelfth-century interpolator. William of Conches is cited in Jacquart and Thomasset, *Sexuality*, p. 88. Lorenz Fries (Phryssen), *Spiegel der Artzney* (1518, 1546), p. 130, says: "Die unfruchbarkeyt wirt auch dardurch geursacht, so die fraw kein lust zu dem mann hat, wie dann die gemeynen frawlin, welche alleyn umb der narung willen also arbeyten." My colleague Elaine Tennent of the German Department at Berkeley suggests that, though the use of "Frawlin" ("Fraulein" in modern German) instead of "Fraw" as in the previous clause, supports reading "gemeynen Frawlin" as prostitutes, it would also support the reading I give in the parentheses in my text. Even if

one were to accept this later reading, Fries's argument still supports my claim that the relationship to production and exchange is marked on the body's capacity to procreate. On heat and religious fervor, see William Bouwsma, *John Calvin* (New York: Oxford University Press, 1988).

108. R. Howard Bloch, *Etymologies and Genealogies: A Literary Anthropology of the Middle Ages* (Chicago: University of Chicago Press, 1983), pp. 173–174. This naturalistic expression of cultural anxiety, in the case of prostitutes and perhaps also of usury, strikes me as an aspect of the new relationship between the sacred and the profane which Peter Brown discusses in his "Society and the Supernatural: A Medieval Change," *Society and the Holy in Late Antiquity* (Berkeley: University of California Press, 1982), pp. 302–322. Indeed, the production of authoritative texts like William of Conches might be construed as evidence for Brown's shift from "consensus to authority." Catherine Gallagher, "George Eliot and *Daniel Deronda*: The Prostitute and the Jewish Question," in *Sex, Politics, and Science in the Nineteenth-Century Novel,* ed. Ruth Yeazell (Baltimore: Johns Hopkins University Press, 1986), pp. 40–41.

109. For a review of this literature up to 1968, see *Journal of the American Psychoanalytic Association,* 16 (July 1968), 405–612, which is made up of a series of articles discussing Mary Jane Sherfey's "The Evolution and Nature of Female Sexuality in Relation to Psychoanalytic Theory" in vol. 14 of the journal. Sherfey's article subsequently came out as a book, *The Nature and Evolution of Female Sexuality* (New York: Vintage, 1973). The view that "equates the occurrence of intercourse with the occurrence of female orgasm," with an adaptationist account of its evolution, is brilliantly criticized in a forthcoming book by Elizabeth A. Lloyd of the Department of Philosophy at Berkeley. Her views are summarized in Stephen J. Gould, "Freudian Slip," *Natural History,* 96 (January 1987), 14–21.

110. Robert Scholes, "Uncoding Mama: The Female Body as Text," in *Semiotics and Interpretations* (New Haven: Yale University Press, 1982), pp. 130–131 and passim.

111. Sigmund Freud, *Three Essays on the Theory of Sexuality* (1905), trans. James Strachey (New York: Avon, 1962), p. 123.

112. Ibid., p. 124.

113. Richard von Krafft-Ebing, *Psychopathia Sexualis,* trans. of 7th enlarged German ed. by Charles Gilbert Chaddock (Philadelphia: F. A. Davis, 1908), p. 31.

114. *Reference Handbook of the Medical Sciences* (New York, 1900–1908), 7.171. Hyrtl taught anatomy in the University of Vienna while Freud was studying there. The Grimms define *Kitzler* as clitoris or female rod, "weibliche Rute," and trace the associations back through a number of earlier forms. *Kitzlerin* is defined as "titillans femina" but the usage given is: "The Emperor Maximillian called one of his blunderbusses the *Kitzlerin.*"

115. Ibid., 7.172. These "endings" take their name from Wilhelm J. F. Krause (1833–1910) and are found not only in the penis and the clitoris but also in the conjunctiva of the eye and in the mucous membranes of the lips and tongue.

116. E. H. Kisch, *The Sexual Life of Women* (English trans., London, 1910), p. 180.

Kisch's *Sterilität des Weibes* (1886) is a major summary of the literature on female sexuality and reproductive biology.

117. *Dictionnaire encyclopédique des sciences médicales,* 18.138; 99.230–288. The vagina, this article reports, is longer in Negro women than in whites, corresponding presumably to the supposedly larger penis of the Negro male.

118. Georg Ludwig Kobelt (1804–1857) was a physician and the eponymous discoverer of Kobelt's network—the junction of the veins of the vestibular bulbs below the clitoris—and several other structures of the genito-urinary system. His *Die Männlichen und Weiblichen Wollusts-Organe des Menschen und verschiedene Saugetiere* (Freiburg, 1844) is the basis for the English text I have generally followed, with slight emendations: Thomas Power Lowry, ed., *The Classic Clitoris* (Chicago: Nelson Hall, 1978).

119. Modern evolutionary biologists would probably not attribute specific purposes to the clitoris but would regard its sensitivities as the female version of the adaptive characteristics of the penis, just as the characteristics of the male nipples are the result of adaptations in the female of the species.

120. *Classic Clitoris,* pp. 38, 43.

121. *Dictionnaire des sciences médicales* (Paris, 1813), 5.373–375; for "clitorisme" see pp. 376–379.

122. Ibid. (Paris, 1821), 56.446–449. "Vagin" began to refer to the organ to which it refers today in the late seventeenth century. As late as 1821, a reference work still found it necessary to note that serious mistakes arose from lexical imprecision.

123. Mauriceau, *Description anatomique des parties de la femme, qui servent a la generation* (Paris, 1662, 1708), pp. 8, 13–14. Mauriceau points out that the clitoris does not emit semen because it has no urethra.

124. Duval, *Traité des hermaphrodites,* p. 68.

125. Freud, "Infantile Sexuality," in *Three Essays,* p. 93.

126. I am indebted in my account of the "aporia of anatomy" in Freud's essay on feminity to Sarah Kofman, *The Enigma of Woman* (Ithaca: Cornell University Press, 1985), esp. pp. 109–114.

127. Rubin, "The Traffic in Women," pp. 179–180, 187.

128. *Civilization and Its Discontents,* trans. James Strachey (New York: Norton, 1962).

129. Rosalind Coward, *Patriarchal Precedents: Sexuality and Social Relations* (London: Routledge and Kegan Paul, 1983), p. 286.

130. Freud to Abraham, November 11, 1917, cited in Peter Gay, *Freud: A Life for Our Times* (New York: Norton, 1988), p. 368.

131. Bonaparte, *Female Sexuality,* p. 203.

Credits

Figs. 10–12. Courtesy of the Bancroft Library, University of California, Berkeley.

Figs. 29, 59. By permission of CIBA Publications.

Figs. 38, 39. By permission of the Frick Collection, New York.

Fig. 55. From R. M. H. McMinn and R. T. Hutchings, *Color Atlas of Human Anatomy* (1975), by permission of Year Book Medical Publishers, Inc.

Fig. 57. From Ernest Gardner, M.D., *Anatomy: A Regional Study of Human Structure,* by permission of W. B. Saunders Company.

Figs. 41, 58. From James E. Anderson, M.D., *Grant's Atlas of Anatomy,* by permission of the Williams and Wilkins Company.

Fig. 61. Courtesy of the Sterling Library, Yale University.

Index

References to illustrations and captions are printed in italics.

Abel, Elizabeth, 249n50, 250n51
Abortion, drugs for, 103
Abraham, Karl, 241
Acton, William, 190, 195
Adelmann, Howard B., 274n90
Adler, L., 220
Adler, Otto, 190
Aeschylus, 57
Aetios of Amida, 49
Alberti, Leon Battista, 168
Albertus Magnus, 45
Albinus, Bernard, 167, 168
Alcoff, Linda, 248n30
Alpers, Paul, 270n57
Alpers, Svetlana, 268n43
Amazons, sexuality of, 225–226
Amenorrhea: resolution of, 36; cures for, 100, 107; among Indians, 105, 106; and hormonal changes, 106; and hysteria, 107; and obesity, 254n29
Anal births, 19
Anal sex, 44, 53
Anatomy: as proof of one-sex model, 70–98; as proof of two-sex model, 157–161, 168–169; and aesthetics, 163–169, 267n38
Androgen-dihydrostestosterone deficiency, 7
Androgyny: in animals, 18–19; in politics, 122–123
Animalculism, 171, 172, 173–174

Animals: ovulation in, 8, 9, 198, 210–212; sexual difference in, 17–18; genital organs of, *18*; testes in, 32; genitals organs as, 109–110; experimental, 161, 174, 180
Anthropology, 68, 105, 225–226
Aquinas, Thomas, 256n47
Araeteus the Cappadocian, 37
Aretino, Pietro, 133
Ariès, Philippe, 245n3
Aristophanes, 52
Aristotle: on one sex, 28–32, 41, 52, 108, 156, 254n25; on fatherhood, 31; on generative organs, 32–33; on lactation in men, 36, 106; on menstrual fluid, 37–38, 41, 42, 107, 214; on male ejaculation, 41–42; on orgasm, 44, 47–48, 67; on conception, 50, 54–55, 58, 99–100, 231, 245–246n6, 255n36; on sex of slaves, 54; in Renaissance, 99–100, 126, 141, 144, 145, 256n46; on status of women, 149, 151; continued influence of, 150, 271n71; and castration, 178, 180; and prostitution, 232; and female ejaculation, 253n23, 255n36
Aristotle's Masterpiece, 150, 245n6, 246n11, 263n5
Artificial insemination, 161, 284n31
Augustine, 59–61
Autopsy, 188, 212, 266n35

Averroës, 67
Avicenna, 40–41, 45, 50, 101

Bachofen, Johann Jakob, 225
Baer, C. E. von, 184
Bakhtin, Mikhail, 121, 122
Barbin, Herculine, 278–279n41
Barclay, John, 168
Barkan, Leonard, 257n54
Barrenness: and sexual pleasure, 49–52, 67,
190–191; cures for, 100–103, 116, 122,
271n73, 281n3; of prostitutes, 107, 230–
232. *See also* Infertility
Barrows, Susanna, 196
Bartholin, Kaspar, 65, *91,* 92–93, *93,* 158,
158, 263n10
Bartisch, Georg, 88
Battey, Robert, 176, 179, 180
Bauhin, Gaspard, 97, 127, 269n48
Baust, Michael, 273n85
Bayfield, Robert, 117–118
Beer, Gilian, 293n36
Belly, as uterus, 27, 94, 251nn3, 4
Bellybutton, bleeding from, 105
Benzo, Antonio, 106
Berengario da Carpi, Jacopo, 74–75, *78, 79–*
82, *79, 80,* 97
Birke, Linda, 250n59
Birthmarks, 104
Bischoff, Theodor von, 178, 179, 207, 211,
213, 214, 222, 223
Blackwell, Elizabeth, 205–206, 286–287n67
Bleeding: hemorrhoidal, 16, 37, 107; nasal,
36, 37, 107; abnormal, 105; by men, 107.
See also Blood; Menstruation
Bleier, Ruth, 13, 250n59
Bloch, R. Howard, 232
Blood: menstrual, 35, 37, 42, 68, 222, 252–
253n18; and food, 35–36, 38, 41, 42–43,
252–253n18; and milk, 36, 68, 104, 105,
106; vomiting of, 37, 105, 254n31; con-
coction of, 38; and male seed, 38, 39, 47,
56, 103; and female seed, 38, 39, 41; and
conception, 121, 143, 145, 146; circulation of, 142, 143. *See also* Bleeding; Menstruation
Blumenbach, Johann Friedrich, 168, 186,
214, 219
Body. *See* Anatomy
Boerhaave, Hermann, 107
Bonaparte, Marie, 242

Bonnet, Charles, 173, 186
Book of the Courtier (Castiglione), sex changes
in, 125–126, 128
Borie, Jean, 215
Bostock, John, 187
Bouchet, Guillaume, 63
Bourgeois, Louise, 68, 108
Bouwsma, William, 300n107
Boylan, Michael, 255n36, 256n44, 257n48,
261n95
Bracken, Henry, 288n81
Breasts: and uterus, 104, 105; bleeding from,
105; Renaissance enjoyment of, 130. *See
also* Lactation
Brown, Peter, 59, 61, 257n50, 262n105,
300n108
Browne, Sir Thomas, 128
Bruhier, Jacques-Jean, 2, 3
Brunst, 218–219
Bryk, Felix, 242
Buckley, Thomas, 265n25, 274n91
Buffon, George Louis Leclerc de, 154, 175
Bull, Emma, 188
Burn, Richard, 161, 162
Burton, Robert, 107
Bylebyl, Jerome, 266n32, 271n71
Bynum, Caroline, 7

Cabinis, P. J. G., 196
Caelius Aurelius, 253n21
Calvin, John, 231
Calvino, Italo, 245n3
Canto, Monique, 260n86
Capellman, Carl, 213
Cardanus, Hieronymous, 106
Carlile, Richard, 229
Cassario, G., 73, *73*
Castiglione, Baldassare, 125–126
Castration, 31, 128, 176–177, 178. *See also*
Ovaries, removal of
Catemenia, 41–42. *See also* Blood, menstrual
Cellini, Benvenuto, 130
Cell theory, and conception, 57, 143, 171–
175
Celsus, 251n3, 253n21, 270n59
Cervix, 90, 92, 96, 97, 239
Charcot, Jean-Martin, 207
Charles I, 145
Chaucer, Geoffrey, 94, 272n79
Chereau, Achille, 175
Christianity, and sexuality, 59–61

Churchill, Fleetwood, 299n105
Churchill, Frederick B., 285n55, 286n60
Cinaedus, 53
Circumcision, 271n75
Clark, Anna, 284n38
Clark, Lorenne M. G., 283n22
Clement of Alexandria, 258n63
Clitoris: as uvula, *37*; as penis, 64, 65, 92–
    93, *93, 98,* 137–138, 140–141, 188, 234,
    301n119; discovery of, 64–65, 66, 98,
    112–113, 233; and orgasm, 65–66, 233–
    237, 240, 242; nomenclature for, 97, 237,
    239, 270n64, 271n68; and hermaphrod-
    ism, 137–138, 161; embryonic, 169; of
    prostitutes, 230; Freud on, 233–237, 240,
    242
Clover, Carol C., 250n58
Codpieces, 94, *95,* 269n54
Cody, Lisa, 275n104, 284n27
Cole, Abadiah, 269n50
Colombo, Matteo Realdo. *See* Columbus,
    Renaldus
Columbus, Renaldus: discovery of clitoris by,
    64, 65, 66, 98, 112–113, 233; and one
    sex, 70; frontispiece to *De re anatomica, 75;*
    and nomenclature for female genitals, 96–
    97; on hermaphrodism, 135–136
Comte, Auguste, 17
Conception: and orgasm, 2–3, 8, 43, 45–46,
    47, 48, 49–52, 60–61, 64, 66–68, 99–
    103, 116, 146–148, 161–163, 181–192;
    and fertility, 9, 49–52; as male "idea," 35,
    42, 59, 116, 142, 147; and nutrition, 51–
    52; and cell theory, 57, 143, 171–175; as
    act of will, 60; scientific control of, 153;
    and rape, 161–162, 284n36; unconscious,
    245n4. *See also* Eggs; Procreation; Semen;
    Sperm
Condorcet, Marquis de, 194
Consanguinity, 56
Constantinius Africanus, 42, 257n51
Contract theory, and sexual difference, 196–
    197, 250n58
Cooper, Sir Astley, 188
Corner, George W., 247n25
Corpus albigans, 15
Corpus luteum: and conception, 184, 211,
    220; role discovered, 184; and virginity,
    184, 185, 185–186
Cott, Nancy F., 246n9, 248n31
Coward, Rosalind, 241

Cowper, William, 159
Crooke, Helkiah, 90, 105
Cruickshank, W. C., 184
Culpepper, Nicholas, 105, 269n50, 276n8
Culture, and sexual difference, 7–8, 9–10,
    11–13, 22, 52–62, 109, 112, 113, 115,
    124–125, 134–135, 142, 149–150, 153,
    155, 248n38
Cushing, Harvey, 70

Dante Alighieri, 42
d'Arconville/Sue skeleton, 167–168
Darwin, Charles, 18, 201, 208, 293n36
Davidoff, Lenore, 204, 250n56, 291n5
Davis, Natalie Z., 264n22, 275n98, 279n43
Death, and sexual pleasure, 1–2
Debay, Auguste, 195
de Graaf, Regnier, 158–159, *159,* 171, 182,
    261n94
De la Motte, Guillaume, 101
Democritus, 46, 175
Descartes, René, 155
Desire. *See* Sexual pleasure
Dickinson, Emily, 181
Diderot, Denis, 26, 200, 235
Dimorphism. *See* Sexual dimorphism
Dissection scenes, 70–79
DNA, 143, 174
Dover, K. J., 260nn81, 84
Dreams, erotic, 297n90
Drugs: and infertility, 102–103, 116,
    273nn84, 85, 281n3; for abortions, 103
Dryander, John, 84, 86, 273n85
Duden, Barbara, 264n22
Duncan, Matthews, 191–192, 213
Durkheim, Emile, 17
Duval, Jacques, 93, 94, 96, 137, 240

Ear, bleeding through, 105
Effeminacy, 123, 124, 125, 128
Eggs: discovery of, 38, 142, 143, 161, 171–
    172, 184, 212, 261n94; unfertilized, 58,
    211. *See also* Conception; Ovism
Ejaculation: and orgasm, 3, 46–47, 48, 49–
    50, 51; premature, 101; and sexual differ-
    ence, 171. *See also* Wet dreams
Elephants, genitals of, 18, 32
Elias, Norbert, 122
Elizabeth I, 122–123
Ellis, Havelock, 189, 221, 226, 293n38

Ellis, Sarah, 201, 204
Elshtain, Jean Bethke, 291nn11, 14
Embryogenesis, 169–171, *170. See also* Fetus
Embryonic sexuality, 10, 169, 170
Endometrium, 220
Enlightenment: views on conception held during, 3; epistemology of, 6–7; and two-sex model, 8, 10–11, 194; political theory of, 11, 197–201
Epigastric vein, 104
Epigenesis, 143, 146, 169, 172, 174–175
Erasmus, Desiderius, 122
Estienne, Charles, 77, *78,* 90, 130–133, *131, 132, 134*
Estrogen, 249n42
Estrous cycle. *See* Heat
Estrus, 219–220
Eunuchs. *See* Castration
Eyes, bleeding from, 105

Fabliaux, 123
Fallopian tubes, 5, 82, 158, 184, 187, 253n20, 270n62
Fallopius, Gabriel (Gabriello Fallopio): discovery of clitoris by, 65; on one sex, 70; and nomenclature for female genitals, 97; discovery of tubes by, 97, 263n9, 270n62; and fertility, 100
Farr, Samuel, 161
Fat, in relation to menstruation, 36, 38, 254n29
Fatherhood, 31, 55–59, 135. *See also* Conception; Paternity
Fausto-Sterling, Anne, 9, 250n59
Fawcett, Millicent, 197, 250n57
Fellatio, 250n54
Feminism, and sexual difference, 12, 22, 197
Fertility, 9, 49–52, 207–227. *See also* Barrenness; Infertility
Fertilization. *See* Conception
Fetus, 10, 36, 105, 144. *See also* Embryogenesis; Embryonic sexuality
Fingers, bleeding from, 105
Fischer-Homberger, Esther, 8
Fleming, George, 220
Fletcher, Angus, 115
Fontanelle, Bernard de, 155
Foreplay, 67, 102, 272n82
Foreskin, male and female, 4, 34, 45, 91, 97–98, 100–101, 236
Foucault, Michel, 5, 10, 12, 13, 21, 51, 124, 151, 228, 275n2, 283n26

Francis I, 123, 130
French Revolution, 11, 194, 196
Freud, Sigmund, 15; and sexuality, 13, 21; and fatherhood, 57–58; on gender, 70; "anatomy is destiny," 189, 233; on one and two sexes, 233, 243; and orgasm, 233, 237–240, 242–243; and patriarchy, 241
Fries, Lorenz, 231, 232
Frigidity, 102. *See also* Barrenness
Frogs, in taffeta trousers, 174

Galen of Pergamum: and inversion of genitals, 4, 9, 25–28, 29, 90, 92, 93, 132–133, 170; nomenclature for genitals, 4–5; and two-seed theory, 40, 45, 58, 255n36; on orgasm, 44, 45, 46, 103, 148, 186; in Renaissance, 62, 70, 145, 146, 271n71; and animal anatomy, 83, *86;* and hysteria, 110; on status of women, 149; continued influence of, 151, 170, 265–266n28, 282n12
Gallagher, Catherine, 201, 232
Gallop, Jane, 250n50
Ganong, W. F., 225
Gardiner, Augustus, 213
Gascoigne, George, 94, 95, 107
Gasking, Elizabeth B., 279n53, 280nn55, 59, 286n60
Gay, Peter, 282n8, 288n77, 290n103
Geddes, Patrick, 6, 41
Gender, as cultural category, 7–8, 9–10, 11–13, 22, 52–62, 109, 112, 113, 115, 124–125, 134–135, 142, 149–150, 153, 155, 248n38
Gender identity, 138
Genitals, female. *See* Cervix; Clitoris; Labia; Ovaries; Uterus; Vagina
Genitals, male. *See* Foreskin; Penis; Scrotum; Testicles
Giles of Rome, 67
Girdwood, G. F., 188, 216–217
Gleason, Maud, 260nn80, 83
Godelier, Maurice, 11
Godwin, William, 201
Gombrich, E. H., 193, 267n38, 268n43
Gottlieb, Alma, 265n25, 274n91
Gouges, Olympe de, 195
Grant, Robert, 188
Gray, Henry, 166
Greenblatt, Stephen, 115
Gregory of Nyssa, 7
Guenther of Andernach, 273n86

Hall, Catherine, 204, 250n56, 291n5, 293n29
Haller, Albrecht von, 107, 173, 174, 183, 184, 214, 286n63
Haraway, Donna, 248n30
Harrison, Frederic, 250n57, 291n12
Hartsoeker, Niklaas, 171, 285n53
Harvey, William: on orgasm and conception, 67; on uterus as scrotum, 94; discovery of egg by, 142, 171; on generation (conception), 142–148, 174; and Aristotle, 144, 256n46, 280n60
Heape, Walter, 220, 221
Heat: vital, 4, 5, 41, 45, 101; and menstruation, 9, 101, 107, 218–220; male and female, 27, 28, 29, 34, 36, 40, 55, 101, 141–142; and orgasm, 44, 45, 46–47, 48, 50, 100, 101–102, 191; Christian view of, 59, 60; insufficient, 102; and sex change, 127; in animals, 157, 218–219; and female receptivity, 157; nomenclature for, 218–219
Heckscher, W. S., 266n33
Hegar, Alfred, 176, 179–181
Henle, Jakob, 87, 165
Henry VIII, 63
Hensen, Victor, 211, 237, 287n72
Herdt, Gilbert, 19
Héritier-Augé, Françoise, 265n25
Hermaphrodism: human, 18, 124, 135–142, 161, 169, 174; in animals, 18–19
Herophilus of Alexandria, 4, 5, 110, 253n19
Hertwig, Oskar, 174
Highmore, Nathaniel, 99, 280n59
Hildegard of Bingen, 120
Hippel, Theodor Gottlieb von, 203, 204
Hippocrates, 35, 36, 39–40, 46, 49, 50, 64, 67, 104, 150, 255nn36, 37, 264n21, 282n12
Hippomanes, 253n19
Hirschman, Albert, 298n102
Hirst, B. C., 192
Hitschmann, H., 220
Hoag, Roger, 287n70
Hobbes, Thomas, 156, 157, 198, 225
Homosexuality, 44, 52, 53, 82, 250n54, 260nn81
Hormones, and ovulation, 9, 153, 213
Horowitz, Mary Cline, 261n89
Hrdy, Sarah Blaffer, 288n78, 293n37, 297–298n91
Hull, Isabel, 298n102

Hunter, John, 161
Hutchinson, Anne, 121, 230
Huxley, Aldous, 207, 293n35
Hyrtl, Joseph, 237, 270n66
Hysteria: causes of, 107, 286n66; and "wandering womb," 108, 110–112, 251n3; Galen on, 110; cures for, 118; and ovaries, 176, 185; and vaginal orgasm, 193; Freud on, 233, 241–243; in men, 251n3

Illegitimacy, 56
Illich, Ivan, 5, 277n20
Impotence, 103, 108, 129–130, 272n81
Indian women, menstruation of, 105, 274n91
Infanticide, 59
Infertility, 36, 49, 50–51, 67, 100–103, 108, 257n49. See also Barrenness; Fertility; Impotence
Innocent III, 61
Isidore of Seville: on uterus and belly, 27; on milk and blood, 36; on male and female seed, 55–56

Jacob, François, 17
Jacobi, Mary Putnam, 222–224, 286–287n67
Jacobus, Mary, 246n7
Jacquart, Danielle, and Claude Thomasset, 256n43, 257n51, 258n58, 261n90, 270n64, 273nn84, 87, 299n107
Jeffreys, Sheila, 282n9, 293n34
Johnson, Barbara, 17
Johnson, Virginia, 234
Joubert, Laurent, 104, 105, 106
Jouissance, 200
Jozé, Victor, 149
Jurisprudential medicine: distinguishing male from female, 132, 134–135, 142; and hermaphrodism, 135–142; and conception, 162; and virginity, 185–186; and sodomy, 279n42; and paternity, 297–298n91
Justinian I, 49, 161

Kaulos (tube), 33–34
Keller, Evelyn Fox, 248n30, 249n46
Kember, O., 262n106
Kent, Susan, 297n89
Keuls, Eva, 252n17
Kisch, E. H., 192, 237
Kleist, Heinrich von, 246n7
Kobelt, Georg Ludwig, 238–239
Kofman, Sarah, 15, 301n126

Krafft-Ebing, Richard von, 236
Krause, Wilhelm J. F., 237, 238
Kristeva, Julia, 12
Kuhn, Thomas, 96, 265n26

Labia, 4, 45, 97–98
Lacan, Jacques, 12, 141
Lactation: in women, 36, 104, 105, 106; in men, 36, 106, 151, 256n42
Landes, Joan, 198
Laqueur, Ernst, 15, 243
Laqueur, Werner, 15
Laurentius, Andreas, 269n48
Lee, Robert, 188, 289n91
Lemnius, 64, 99
Leonardo da Vinci, *85,* 104, 170, 264n13
Leuwenhoek, Anton van, 171, 261n94
Lévi-Strauss, Claude, 19
Liberalism, and sexual difference, 197, 250n58
Libido, 43, 44, 190. *See also* Sexual pleasure
Linnaeus, Carolus, 172–173
Lister, Joseph, 176, 180
Lloyd, Elizabeth A., 300n109
Lloyd, G. E. R., 34, 254n30, 262n106
Lloyd, Lisa, 264n20
Locke, John, 156, 194
Louis, Antoine, 2, 3
Lucretius, 47, 60, 231
Lust, and gluttony, 51
Luther, Martin, 231

MacCormack, Carol P., 262n106
MacKinnon, Catharine, 12
McLaren, Angus, 8, 291n6
Maclean, Ian, 108, 271n71, 275n99, 277n21
Malthus, Thomas, 201
Marche, Madame de la, 68
Marcis, Marie de, 136–137, 138
Maria Theresa, 150
Marshall, F. H. A., 295n57
Martin, Emily, 265n23
Massa, Niccolo, 269n50
Masters, William H., 234
Masturbation, 185, 190, 196, 227–230, 232, 272–273n83, 297n90
Maubray, John, 245n3
Maupertuis, Pierre de, 175
Mauriceau, François, 239–240
Medicine, modern: and sexual difference, 5–6, 9–10, 14–15, 154–163; and society, 11. *See also* Jurisprudential medicine

Mendelsohn, Everett, 251n7
Menses. *See* Heat; Menstruation
Menstruation: and timing of conception, 9; seen as heat, 9, 101, 107, 218–220; as plethora, 35, 36, 107, 109, 213; disturbances of (amenorrhea), 36, 101; and lactation, 36, 104, 105, 106; male, 107; and uterus, 152, 220–221; and ovaries, 176, 178, 179. *See also* Bleeding; Blood
Merchant, Carolyn, 266n34, 280n56
Metchnikoff, Elie, 222
Michelangelo, 77
Michelet, Jules, 211, 217, 221–222
Microscope, and discovery of sperm, 171
Milk. *See* Blood, and milk; Lactation
Mill, James, 202
Millar, John, 200–201, 206
Minnow, Martha, 197
Minorities, sexuality of, 152, 283n16
Misogyny, 21–22, 125
Moerloose, Isabella De, 68, 272–273n83
Mohammed, 7
Mola, 58–59, 278n32
Mollis, 53, 258n58
Mondino de' Luzzi (Mundinus), 65, 71, *72,* 97
Montaigne, Michel de, 7, 113, 126–127, 128–130, 139, 246n7
Moravia, Alberto, 258n63
More, Hannah, 201
Moreau, Jacques-Louis, 5, 149, 196
Morgan, Lewis Henry, 225
Morton, Peter, 293nn35, 36
Moses, Bessie, 295n53
Mosher, Clelia Duel, 191
Motherhood, 56, 57. *See also* Conception
Müller, Johannes, 169, 187, 215

Nägele, Carl Franz, 219
Napheys, George, 212
Necrophilia, 1
Nehamas, Alexander, 249n39
Nemesius of Emesa, 4
Netter, Frank, *87, 170*
Nietzsche, Friedrich, 13, 248–249n39
Nightingale, Florence, 233
Nihell, Elizabeth, 112
Nosebleed, 36, 37, 107

One-sex model, 8, 19–20, 21, 25, 26–62 passim, 63–113 passim, 114–142 passim, 150–151, 153–154

Orgasm, 43–52; and conception, 2–3, 8, 43, 45–46, 47, 48, 49–52, 60–61, 64, 66–68, 99–103, 116, 146–148, 161–163; and heat, 44, 45, 46–47, 48, 50, 100, 101–102; and sexual difference, 150, 181–192. *See also* Sexual pleasure

Origen, 7

Ortner, Sherry, 12

Outram, Dorinda, 291n2

Ovaries: as female testicles, 4, 9, 26, 82, 149, 158, *159,* 161, 176–177, 236, 270n61; view of in antiquity, 4–5; nomenclature for, 4–5, 172, 253n19; view of in eighteenth century, 5, 161, 252n14; removal of, 31, 176–181; and status of women, 149, 175–181, 213–214, 216; and menstruation, 176, 178, 179

Ovid, 43–44, 129, 139

Ovism, 173–174

Ovulation: and sexual pleasure, 8, 163, 182, 183–184, 211; coitally induced, 8, 288n79; spontaneous, 8–9, 178, 184, 187, 211, 213, 214, 223, 288n79; hormonal control of, 9, 153, 213; timing of, 68, 212–213, 222, 294n48; hidden, 297–298n91

Owen, R. D., 228

Owen, Richard, 173, 190

Paget, Sir James, 212

Pangenesis, 39, 46, 175, 255–256n40, 258n63, 286n63

Papanicolaou, George N., 213

Paracelsus, 22, 110

Paré, Ambroise, 7, 102, 104, 126, 127

Parent-Duchâtelet, Jean-Baptiste, 230

Park, Katherine, 226n35, 279n48

Parr, Bartholomew, 219

Parthenogenesis, 18, 58, 144, 173

Passionlessness, 3, 8, 150, 161, 182, 189–191, 195–196, 206, 246n9

Pasteur, Louis, 294n43

Pateman, Carole, 196, 250n58, 283n21, 291n10

Paternity, 297–298n91

Pathicus, 53

Patriarchy, 20, 62, 100, 157, 194–195, 241. *See also* Subordination, of women

Paulus Aegineta, 254n29

Pechy, John, 98

Peck, A. L., 252n10

Penis: and vagina, 4, 16, 21, 26, 28, 33–34, 35, 63, 65, *78–93* passim, 79–88, 90–92, 109, 160; Aristotle's nomenclature for, 30; size of, 31–32, 50, 101, 130; and sexual pleasure, 44, 45; and clitoris, 64, 65, 92–93, *93,* 137–138, 140–141, 188, 234, 301n119; and sex changes, 123, 126, 127, 128, 129, 136–142; and male status, 134, 135, 137, 139, 140

Penis envy, 129, 240

Perry, R. L., 228

Peter of Spain, 273n84

Pflüger, E. F., 222

Phagocytosis, 222

Phlebotomy, 107

Phlegm, and semen, 35, 42, 254n26

Phrenology, 208–210

Phusis/nomos, 7

Plants, sex among, 172–173

Plato, 54, 109, 112, 199, 275n100

Pleasure. *See* Orgasm; Sexual pleasure

Plethora: and menstruation, 35, 36, 107, 109, 213; and fat, 36

Pliny, 128, 129, 214, 270n59

Plutarch, 56, 58

Pollux, Julius, 34, 253n22

Pontormo, Jacobo, *95*

Poovey, Mary, 203, 248n34

Positivism, 17, 249n48

Pott, Percival, 178, 179

Pouchet, F. A., 211, 213, 215–216, 222, 223

Poullain de la Barre, François, 155–156

Preformationism, 173–175, 282–283n14

Procreation, devaluation of, 8, 247n24. *See also* Conception

Prostate, 15, 92, 249n43

Prostitution, 107, 227, 230–233

Pseudo-Albertus Magnus, 43, 64, 151, 256n47, 257n53

Pseudo-Aristotle, 34, 50, 245–246n6

Pufendorf, Samuel von, 198

Pulley, John, 185

Quaife, G. R., 273n85

Quine, W. V., 69

Quine-Duhem thesis, 69, 280n61

Rabelais, François, 94, 108, 109

Races, scientific basis for distinguishing among, 155, 208

Raciborski, Adam, 190, 217–218, 222, 223

Rape, and conception, 161–162, 260n78, 284n36

Remak, Robert, 215
Renal vein, and sexual pleasure, 103, *104*
Reproduction, 155. *See also* Conception
Restitutus, 60
Reynolds, Sir Joshua, 167
Rhazes, 50
Riolan, John, 73–74, *74*
Roberts, Dr. G., 178, 179
Rokitansky, Karl, 188
Rose, Jacqueline, 264n13
Rosenberg, Rosalind, 191
Rouselle, Aline, 39, 255n37
Rousseau, Jean-Jacques, 157, 197, 198–200, 225, 228, 235
Roussel, Pierre, 6, 152, 196, 281n2
Rubin, Gayle, 12, 241
Rueff, Jacob, *89,* 102
Ruysch, Frederik, 14
Ryan, Michael, 3, 8
Ryff, Walther, 88, *89*

Sanchez, Regina Marantz, 287n67
Satyriasis, 52
Sayers, Dorothy L., 1, 10
Schiebinger, Londa, 247n27, 283n25, 285nn46, 48, 49
Schmitt, Charles B., 265–266n28
Scholes, Robert, 234
Schuria, Henrika, 137
Schwartz, Joel, 291n14, 292n16, 298n96
Schwartz, Vanessa, 279n41
Scott, Joan, 12, 248n34
Scrotum, uterus as, 4, 25, 28, 33, 35, 63, 64, 79, *85, 88, 89,* 94, 160, 236, 257n55
Secrets of Women, The (Pseudo-Albertus Magnus), 43, 64, 151, 256–257n47
Semen, 35, 40, 41, 42, 44, 45, 46, 53, 56, 66, 68, 99, 101, 103, 106, 120, 144. *See also* Sperm
Seminal vesicles, 85
Sex changes: human, 7, 123, 124, 125–130, 136, 151; in animals, 18–19, 128
Sexual dimorphism, 6, 163. *See also* Gender
Sexual pleasure: and conception, 2–3, 8, 43, 45–46, 47, 48, 49, 60–61, 64, 66–68, 99–103, 116, 146–148, 161–163; male and female, 4, 43–44, 46, 150, 181–192; and warmth, 40, 46, 48, 49, 102, 107. *See also* Orgasm
Sexual selection, Darwinian, 208
Shakespeare, William, 24, 113, 114–115, 123, 243

Sharp, Jane, 65, 68
Sherfey, Mary Jane, 300n109
Siraisi, Nancy G., 261n97, 265n28
Sissa, Giulia, 252n11
Skeleton, female, 166–169
Skene, A. J. C., 171
Smellie, William, 112, 182, 288n81
Smith, Adam, 233
Smith-Rosenberg, Carol, 286n65
Smith, J. G., 185
Smollett, Tobias, 112
Snakes, sexuality of, 44
Sociability, and sex, 124
Social-contract theory. *See* Contract theory
Sociobiology, 21
Soemmerring, Samuel Thomas von, 167–168
Soranus, 34, 37, 51, 59, 150, 161
Spallanzani, Lazzaro, 161, 174
Speculums, 206
Spellman, Vicky, 54
Spenser, Edmund, 118–119
Sperm: discovery of, 38, 143, 161, 171, 172, 261n94; strength of, 39–40; and sexual pleasure, 45; as parasite, 57, 174; role of, 145, 146–147, 187; economy of, 196; nomenclature for, 285n53. *See also* Animalculism; Conception; Semen
Spittle, and blood, 105
Spontaneous ovulation. *See* Ovulation
Sterility. *See* Barrenness; Infertility
Stoller, Robert, 252n15
Stone, Lawrence, 5
Stopes, Mary, 213
Storch, Johann, 264–265n22
Stubbs, George, 168
Subordination, of women, 156–157. *See also* Patriarchy
Sweating, 35, 37, 255n32
Sylvius, Jacobus, 70
Symons, Donald, 288n78

Tanner, John, 117
Tardieu, Ambroise, 136
Taylor, Barbara, 292n22
Temkin, Oswei, 282n12
Tertullian, 46
Testicles: male and female, 4, 16, 25, 26, 32, 63, 82, 85–86, 149, 160–161, 176, 236, 270n61; like weights on loom, 32; removal of, 178
Thomas, Joseph, 219
Thomas, Keith, 275n4

Thomasen, Anne-Liese, 257n47
Thompson, J. Arthur, 6
Thompson, William, 202–203
Throat, and vagina, 36, *37,* 50, 110, 254n30, 258n59
Tilt, Edward John, 287n71
Tissot, Samuel August, 228
Tocqueville, Alexis de, 157
Todd, Mabel Loomis, 181–182
Transvestites, 136–139
Trembley, Abraham, 173
Tribade, 53, 136, 137
Trotula of Salerno, 272n79
Two-sex model, 8, 16, 20–21, 69, 148, 149–192 passim, 193–243 passim

Uterus: as scrotum, 4, 25, 28, 33, 35, 63, 64, 79, *85,* 88, *89,* 94, 160, 236, 256n43, 257n55; as archetype for other organs, 27; and childbearing, 27, 86–88; belly as, 27, 94, 251nn3, 4; Aristotle's interpretation of, 31, 33; as purse, 64, 263nn3, 4; nomenclature for, 64, 96; cells in, 65; links to breasts, 104, 105; and hysteria, 108, 110–112, 251n3; as vessel, 131, 251n2; as brain, 147; and status of women, 149, 152, 155, 225; and menstruation, 152, 220–221
Uterus masculinus (prostatic utricle), 15, 249n43

Vaga, Perino del, 130–131
Vagina: as penis, 4, 16, 21, 26, 28, 33–34, 35, 63, 65, *78–93* passim, 79–88, 90–92, 109, 160; and throat, 36, *37,* 50, 254n30, 258n59; nomenclature for, 96–97, 149, 158, 159, 239, 270n60, 301n122; and orgasm, 45, 92, 233, 236–237, 238, 239–240; vestigial, 249n43
Valverde, Juan de, 75–77, *76, 77, 84*
Varicose veins, 37

Venette, Nicholas, 93, 150–151, 245n5
Veronese, Paolo, *75*
Vesalius, 63, 70, *71,* 71–72, 73, *73,* 75, 77, *77, 80, 81,* 82, *82,* 83, *84,* 103, *111,* 115–116, 164, 170
Vicary, Thomas, 98
Vickers, Nancy J., 278n33
Vidius, Vidus, *83*
Virchow, Rudolf, 207, 221
Virgin births, 119
Virginity, 59, 184, 185, 188
Vlastos, Gregory, 260–261n86
Vogt, Karl, 208
Vulva: medieval meaning of, 27; and female genitals, 94, 98, 160, 236, 269n52, 270n59

Wagner, Richard, 219
Wandering womb. *See* Hysteria
Warner, John Harley, 282n12
Weeks, Jeffrey, 13
Westermarck, Edward, 225, 226, 227
Wet dreams, 48, 101, 259nn68, 71
Wheeler, Anna, 201, 202–203
Whitehead, Harriet, 12
William of Conches, 231, 300n107
Wind eggs, 58
Winthrop, John, 121
Wirsung, Christopher, 105
Wolff, Kaspar Friedrich, 169, 272n79
Wollstonecraft, Mary, 201, 203–204
Wolstenholme, Elizabeth, 152, 226, 227
Womb. *See* Uterus
Woolf, Christian, 174
Woolf, Virginia, 5

Xeuxis, 168

Zacchia, Paolo, 140–141, 142, 161
Zapperi, Roberto, 277n18
Zodiacal man, 116, *117*